T0189623

Protein Phosphatase Protocols

METHODS IN MOLECULAR BIOLOGY™

John M. Walker, SERIES EDITOR

Protein Phosphatase Protocols

Edited by

John W. Ludlow

University of Rochester Cancer Center, Rochester, NY

Springer Science+Business Media, LLC

Cover illustration: Fig. 2 from Chapter 22, "Identifying Protein Phophatase 2A Interacting Proteins Using the Yeast Two-Hybrid Method," by Brent McCright and David M Virshup.

Cover design by Patricia F. Cleary.

Library of Congress Cataloging in Publication Data

Main entry under title:

Methods in molecular biology™.

Protein phosphatase protocols/edited by John W. Ludlow.
        p.   cm.—(Methods in Molecular Biology; 93)
    Includes index.
    ISBN 978-1-4899-4276-0   ISBN 978-1-59259-267-8 (eBook)
    DOI 10.1007/978-1-59259-267-8

    1. Phosphoprotein phosphatases—Laboratory manuals. I. Ludlow, John W. II. Series: Methods in molecular biology (Totowa, NJ); 93
QP609.P56P76 1998
572'.76—dc21
DNLM/DLC   98—15957
for Library of Congress   CIP

Preface

A major mechanism by which cells regulate protein function is to place phosphate groups on serine and threonine residues. Though the steady-state level of protein phosphorylation depends on the relative activities of both kinases and phosphatases, a much greater effort has previously gone into the study of the former that the latter. Today, however, there is an increasing appreciation for the role that protein phosphatases play in the dynamic process of protein phosphorylation. To date, there are four major types of protein serine/threonine phosphatase catalytic subunits, designated protein phosphatase type 1, 2A, 2B, and 2C. Each has been identified by the techniques of protein chemistry and enzymology and can be distinguished from one another by their preference for specific substrates as well as their sensitivity to certain activators and inhibitors.

Protein Phosphatase Protocols has been assembled in response to the growing interest these enzymes are receiving. The goal of this compilation is to provide a "how-to" experimental guide to aid newcomers as well as seasoned veterans in their research endeavors, thus further contributing towards our ever increasing knowledge of serine/threonine phosphatases.

What you have before you contains contributions by many of the current and emerging leaders in the field. To highlight just a few, these chapters contain step-by-step information on how to isolate novel phosphatases and regulatory subunits, assay for activity, and generate immunological reagents for both biochemical and biological characterization of these enzymes. Though it is obviously not possible to include contributions by each and every researcher in this field, every effort was made to be inclusive, and avoid being exclusive, regarding the methods used to investigate these phosphatases. We hope that you find our work both informative and thought provoking.

John W. Ludlow

Contents

Contributors

JOAQUIN ARIÑO • *Department de Bioquimica i Biologia Molecular, Universitat Autonoma, Barcelona, Spain*

ARND BAUMANN • *Institut für Biologische Informationsverarbeitung, Forschungszentrum Jülich, Germany*

NAJMA BEGUM • *Diabetes Research Laboratory, Winthrop University Hospital, Mineola, NY*

NORBERT BERNDT • *Department of Pediatrics, Children's Hospital of Los Angeles, CA*

MONIQUE BEULLENS • *Afdeling Biochemie, Faculteit Geneeskunde, Katholieke Universiteit Leuven, Belgium*

MATHIEU BOLLEN • *Afdeling Biochemie, Faculteit Geneeskunde, Katholieke Universiteit Leuven, Belgium*

DAVID L. BRAUTIGAN • *Center for Cell Signalling, Hospital West, University of Virginia, Charlottesville, VA*

JOHN F. CANNON • *Department of Molecular Microbiology and Immunology, University of Missouri, Columbia, MO*

QI CHENG • *Department of Biochemistry, North Dakota State University, Fargo, ND*

NAOKI CHIDA • *Department of Biochemistry, Tohoku University, Sendai, Japan*

JOSEP CLOTET • *Department de Bioquimica i Biologia Molecular, Universitat Autonoma, Barcelona, Spain*

ELIZABETH COLLINS • *Medical Biochemistry Faculty of Medicine, University of Newcastle, Australia*

JOHN H. CONNOR • *Department of Pharmacology and Cancer Biology, Duke University Medical Center, Durham, NC*

BRIAN K. DALLEY • *Department of Molecular Microbiology and Immunology, University of Missouri, Columbia, MO*

ZAHI DAMUNI • *Cellular and Molecular Physiology, College of Medicine, Pennsylvania State University, Hershey, PA*

MARIAM DOHADWALA • *Department of Pediatrics, Children's Hospital of Los Angeles, CA*

SAMUEL C. EDWARDS • *Department of Pharmacology and Therapeutics, University of South Florida College of Medicine, Tampa, FL*

ALBERT FERRER • *Departament de Bioquimica i Biologia Molecular, Universitat de Barcelona, Spain*

JOZEF GORIS • *Afdeling Biochemie, Faculteit Geneeskunde, Campus Gasthuisberg, Leuven, Belgium*

VEERLE JANSSENS • *Afdeling Biochemie, Faculteit Geneeskunde, Campus Gasthuisberg, Leuven, Belgium*

ULRICH B. KAUPP • *Institut für Biologische Informationsverarbeitung, Forschungszentrum Jülich, Germany*

PETER J. KENNELLY • *Department of Biochemistry and Anaerobic Microbiology, Virginia Polytechnic Institute and State University, Blacksburg, VA*

S. DEREK KILLILEA • *Department of Biochemistry, North Dakota State University, Fargo, ND*

SUSANNE KLUMPP • *Universität Pharmazeutische Chemie, Marburg, Germany*

TAKAYASU KOBAYASHI • *Department of Biochemistry, Tohoku University, Sendai, Japan*

NANCY A. KRUCHER • *University of Rochester Cancer Center, Rochester, NY*

KAZUYUKI KUSUDA • *Department of Biochemistry, Tohoku University, Sendai, Japan*

NED J. C. LAMB • *Cell Biology Unit, Centre de Recherche de Biochimie Macromoleculaire, Centre National de la Recherche Scientifique, Montpellier, France*

MEI LI • *Cellular and Molecular Physiology, College of Medicine, Pennsylvania State University, Hershey, PA*

JOHN W. LUDLOW • *Department of Biochemistry and Biophysics, University of Rochester Cancer Center, Rochester, NY*

BRENT MCCRIGHT • *Department of Oncological Sciences, University of Utah, Salt Lake City, UT*

WILFRIED MERLEVEDE • *Afdeling Biochemie, Faculteit Geneeskunde, Campus Gasthuisberg, Leuven, Belgium*

DEIRDRE A. NELSON • *Department of Biochemistry and Biophysics, University of Rochester Cancer Center, Rochester, NY*

MOTOKO OHNISHI • *Department of Biochemistry, Tohoku University, Sendai, Japan*

CAREY J. OLIVER • *Department of Pharmacology and Cancer Biology, Duke University Medical Center, Durham, NC*

ERIC M. PHIZICKY • *Department of Biochemistry, University of Rochester School of Medicine and Dentistry, NY*

FRANCESC POSAS • *Departament de Bioquimica i Biologia Molecular, Universitat de Barcelona, Spain*

GEMMA PUJOL • *Departament de Bioquimica i Biologia Molecular, Universitat de Barcelona, Spain*

HAI QUAN • *Department of Pharmacology and Cancer Biology, Duke University Medical Center, Durham, NC*

LOUIS RAGOLIA • *Diabetes Research Laboratory, Winthrop University Hospital, Mineola, NY*

NADJA T. RAMASWAMMY • *Department of Molecular Microbiology and Immunology, University of Missouri, Columbia, MO*

MATTHEW K. ROBINSON • *Program in Genetics, University of Rochester School of Medicine and Dentistry, NY*

AXEL H. SCHÖNTHAL • *Department of Molecular Microbiology and Immunology, K. Norris Comprehensive Cancer Center, University of Southern California, Los Angeles, CA*

DAGMAR SELKE • *Universität, Pharmazeutische Chemie, Marburg, Germany*

SHIRISH SHENOLIKAR • *Department of Pharmacology and Cancer Biology, Duke University Medical Center, Durham, NC*

ALISTAIR T. R. SIM • *Medical Biochemistry Faculty of Medicine, University of Newcastle, Australia*

WILLY STALMANS • *Afdeling Biochemie, Faculteit Geneeskunde, Katholieke Universiteit Leuven, Belgium*

SHINRI TAMURA • *Department of Biochemistry, Tohoku University, Sendai, Japan*

MARGHERITA TOGNARINI • *Department of Biomedicine, University of Pisa, Italy*

PATRIC TUROWSKI • *Cell Biology Unit, Centre de Recherche de Biochimie Macromoleculaire, Centre National de la Recherche Scientifique, Montpellier, France*

TIMOTHY H. VAN DYKE • *Department of Pharmacology and Therapeutics, University of South Florida College of Medicine, Tampa, FL*

TRAVIS B. VAN DYKE • *Department of Pharmacology and Therapeutics, University of South Florida College of Medicine, Tampa, FL*

CHRISTINE VAN HOOF • *Afdeling Biochemie, Faculteit Geneeskunde, Campus Gasthuisberg, Leuven, Belgium*

EMMA VILLA-MORUZZI • *Department of Biomedicine, University of Pisa, Italy*

DAVID VIRSHUP • *Department of Oncological Sciences, University of Utah, Salt Lake City, UT*

BRIAN E. WADZINSKI • *Department of Pharmacology, Vanderbilt University, Nashville, TN*

ZHI-XIN WANG • *Department of Biochemistry, North Dakota State University, Fargo, ND*

RYAN S. WESTPHAL • *Department of Pharmacology, Vanderbilt University, Nashville, TN*

JULIE A. ZAUCHA • *Department of Pharmacology, Vanderbilt University, Nashville, TN*

JIANHONG ZHENG • *Department of Molecular Microbiology and Immunology, University of Missouri, Columbia, MO*

1

Prokaryotic Protein-Serine/Threonine Phosphatases

Peter J. Kennelly

1. Introduction

1.1. Prokaryotic Protein-Serine/Threonine Phosphatases: A Brief Review

1.1.1. Why Study Protein Phosphorylation Events in Prokaryotes?

As this chapter deals with the protein-serine/threonine phosphatases of prokaryotic organisms, some comments on the role of prokaryotes in the study of these important enzymes would appear to be in order. Prokaryotic organisms dominate the living world. They represent by the largest source of biomass on the planet, forming the indispensable foundation of the food chain upon which all other living organisms depend. They are the exclusive agents for carrying out biological nitrogen fixation, and are responsible for the majority of the photosynthetic activity that generates the oxygen we breath. In absolute numbers, in number of species, in range of habitat, and in the spectrum of their metabolic activities, the prokaryotes far outpace their eukaryotic brethren. More immediately, in humans prokaryotes perform essential functions in the digestion and assimilation of nutrients, whereas infection by bacterial pathogens can lead to illness or death.

The intrinsic biological importance of prokaryotic organisms in the biosphere renders them important and interesting objects of study *(1)*. Be that as it may, the question remains as to why protein phosphorylation in prokaryotes should be of interest to "mainstream" signal transduction researchers whose attention has long been fixed on humans and other higher eukaryotes. At least part of the answer lies in the recent realization that prokaryotes and eukaryotes employ many of the same molecular themes for the construction and operation of their protein phosphorylation networks *(2,3)*. Virtually every major family

From: *Methods in Molecular Biology, Vol. 93: Protein Phosphatase Protocols*
Edited by: J. W. Ludlow © Humana Press Inc., Totowa, NJ

of protein kinases or protein phosphatases identified in eukaryotic organisms possesses a prokaryotic homolog(s), and vice versa. Consequently, the prokaryotes represent a voluminous library of fundamentally important, universally applicable information concerning the structure, function, origins, and evolution of protein kinases, protein phosphatases, and their target phosphoproteins. In addition, prokaryotes offer significant advantages as venues for the study of protein kinases and protein phosphatases, particularly with regard to dissecting their physiological functions and the factors that influence them. Prokaryotes carry out their life functions and the regulation thereof utilizing a many-fold smaller suite of genes and gene products than does the typical eukaryote. Although they employ molecular mechanisms as subtle and sophisticated as any found in "higher" organisms, the fewer "moving parts" in the prokaryotes materially facilitates the design, execution, and analysis of molecular genetic experiments. In addition, their robustness in the face of a wide range of nutritional and environmental challenges greatly facilitates the identification and analysis of resulting phenotypes. The prokaryotes thus represent a rich and presently underutilized tool for understanding the fundamental principles governing the form and function of protein phosphorylation networks.

1.1.2. Not All Prokaryotes Are Created Equal: A Brief Outline of Phylogeny

Most readers of this chapter were taught that all living organisms could be grouped into two phylogenetic domains whose names were often given as the eukaryotes and the prokaryotes *(4)*. However, these latter terms actually refer to a morphological classification, not a genetic/hereditary one *(5)*. The term eukaryote describes those organisms whose cells manifest internal compartmentation, more precisely the presence of a nuclear membrane. The prokaryotes include all organisms lacking such intracellular organization, in other words everything that is not a eukaryote. Early studies of phylogeny based on the first protein sequences, the gross structural and functional characteristics of key macromolecules, the architecture of common metabolic pathways, and so forth, suggested that this morphological classification of living organisms paralleled their hereditary relationships. However, as researchers gained facility with the isolation, sequencing, and analysis of DNA, a truly genetic outline of phylogeny has emerged, one that groups living organisms into three distinct phylogenetic domains—the *Eucarya, Bacteria,* and *Archaea (Archaebacteria) (6)*.

Whereas the prior supposition that the eukaryote morphological phenotype characterized members of a coherent phylogenetic domain—the *Eucarya*—proved correct, molecular genetic analysis revealed that the prokaryotes segregated into two distinct and very different domains: the *Bacteria* and the

Archaea. The domain *Bacteria* includes essentially all of the prokaryotic organisms one encounters in a typical microbiology course: *E. coli, Salmonella, Pseudomonas, Rhizobium, Clostridia, Staphylococcus, Bacillus, Anabaena,* and so on. The domain *Archaea,* on the other hand, is populated largely by extremophiles that occupy habitats whose heat, acidity, salinity, or oxygen tension render them hostile, if not deadly to other living organisms. However, it would be wrong to suppose that the *Archaea* are simply a set of unusual bacteria. Examination of the genes encoding their most fundamentally important macromolecules, ranging from DNA polymerase to ribosomal RNAs, make it clear that the *Archaea* have much more in common with the *Eucarya* than they do with the superficially-similar *Bacteria (6,7).* The earliest detectable branch point in the evolutionary time line resulted in the segregation of the *Bacteria* away from the organism that eventually gave rise to both the *Eucarya* and the *Archaea.* The common progenitor of these latter domains then evolved for a considerable period following this first bifurcation. As a consequence, many investigators believe that present day archaeons still possess numerous features reflective of ancient proto-eukaryotes *(7).* This combination of prokaryotic "simplicity" with high relatedness to medically relevant eukaryotes render the *Archaea* a particularly intriguing target for the study of protein phosphorylation phenomena.

1.1.3. Prokaryotic Protein-Serine/Threonine Phosphatases Identified to Date

When one considers that the modification of prokaryotic proteins by phosphorylation-dephosphorylation first was reported nearly 20 yr ago *(8–10),* surprisingly little is known about the enzymes responsible for the hydrolysis of phosphoserine and phosphothreonine residues in these organisms. The first prokaryotic protein-serine/threonine phosphatase to be identified and characterized was the product of the *aceK* gene in *E. coli (11).* This gene encodes a polypeptide that contains both the protein kinase and protein phosphatase activities responsible for the phosphorylation-dephosphorylation of isocitrate dehydrogenase. Today, AceK remains an anachronism by virtue of its hermaphroditic structure, and because the sequences of its protein kinase and protein phosphatase domains are unique, exhibiting no significant resemblance to other protein kinases or protein phosphatases *(12).*

The next prokaryote-associated protein-serine/threonine phosphatase to be discovered was ORF 221 encoded by bacteriophage λ *(13,14).* This enzyme, and a potential protein encoded by an open reading frame in bacteriophage φ80, exhibit significant sequence homology with the members of the PP1/2A/2B superfamily, one of the two major families of eukaryotic protein-serine/threonine phosphatases *(15).* Whereas this represented the first discovery of a eukaryote-like protein phosphorylation network component having any association with a prokaryotic organism, the mobility and malleability of viral vectors begged the question of whether the genes for these protein phosphatases were bacterial in origin. Moreover, it remains unclear to what degree a protein phosphatase from a pathogen can shed light on how bacterial proteins are dephosphorylated under normal physiological circumstances.

More recently, two unambiguously bacterial enzymes have been described that possess protein-serine/threonine phosphatase activity. The first, IphP from the cyanobacterium *Nostoc commune (16)*, is a dual-specificity protein phosphatase that acts on phosphoseryl, phosphothreonyl, and phosphotyrosyl proteins in vitro *(17)*. Like other dual-specific protein phosphatases, IphP contains the characteristic HAT (His-Cys-Xaa$_5$-Arg, or His-Arg-Thiolate) active site signature motif characteristic of protein phosphatases capable of hydrolyzing phosphotyrosine *(18)*. The second is SpoIIE from *Bacillus subtilis*, a bacterial homolog of the second major family of "eukaryotic" protein-serine/threonine phosphatases, the PP2C family *(19,20)*.

"Eukaryotic" protein-serine/threonine phosphatases have been uncovered in the *Archaea* as well. In the author's laboratory a protein-serine/threonine phosphatase, PP1-arch, has been purified, characterized, cloned, and expressed from the extreme acidothermophilic archaeon *Sulfolobus solfataricus (21,22)*. This protein is a member of the PP1/2A/2B superfamily, with whose eukaryotic members it shares nearly 30% sequence identity *(22)*. Surveys of two other archaeons, which are phylogenetically and phenotypically distinct from *S. solfataricus*, the halophile *Haloferax volcanii* and the methanogen *Methanosarcina thermophila* TM-1, indicate that PP1-arch from *S. solfataricus* is the first representative of what may prove to be a widely distributed family of archaeal protein-serine/threonine phosphatases *(23,24)*. This recently has been confirmed at the sequence level through the cloning of a second form of PP1-arch from *M. thermophila* via the polymerase chain reaction (PCR).

1.1.4. Limited Applicability of Cohen's Scheme to the Classification Prokaryotic Protein-Serine/Threonine Phosphatases

Recent experience with prokaryotic protein phosphatases has revealed that Cohen's criteria for classifying the protein-serine/threonine phosphatases cannot be extrapolated with confidence to prokaryotic enzymes. To briefly review, in the early 1980s, Cohen and coworkers compiled a set of functional attributes characteristic of each of the major protein-serine/threonine phosphatases found in eukaryotes *(25)*. These attributes included their preference for dephosphorylating the α- vs the β-subunit of phosphorylase kinase, their sensitivity to the heat-stable inhibitor proteins I-1 and I-2, and the (in)dependence of their catalytic activity on the presence of divalent metal ions such as Mg^{2+}, Mn^{2+}, or Ca^{2+}. In later years sensitivity to potent microbial toxins—such as microcystin-

LR, okadaic acid, and tautomycin—that inhibited the activity of PP1 and PP2A were added to the list *(26)*. While this scheme soon was adopted as standard for the classification of eukaryotic protein-serine/threonine phosphatases, attempts to apply it to prokaryotic enzymes have met with mixed success. For example, PP1-arch from *S. solfataricus* is okadaic acid-insensitive and requires exogenous divalent metal ions for activity *(21)*. Under Cohen's scheme, this would classify it as a member of the PP2C family. However, the amino acid sequence of PP1-arch clearly places it in the PP1/2A/2B superfamily *(22)*. The same holds true for another divalent metal ion-dependent, okadaic acid-insensitive PP1/2A homolog, ORF 221 from bacteriophage λ *(14)*.

1.2. An Overview of Methods for Assaying, Purifying, and Identifying Clones of a Prokaryotic Protein-Serine/Threonine Phosphatase, PP1-Arch

We use [^{32}P]phosphocasein that has been phosphorylated using the catalytic subunit of the cAMP-dependent protein kinase *(27)* as a general-purpose substrate for the assay of protein-serine/threonine phosphatase activity in prokaryotic organisms. Although it is a eukaryotic phosphoprotein, all of the prokaryotic protein-serine/threonine phosphatases that have been studied *(16,17,21–24)*, as well as the ORF 221 protein-serine/threonine phosphatase from bacteriophage λ *(14)*, hydrolyze phosphocasein at a usefully high rate in vitro. Its major virtue resides in the fact that it is readily prepared in quantity by procedures that are simple and economical with regard to both effort and expense. The major drawback of phosphocasein is the very high quantity of unlabeled phosphate that is already bound to it, which renders it unsuitable for determining kinetic parameters. However, for routine applications—those requiring knowledge of the relative protein phosphatase activity present in a sample such as surveying cell homogenates or column fractions, screening potential activators or inhibitors, and so on—phosphocasein is entirely suitable.

For the assay of PP1-arch, a sample of protein phosphatase is incubated with [^{32}P]phosphocasein in the presence of a divalent metal ion cofactor and a protein carrier, bovine serum albumin (BSA). Inclusion of the divalent metal ion cofactor is very important. Every PP1/2A homolog characterized to date in both the *Archaea (21,23,24)* and bacteriophage λ *(14)* requires divalent metal ions for activity, as does the bacterial PP2C homolog SpoIIE *(20)*. (Eukaryotic PP1 is a metalloenzyme *(28)*, but it normally binds divalent metal ions in a sufficiently tenacious manner to render the addition of exogenous cofactors unnecessary.) In the author's experience, Mn^{2+} has proven the most efficacious and general cofactor. However, activation by Co^{2+}, Ni^{2+}, or Mg^{2+} has been observed on occasion *(21,23,24)*. The assay is terminated by adding trichloroacetic acid (TCA) and centrifuging. With the assistance of the BSA carrier, the

TCA quantitatively precipitates the unreacted [^{32}P]phosphocasein whereas the inorganic [^{32}P]phosphate that was released by the action of the protein phosphatase remains in the supernatant liquid. An aliquot of the supernatant is then removed and analyzed for ^{32}P content by liquid scintillation counting. (Methods for verifying that the radioactivity detected is derived from inorganic phosphate and not small, TCA-soluble phosphopeptides produced by the action of proteolytic enzymes can be found in **ref. *21***)

Purification of PP1-arch from *S. solfataricus* is a relatively straightforward process involving ion-exchange chromatography, gel filtration chromatography, and absorption onto and elution from hydroxylapatite. As with many prokaryotic organisms, breaking the cells themselves is a much more arduous task than is typical for most animal cells. In the case of *S. solfataricus,* repeated sonication is sufficient, but other organisms may require repeated passage through a French Press or similarly severe methods. Advantage is taken of the fact that *S. solfataricus* releases a soluble, pea-green pigment upon cell rupture. By monitoring the release of pigment at 400 nm after each sonication cycle, the point at which the majority of the cells have been broken open can readily be determined.

The PP1-arch obtained by the procedure described herein is ≈1000-fold purified over the Soluble Extract. Although this preparation falls somewhat short of absolute homogeneity, the major protein species is PP1-arch, which constitutes 40–70% of the total protein present. The unambiguous identification of the PP1-arch polypeptide, and subsequent determination of its relative abundance, was accomplished by assaying its catalytic activity in gel slices following polyacrylamide gel electrophoresis in the presence of sodium dodecyl sulfate (SDS-PAGE) *(22)*. The key to recovering at least a portion of the PP1-arch in an active state following electrophoresis is the selection of a much lower temperature, 65°C vs the usual 100°C, for incubating of the protein with SDS Sample Buffer.

In addition to identifying and characterizing archaeal protein phosphatases by classic purification, sequencing, and cloning techniques *(22)* the gene encoding a second member of the PP1-arch family from *M. thermophila* has been identified using PCR. This was accomplished using primers modeled after regions of PP1-arch from *S. solfataricus,* that are highly conserved with homologous eukaryotic protein phosphatases (*see* **Fig. 1**). Included in these primers are 5' extensions containing nucleotide sequences suitable for annealing the ends of the primers to sites cut by the endonucleases *Eco*RI or *Bam*HI. This permits the direct cloning of the PCR product(s) into a variety of plasmid vectors. The selectivity of PCR amplification is enhanced by using the "touchdown" method *(29)*. The touchdown method is essentially a PCR titration in which the annealing temperature is lowered by one degree every few, usually three, cycles. Under these conditions, the region of DNA that most tightly binds

```
PP1 S. solf.    GDYVDREPQTGVENLSLIL-KKLIESDENKGKTKIVVLRGNHE

PP1 rabbit      GDYVDRGKQS-LETICLLLAYKI-KYPEN-----FFLLRGNHE

PP2A yeast      GDYVDRGYYS-VETVSYLVAMKV-RYPHR-----ITILRGNHE

PP2B rat        GDYVDRGYFS-IECVLYLWALKIL-YPKT-----LFLLRGNHE


Primer 1        5'GGAATTCCGGNGA(T/C)TA(T/C)GTNGA(T/C)(A/C)G 3'
                EcoRI


Primer 2        5'CGGGATCCG(T/C)TC(A/G)TG(A/G)TTNCCNG(T/G)NA 3'
                BamHI
```

Fig. 1. Design of degenerate oligonucleotide primers cloning of PP1-arch homologs by PCR. At top is shown the sequence of amino acids 63–104 of PP1-arch from *S. solfataricus* aligned with the corresponding regions of a rabbit PP1, a yeast PP2A, and a rat PP2B (Adapted from **ref. 22**). The areas of highly conserved amino acid sequence used to design the primers are designated with bold lettering. The conserved GDYVDR sequence was used to design primer 1 and the conserved LRGNHE sequence was used to design primer 2. Below these protein sequences are given the nucleotide sequences of each primer. The underlined portions represent the extensions added to enable primer 1 to anneal to restriction sites for *Eco*RI and primer 2 to anneal to restriction sites for *Bam*HI. Positions where two bases are enclosed in parentheses indicate that both of the indicated nucleotide bases were incorporated at that position in the oligonucleotide sequence, whereas N indicates positions where all four possible nucleotide bases were included.

the primers is amplified first, and, therefore, constitutes the predominant end product because it is amplified through several-fold more cycles than the next best match. If three cycles are performed at each temperature and two sequences differ by 2°C in annealing temperature, the higher annealing product will be amplified $(2)^6$-, or 64-, fold more than the lower annealing product. By scanning through a range of temperatures, the experimentalist gains the selectivity of using the highest possible annealing temperature without the risk of overshooting it completely. It should be noted, however, that PCR is not a panacea. Despite biochemical evidence for the existence of a PP1-arch homolog in the archaeon *H. volcanii* **(23)**, PCR reactions have failed to yield an oligonucleotide product derived from its gene.

2. Materials

2.1. Assay of PP1-Arch

2.1.1. Preparation of [³²P]Phosphoseryl Casein

1. Catalytic subunit of cAMP-dependent protein kinase: 1000 U, from Sigma (St. Louis, MO, cat. no. P 2645).

2. Casein solution: Autoclaved, hydrolyzed, and partially dephosphorylated casein (5% w/v) from bovine milk from Sigma (cat. no. C 4765).
3. ATP, 10 m*M*, pH 7.5.
4. [γ-^{32}P]ATP, 0.8 mCi (*see* **Note 1**).
5. Buffer A: 50 m*M* Tris-HCl, pH 7.0, 1 m*M* dithiothreitol (DTT), 0.1 m*M* EGTA (*see* **Note 2**).
6. Buffer B: 60 m*M* magnesium acetate in Buffer A.
7. Buffer C: 5% (v/v) glycerol in Buffer A.
8. Stop solution: 100 m*M* sodium pyrophosphate, pH 7.0, 100 m*M* EDTA.
9. A 1.0 × 17 cm column of Sephadex G-25 fine (Pharmacia, Uppsala, Sweden) equilibrated in Buffer C (*see* **Note 3**).

2.1.2. Assay of Phosphocasein Phosphatase Activity in Soluble Samples of Protein Phosphatase

1. Buffer D: 50 m*M* MES, pH 6.5.
2. Buffer E: 120 m*M* MnCl$_2$ in Buffer D.
3. Buffer F: 2 mg/mL BSA in Buffer D.
4. TCA, 20% (w/v).

2.1.3. Assay of Phosphocasein Phosphatase Activity in Slices from SDS-Polyacrylamide Gels

1. SDS Sample Buffer: 5% (w/v) SDS, 40% (v/v) glycerol, 0.1% (w/v) bromophenol blue.
2. Buffer D: 50 m*M* MES, pH 6.5.
3. Buffer G: 0.5 m*M* EDTA in Buffer D.
4. Buffer H: 100 m*M* MES, pH 6.5, 0.66 mg/mL BSA, 40 m*M* MnCl$_2$.
5. Buffer I: 100 m*M* MES, pH 6.5, 0.66 mg/mL BSA, 10 m*M* EDTA.

2.2. Purification of PP1-Arch from Sulfolobus Solfataricus

1. Buffer J: 20 m*M* MES, pH 6.5, 100 m*M* NaCl, 1 m*M* EDTA, 1 m*M* EGTA, 1 m*M* DTT, 5 μg/mL leupeptin, 5 μg/mL soybean trypsin inhibitor, 0.5 m*M* phenylmethylsulfonyl fluoride (PMSF), 0.5 m*M* tosyllysyl chloromethylketone (TLCK), 0.5 m*M* tosylphenylalanyl chloromethylketone (TPCK) (*see* **Note 4**).
2. Buffer K: 10 m*M* MES, pH 6.5, 0.5 m*M* EDTA, 0.5 μg/mL leupeptin, 0.2 m*M* PMSF.
3. 150 m*M* NaCl in Buffer K.
4. 400 m*M* NaCl in Buffer K.
5. Buffer L: 1 m*M* sodium phosphate, pH 6.5, 0.5 m*M* EDTA, 0.5 μg/mL leupeptin, 0.2 m*M* PMSF.
6. Buffer M: 400 m*M* sodium phosphate, pH 6.5, 0.5 m*M* EDTA, 0.5 μg/mL leupeptin, 0.2 m*M* PMSF.
7. Buffer N: 20 m*M* MES, pH 6.5, 10 m*M* NaCl, 0.5 m*M* EDTA, 0.5 μg/mL leupeptin, 0.2 m*M* PMSF.
8. A 10 × 4 cm column (*see* **Note 3**) of CM-Trisacryl (Sepracor, Marlborogh, MA) equilibrated in Buffer K.

9. A 6.25 × 30 cm column of DE-52 cellulose (Whatman, Clifton, NJ) equilibrated in Buffer K.
10. A 2.5 × 40 cm column of DE-52 cellulose equilibrated in Buffer K.
11. A 2.5 × 12 cm column of Hydroxylapatite HT (Bio-Rad, Richmond, CA) equilibrated in Buffer L.
12. A 5.0 × 100 cm column of Sephadex G-100 fine (Pharmacia) equilibrated in Buffer M.
13. An FPLC system (Pharmacia) equipped with a 0.5 × 7 cm column of Mono Q that has been equilibrated in Buffer K.

2.3. Cloning of Phosphatase Genes by PCR

1. The enzymes and buffers of the Perkin-Elmer Cetus *GenAmp*™ PCR system were used, although PCR reagents from other commercial sources presumably can be substituted without prejudice to the ultimate results.
2. Oligonucleotide primers 1 and 2 as shown in **Fig. 1**.

3. Methods

3.1. Preparation of [^{32}P]Phosphocasein

1. Combine the following in a 1.5 mL Eppendorff tube: 100 μL of 5% (w/v) casein (*see* **Note 5**), 85 μL of buffer B, 10 μL of 10 mM ATP, and ≈ 0.8 mCi of [γ-^{32}P]ATP. The precise volume of [γ-^{32}P]ATP added will depend on the concentration of the solution as supplied by the manufacturer as well as the age of the preparation, since ^{32}P has a relatively short half-life of 13 d (*see* **Note 6**). Make up the total volume to 325 μL with distilled water.
2. Take a vial containing 1000 U of lyophilized catalytic subunit of the cAMP-dependent protein kinase. Remove the septum cap. Add 87.5 μL of Buffer A. Agitate gently by hand to dissolve the solid. Let stand for a moment to permit the liquid to drain and collect in the bottom. Transfer to the Eppendorff tube from **step 1**.
3. Rinse residual catalytic subunit from its container by adding another 87.5 μL of Buffer A and repeating **step 2**. Securely cap the Eppendorff tube and mix briefly on a Vortex mixer.
4. Incubate for 8–12 h in a 30°C water bath.
5. At the conclusion of the incubation period, add 50 μL of Stop Solution. Mix briefly on a Vortex mixer. You can store at –20°C or proceed immediately with the remaining steps.
6. Remove 5 μL of the incubation mixture and add to 995 μL of distilled water in a 1.5 mL Eppendorff tube. Mix vigorously on a Vortex mixer. Remove three 5 μL portions of the 1:200 diluted incubation mixture, place in individual scintillation vials, then add 1 mL of a water-compatible liquid scintillation fluid such as Eco-Lume (WestChem, Irvine, CA) or Econo-Safe (RPI, Mount Prospect, IL). Measure the radioactivity present in a liquid scintillation counter (*see* **Note 7**). This information is then used to calculate the specific radioactivity of the ATP used to phosphorylate the casein (Assume the cold ATP you added completely accounts for the concentration of total ATP.) Typical specific activities range from 1–3 ×

10^{16} cpm/mole. Please note that it not necessary to try and convert cpm to dpm as long as you use the same scintillation counter and scintillation fluid for all measurements of radioactivity. Under these circumstances, efficiency is a constant that cancels itself in all subsequent calculations of moles of product produced, percent substrate turnover, and so on.

7. Apply the incubation mixture to a 1.0 × 17 cm column of Sephadex G-25 fine that has been equilibrated in Buffer C.

8. Develop the column with Buffer C. Collect 1.0 mL fractions in numbered 1.5 mL Eppendorff tubes.

9. Remove 5 µL aliquots from each fraction, place each in a separate, numbered scintillation vial, add 1.0 mL of scintillation fluid, and count for radioactivity.

10. Graph the radioactivity present in the aliquots as a function of fraction number. Two peaks of [^{32}P]radioactivity should be apparent on the chromatogram. The first peak is the [^{32}P]phosphocasein (*see* **Note 8**) and the second is the unreacted [γ-^{32}P]ATP.

11. Store the two or three peak fractions of [^{32}P]phosphocasein at –20°C. The concentration of casein-bound [^{32}P]phosphate in peak fractions generally ranges from 5–25 µ*M*. Discard the remaining fractions as radioactive waste. Store the column in a shielded location until needed again (*see* **Note 9**).

3.2 Assay of PP1-Arch Activity

1. Thaw a tube of [^{32}P]phosphocasein solution. Mix the contents using a Vortex mixer. Spin briefly in a microcentrifuge to centrifuge the contents into the bottom of the tube. This represents an important precaution designed to minimize the chances of inadvertently contacting radioactive material that might otherwise be clinging to the bottomside of the cap, or scattering it about the lab while opening the tube (*see* **Note 10**). Remove 10% of the volume of [^{32}P]phosphocasein required to perform the planned number of assays, and place in an Eppendorff tube. Return the rest of the [^{32}P]phosphocasein stock to the freezer.

2. For each assay, combine 5 µL of Buffer E and 5 µL of Buffer F in a 1.5 mL Eppendorff tube.

3. Add the protein phosphatase sample to be assayed, plus any additional compounds (activators, inhibitors, and so on) you might wish to test, to the Eppendorff tube. The volume of the sample plus other additions should be ≤ 15 µL. Make up any unutilized portion of this 15 µL volume, if necessary, with Buffer D. Control assays should substitute an equal volume of a suitable buffer in place of the protein phosphatase sample.

4. Initiate the assay by adding 5 µL of [^{32}P]phosphocasein solution, mixing briefly on a Vortex mixer, then place in a 25°C water bath (*see* **Notes 11** and **12**). This quantity of phosphocasein solution generally yields a final concentration of casein-bound [^{32}P]phosphate of 1–4 µ*M*.

5. Terminate reaction, generally after a period of 10–90 min, by adding 100 µL of 20% (w/v) TCA and mixing briefly on a Vortex mixer.

6. Pellet precipitated protein by centrifuging at 12,000*g* for 3 min in a microcentrifuge (*see* **Note 10**).

7. Remove a 50 µL aliquot of the supernatant liquid and determine the amount of [^{32}P]phosphate present by liquid scintillation counting. At the same time, count a 5 µL aliquot of the unused portion of the [^{32}P]phosphocasein stock solution.

8. To calculate the number of moles of casein-bound phosphate hydrolyzed, first subtract the number of cpm of radioactivity present in a control lacking the protein phosphatase from the number of cpm present in the assays where the protein phosphatase was present. This minus-enzyme control accounts for any traces of residual [^{32}P]ATP that may be contaminating the [^{32}P]phosphocasein stock solution as well as any [^{32}P]P$_i$ produced by chemical hydrolysis during storage and/or assay. Next, multiply this difference by 130 µL/50 µL = 2.6 to translate the net cpm of [^{32}P]phosphate present in the 50 µL aliquot into the total number of cpm of [^{32}P]phosphate hydrolyzed in the entire assay, whose final volume after TCA addition is 130 µL. Dividing the total amount of radiophosphate produced, in cpm, by the cpm of radioactive phosphate present in the 5 µL aliquot of the phosphocasein stock solution (assuming the level of any contaminating [^{32}P]ATP and/or [^{32}P]phosphate represents only a few percent or less of the total radioactivity in the [^{32}P]phosphocasein stock) yields the fraction of the total casein-bound [^{32}P]phosphate hydrolyzed. Knowing the molar concentration of casein-bound [^{32}P]phosphate present in the substrate as synthesized, the number of moles of product released can be calculated without the need to perform complex calculations of radioactive decay. In the author's experience, the assay is linear with time up to 30% turnover of casein-bound [^{32}P]phosphate.

3.3. Assay of PP1-Arch Activity Following SDS-PAGE

1. Take samples of Mono Q Fraction, 15-20 µL containing 4–8 µg of protein, and add 5 µL of SDS Sample Buffer. Heat for 5 min at 65°C. (Although in theory this procedure should work with any sample of PP1-arch, regardless of purity, to date it has only been tested on Mono Q fraction.)

2. Apply the samples to parallel lanes of a standard 15% SDS-polyacrylamide gel *(30)* that is 0.1 cm thick and approx 7–8 cm in length. The aim is to create a gel with two halves that are mirror images of each other, one half that will be stained for protein and the other that will be assayed for activity.

3. After electrophoresis, separate the two identical halves of the gel with a razor blade. Take one of the halves and stain for protein as usual.

4. Take the other half and soak it for 30 min in enough Buffer G to completely immerse the gel.

5. Decant the Buffer G, then soak the half gel for 30 min in Buffer D.

6. Decant the liquid. Place the gel on a clean glass plate. Using a clean, sharp razor blade and a clean, clear plastic ruler or other straightedge, excise the individual lane(s) containing the sample(s) to be assayed for phosphocasein phosphatase activity. Next, use the razor blade to divide this gel section lengthwise into 0.2 cm slices, approx 35–40 total.

7. With a clean pair of forceps, place each gel slice in the bottom of its own individually numbered Eppendorff tube. Add 30 µL of Buffer J to each tube and

macerate the gel slices with a plastic pestle. The objective is to expose the greatest possible surface area of the gel to the buffer. If you wish to determine whether any phosphatase activity you might detect is divalent metal ion dependent, take a second, identical lane and repeat **steps 6** and **7**. Substitute Buffer K, which contains EDTA, for Buffer J, which contains Mn^{2+}.

8. Let stand overnight at room temperature.
9. Initiate the assay for protein phosphatase activity by adding 10 μL of [^{32}P]phosphocasein to each tube. Leave the gel slice in the tube during the assay.
10. Incubate 30–120 min in a 25°C water bath, then terminate the reaction by adding 100 μL of 20% (w/v) TCA and work up the assays as described in **steps 5–8** of **Subheading 3.2.**

3.4. Purification of PP1-Arch from Sulfolobus Solfataricus

Steps 1–4 are performed at 4°C. All other operations are performed at room temperature.

1. Thaw frozen cell pellet, containing 100–200 g wet weight of *Sulfolobus solfataricus,* at room temperature until soft, then suspend in 5 vol of Buffer A in a large glass beaker. Disperse the cell paste using a stirring rod or spatula to make a lump-free soup.
2. Break the cells by sonic disruption. Immerse the beaker holding suspended cell paste in ice water to insure efficient cooling. Sonicate for periods of 1 min using the maximum power setting of a sonicator fitted with a large probe. After each sonication cycle, let the mixture sit and cool for at least 1 min while you check progress as described in **step 3**.
3. After each sonication cycle, take an ≈ 1 mL aliquot from the mixture and centrifuge it for 3 min at 12,000g. Measure the absorbance of the supernatant liquid at 400 nm. A plot of absorbance at 400 nm as a function of sonication cycle should resemble a square hyperbola. When two consecutive periods of sonication yield closely comparable values—within 0.1 or 0.2 absorbance units of each other—for absorbance at 400 nm, proceed to the **step 4**. If you are unsure you have reached a plateau, conduct one more cycle of sonic disruption and check the absorbance to be sure. Complete cell rupture generally requires 6–8 periods of sonic disruption and yields a supernatant fraction with a final absorbance in the range of 1.0–1.6.
4. Centrifuge the entire sonication mixture at 12,000g for 30 min. Save the supernatant fraction as the Soluble Extract. The Soluble Extract may be stored for several months at –20°C without significant loss of protein phosphatase activity.
5. Pass the Soluble Extract through a 10 × 4 cm precolumn of CM-Trisacryl and load onto a 6.25 × 30 cm column of DE-52 Cellulose, both of which have been equilibrated in Buffer K. Simply place the two columns in series, the CM-Trisacryl column above and feeding into the DE-52 column.
6. After loading is completed, remove the CM-Trisacryl precolumn and wash the DE-52 column with Buffer K until little or no protein is detectable in the eluate. This generally requires 500–1000 mL of Buffer K.

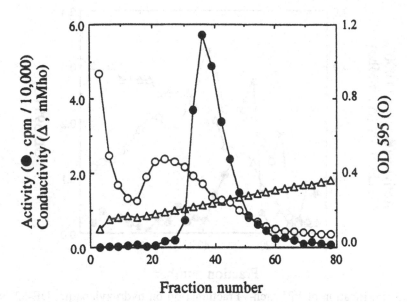

Fig. 2. Purification of PP1-arch—Ion-exchange chromatography on DE-52 cellulose using a linear salt gradient. DE-52 Fraction I was dialyzed, and applied to a second column of DE-52 cellulose, which was then washed and developed with a linear salt gradient as described in the text (*see* **Subheading 3.4., steps 9–11**). This figure shows the results of the gradient elution. Shown are the relative protein phosphatase activity as measured in 20 μL aliquots of selected fractions (●), the relative amount of protein in 20 μL aliquots of selected fractions as determined using the Bradford protein assay (O), and the salt gradient (Δ). Fractions 30–54 were then pooled as DE-52 Fraction II.

7. Wash the DE-52 column with Buffer K containing 150 m*M* NaCl until little or no protein can be detected in the eluate. This generally requires 1500–2000 mL of Buffer K containing 150 m*M* NaCl.
8. Discard both washes. Elute PP1-arch from the column by applying Buffer K containing 400 m*M* NaCl. Pool that portion of the high salt batch eluate that contains detectable levels of protein, generally about 1 *l* in volume, as DE-52 Fraction I.
9. Dialyze DE-52 Fraction I vs Buffer K. Apply the dialyzed material to a 2.5 × 40 cm column of DE-52 Cellulose that has been equilibrated in Buffer K.
10. Wash the column with 250 mL of Buffer K containing 150 m*M* NaCl.
11. Develop the column with an linear salt gradient consisting of 400 mL of 150 m*M* NaCl in Buffer K and 400 mL of 400 m*M* NaCl in Buffer K. Collect fractions, 10 mL, and assay for the presence of protein and protein phosphatase activity (*see* **Subheading 3.2.**). A single peak of protein phosphatase activity that elutes near the midpoint of the gradient should be detected. Pool the most active fractions as DE-52 Fraction II (*see* **Fig. 2**).
12. Dialyze DE-52 Fraction II vs Buffer L, then apply to a 2.5 × 12 cm column of hydroxylapatite HT that has been equilibrated in Buffer L.

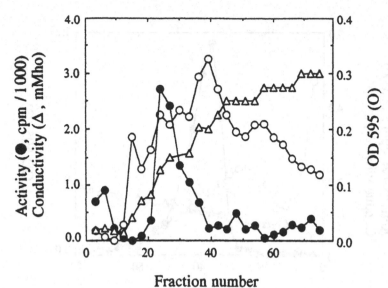

Fig. 3. Purification of PP1-arch—Fractionation on hydroxylapatite. DE-52 Fraction II (*see* **Fig. 2**) was dialyzed and applied to a column of hydroxylapatite, which was then washed and eluted with a linear gradient of sodium phosphate as described in the text (*see* **Subheading 3.4., steps 12–15**). This figure summarizes the results of the gradient elution. Shown are the relative protein phosphatase activity as measured in 10-μL aliquots of selected fractions (●), the relative amount of protein in 100-μL aliquots of selected fractions as determined using the Bradford protein assay (O), and the sodium phosphate gradient (Δ). Fractions 23–33 were pooled as the Hydroxylapatite Fraction.

13. Wash the column with 50 mL of Buffer L, then develop with a linear gradient consisting of 400 mL of Buffer L and 400 mL of Buffer M.
14. Fractions, 10 mL, are collected and assayed for protein and protein phosphatase activity (*see* **Subheading 3.2.**). Although the inorganic phosphate in Buffers L and M inhibits PP1-arch, sufficient residual activity can be detected to permit the identification of active fractions (*see* **Fig. 3**).
15. Pool the active fractions as the Hydroxylapatite Fraction and concentrate to a volume of approx 2 mL via centrifugal ultrafiltration at 3000g using a Centriprep 10 (Amicon, Danvers, MA).
16. Apply the concentrated Hydroxylapatite Fraction to a 5.0 × 100 cm column of Sephadex G-100 that has been equilibrated in Buffer M. Develop the column with this same buffer. Collect 2 ml fractions and assay for protein and protein phosphatase activity (*see* **Subheading 3.2.**). A single, somewhat broad peak of activity should be observed (*see* **Fig. 4**). Pool active fractions as the G-100 Fraction.
17. Apply the G-100 Fraction, using an FPLC system (Pharmacia) and following the manufacturer's instructions, to a 0.5 × 7 cm column of Mono Q that has been equilibrated in Buffer K.

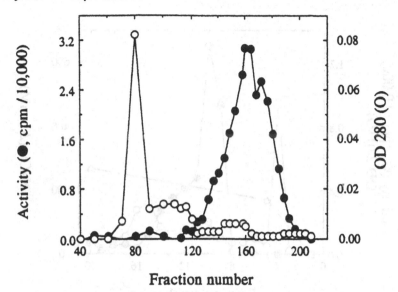

Fig. 4. Purification of PP1-arch—Gel filtration chromatography on Sephadex G-100. Hydroxylapatite fraction (*see* **Fig. 3**) was concentrated, then applied to and eluted from a column of Sephadex G-100 as described in the text (*see* **Subheading 3.4., step 16**). This figure summarizes the results of the elution. Shown are the relative protein phosphatase activity as measured in 10 μL aliquots of selected fractions (●) and the concentration of protein as measured by OD at 280 nm (O). Fractions 136–184 were pooled to give the G-100 Fraction.

18. Wash the column with 5 mL of Buffer K, then develop with a two-stage gradient. The first stage is a linear gradient consisting of 1 mL of Buffer K and 1 mL of Buffer K containing 170 mM NaCl. The second stage is a linear gradient consisting of 17.5 mL of Buffer K containing 170 mM NaCl and 17.5 mL of Buffer K containing 270 mM NaCl. The flow rate should be 1.0 mL/min throughout. Collect 1 mL fractions and assay for protein and protein phosphatase activity (*see* **Subheading 3.2.**). PP1-arch elutes as a single peak during the second stage of the gradient (*see* **Fig. 5**). Active fractions are pooled to give the Mono Q Fraction, which was purified roughly 1000-fold from the Soluble Extract and contains 40–70% of its total protein as PP1-arch (*see* **Table 1**).

3.5. Cloning of PP1-Arch
from Methanosarcina thermophila by PCR

1. Combine the following in a PCR tube: 10 μL PCR buffer II; 2 μL each of dATP, dCTP, dGTP, dTTP solutions; 5 μL each of 20 μM oligonucleotide primers 1 and 2 (*see* **Fig. 1**); 0.5 μL of *AmpliTaq®* DNA polymerase; 16 μL of 25 mM MgCl$_2$; and 0.25, 0.5, or 1.0 μg of DNA (The author generally tries all three concentrations in parallel reactions).

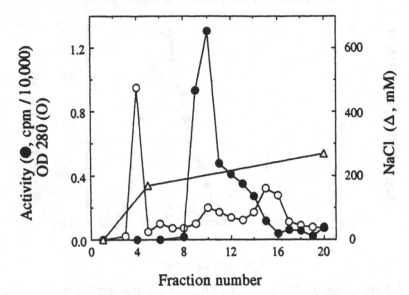

Fig. 5. Purification of PP1-arch—Ion-exchange chromatography on Mono Q. G-100 Fraction (*see* **Fig. 4**) was applied to Mono Q FPLC column, washed, and developed with a two-stage salt gradient as described in the text (*see* **Subheading 3.4., steps 17 and 18**). This figure summarizes the results of the gradient elution. Shown are the relative protein phosphatase activity as measured in 10 μL aliquots of selected fractions (●), the concentration of protein as measured by OD at 280 nm (O), and the salt gradient (△). Fractions 9–13 were then pooled as the Mono Q Fraction.

2. Program the thermal cycler with the temperature cycling protocol described below, then execute. Begin with three cycles each consisting of a 1-min denaturation at 94°C, followed by annealing for 1 min at 55°C, followed by extension for 1 min at 72°C. Follow the first three cycles with 10 more sets of 3 in which the annealing temperature is lowered by one degree each set, covering the range of 55–45°C.
3. Visualize the product by horizontal electrophoresis in agarose and clone the PCR product(s) into a pUC or other plasmid vector cut with *Eco*RI or *Bam*HI for subsequent sequencing, and so on, using standard procedures.

4. Notes

1. Any commercial preparation of [γ-^{32}P]ATP will suffice. The precise specific radioactivity is unimportant since the radiolabeled stock will be diluted into a large excess of unlabeled (i.e., cold) ATP.
2. DTT is included as a protectant against oxidative damage to proteins. Only the reduced form of DTT has protective qualities, and DTT in free solution will be completely oxidized by molecular oxygen within a few hours to days, depending upon the temperature. Therefore, DTT always should be added to buffers on the day of use.

Table 1
Summary of the Purification
of PP1-Arch from 200 g of Sulfolobus Solfataricus

Fraction	Protein	Activity	Specific act.	Recovery	Enrichment
Soluble Extract	1660 mg	380 nmol/min	0.2 nmol/min mg	100%	one-fold
DE-52 fraction I	290	580	2.0	152	10
DE-52 fraction II	60	304	5.1	80	26
Hydroxylapatite	7.5	nd[a]	nd	nd	nd
G-100 fraction	1.9	65	34.2	17	171
Mono Q fraction	0.5	62	206	17	1033

[a]nd, not determined due to inhibition by phosphate in the buffers used for this step.

3. The column dimensions listed reflect, in part, the choice of columns available in the author's laboratory. These dimensions are intended to serve as a guide to aid in the selection of suitable columns possessing similar, but not necessarily identical, dimensions.

4. PMSF, TLCK, and TPCK are slowly hydrolyzed to inactive products in water. Therefore, these compounds should be added to buffers fresh on the day of use. In addition, PMSF and TPCK both possess limited solubility in water. Therefore, they first are dissolved in an anhydrous alcohol such as isopropanol to a concentration of 20 mM, then an aliquot of this alcohol stock is added to the aqueous buffer with vigorous stirring. A blender works very nicely to provide quick and vigorous agitation. Even so, you will likely notice a few grains of undissolved material remaining. They can generally be ignored without compromising the quality of the preparation.

5. The casein solution comes in a bottle with a septum cap. Because this nutrient-rich solution makes an excellent media for the growth of bacteria, mold, and so forth, it is important to leave the cap and its seal intact. Remove approx 200 μL of the solution using a sterile 1 cc disposable tuberculin syringe with needle. Empty the contents of the syringe into an Eppendorff tube until needed. After removing the 100 μL portion needed for making phosphocasein, discard the remainder of the contents of the Eppendorff tube.

6. Since the final specific radioactivity of the [γ-^{32}P]ATP will be determined experimentally, it is unnecessary to engage in detailed calculations of radioactive decay. A quick way to estimate the specific radioactivity of a sample of ^{32}P on a particular day is to multiply the original figure by a factor of 0.5 for every 2 wk that have passed since the reference or calibration date, by a factor of 0.7 for any remaining week, and 0.84 for a half week. If a stock solution is 5 1/2 wk old, the specific radioactivity will be 0.5 × 0.5 × 0.7 × 0.84 ≈ 15% that listed for the reference date.

7. The upper limit of the dynamic range of most liquid scintillation counters is roughly 1–5 × 10^6 cpm. A 5 μL aliquot of the raw incubation mixture contains levels of ^{32}P approaching or exceeding this limit. Therefore, a dilution step is used to reduce the amount of radioactivity present in an accurately measurable volume of sample to a level that the scintillation counter can handle.

8. If there is any confusion as to which fractions contain [^{32}P]phosphocasein, take a 5–10 μL aliquot from each fraction, place each in a disposable plastic tube, and add Bradford protein assay reagent (31). In the author's laboratory 1.0 mL of Commassie Protein Reagent from Pierce (cat. #23200) is used. The casein will turn the color of the reagent from rust brown to blue. Examine the relative intensity of the color in each tube by eye through a plexiglass shield (Do not risk contaminating your lab's spectrophotometer) in order to determine which fractions contain the highest concentrations of casein.

9. Do **not** try to decontaminate the column after use. The short half-life of ^{32}P dictates that you will prepare [^{32}P]phosphocasein on a regular basis. Therefore, trying to scrub the column free of every count of radioactive contamination

represents an unnecessary hazard to laboratory personnel. Simply flush it with a couple of column volumes of Buffer C to remove most of the "mobile" radioactivity. Then seal the column (it is best to have valves on both ends), put it in a plastic bag, and stand the bagged column in a reasonably upright position in a 4 L plastic beaker. Store in a safe, authorized location until needed again.

10. The preparation of [^{32}P]phosphocasein and the assay of protein phosphatase activity require the frequent use of a microcentrifuge with radioactive materials. Thus, the potential for radioactively contaminating this device is quite high. If your lab possesses more than one microcentrifuge, it is best to confine work with radioactive samples exclusively to one of them. Instruments whose lids are equipped with a plastic gasket and whose rotors possess a screw-on lid are highly recommended. If at all possible, use disposable plastic liners to simplify the task of decontamination. Coating the inside of the rotor with a fine layer of mineral oil can be helpful, since phosphate-containing compounds tend to bind quite tenaciously to the metal rotors found in most microcentrifuges. The oil film traps droplets containing the radioactive contaminants before they directly contact the rotor, allowing them to be washed away with the oil. Since the centrifugal forces generated by the instrument will eventually drive the oil to the bottom of the rotor, re-exposing the bare metal, regular renewal of the protective coating is necessary.

11. When working with small volumes in an Eppendorff tube, it is preferable to mix the contents on the Vortex mixer without securing the cap. Simply deposit all of the material to be added at points within the lower, conical portion of the tube. Hold the tube by the stem joining the tube to its cap and then lightly touch the lower side portion of the tube to the side of the plastic head of a running vortex mixer. As soon as you see the contents begin to swirl together under this agitation, remove the tube and place it in the water bath. The opening and closing of the caps on Eppendorff tubes often leads to torn gloves and the potential for direct contact of radioactive materials with fingers or hands. Also, opening a tightly sealed cap often results in the newly opened tube flipping out of your hands and onto the bench, the floor, or worse. To minimize the risk of such incidents, tubes should be capped only when necessary, i.e., when they contain greater than 100 µL of material.

12. PP1-arch from *S. solfataricus* is stable for 30 min or more at temperatures as high as 80°C. Its phosphocasein phosphatase activity has been assayed at temperatures as high as 45°C with a concomitant thermodynamic enhancement of the rate of reaction. Although it is theoretically possible to assay the enzyme at even higher temperatures, the casein and other eukaryotic phosphoproteins we use as substrates generally denature at 50–60°C.

References

1. Woese, C. R. (1994) There must be a prokaryote somewhere: microbiology's search for itself. *Microbiol. Rev.* **58**, 1–9.
2. Kennelly, P. J. and Potts, M. (1994) Protein phosphatases in prokaryotes: reflections of the past. Windows to the future? *Adv. Prot. Phosphatases* **8**, 53–68.

3. Kennelly, P. J. and Potts, M. (1996) Fancy meeting you here! A fresh look at 'prokaryotic' protein phosphorylation. *J. Bacteriol.* **178**, 4759–4764.

4. Stanier, R. Y. and van Niel, C. B. (1962) The concept of a bacterium. *Arch. Microbiol.* **42**, 17–35.

5. Chatton, E. (1937) Titres et travoux scientifiques. Sete, Sottano, Italy.

6. Olsen, G. J. and Woese, C. R. (1993) Ribosomal RNA: a key to phylogeny. *FASEB J.* **7**, 113–123.

7. Keeling, P. J., Charlebois, R. L., and Doolittle, W. F. (1994) Archaebacterial genomes: eubacterial form and eukaryotic content. *Curr. Opin. Genet. Dev.* **4**, 816–822.

8. Garnak, M. and Reeves, H. C. (1978) Phosphorylation of isocitrate dehydrogenase of *Escherichia coli*. *Science* **203**, 1111,1112.

9. Wang, J. Y. J. and Koshland, D. E., Jr. (1978) Evidence for protein kinase activities in the prokaryote *Salmonella typhimurium*. *J. Biol. Chem.* **253**, 7605–7608.

10. Manai, M. and Cozzone, A. J. (1979) Analysis of the protein kinase activity of *Escherichia coli* cells. *Biochem. Biophys. Res. Commun.* **91**, 819–826.

11. LaPorte, D. C. and Koshland, D. E., Jr. (1982) A protein with kinase and phosphatase activities involved in regulation of the tricarboxylic acid cycle. *Nature* **300**, 458–460.

12. Klumpp, D. J., Plank, D. W., Bowdin, L. J., Stueland, C. S., Chung, T., and LaPorte, D. C. (1988) Nucleotide sequence of *aceK,* the gene encoding isocitrate dehydrogenase kinase/phosphatase. *J. Bacteriol.* **170**, 2763–2769.

13. Cohen, P. T. W., Collins, J. F., Coulson, A. F. W., Berndt, N., and da Cruz e Silva, O. B. (1988) Segments of bacteriophage λ (orf 221) and ɸ80 are homologous to genes encoding for mammalian protein phosphatases. *Gene* **69**, 131–134.

14. Cohen, P. T. W. and Cohen, P. (1989) Discovery of a protein phosphatase activity encoded in the genome of bacteriophage λ. Probable identity with open reading frame 221. *Biochem. J.* **260**, 931–934.

15. Barton, G. J., Cohen, P. T. W., and Barford, D. (1994) Conservation analysis and structure prediction of the protein serine / threonine phosphatases. Sequence similarity with diadenosine tetraphosphatase from *Escherichia coli* suggests homology to the protein phosphatases. *Eur. J. Biochem.* **220**, 225–237.

16. Potts, M., Sun, H., Mockaitis, K., Kennelly, P. J., Reed, D., and Tonks, N. K. (1993) A protein-serine/tyrosine phosphatase encoded by the genome of the cyanobacterium *Nostoc commune* UTEX 584. *J. Biol. Chem.* **268**, 7632–7635.

17. Howell., L. D., Griffiths, C., Slade, L. W., Potts, M., and Kennelly, P. J. (1996) Substrate specificity of IphP, a cyanobacterial dual-specificity protein phosphatase with MAP kinase phosphatase activity. *Biochemistry* **35**, 7566–7572.

18. Guan, K. and Dixon, J. E. (1991) Evidence for protein-tyrosine-phosphatase catalysis proceeding via a cysteine-phosphate intermediate. *J. Biol. Chem.* **266**, 17,026–17,030.

19. Bork, P., Brown, N. P., Hegyi, H., and Schultz, G. (1996) The protein phosphatase 2C (PP2C) superfamily: detection of bacterial homologs. *Protein Sci.* **5**, 1421–1425.

20. Duncan, L., Alper, S., Argioni, F., Losick, R., and Stragier, P. (1995) Activation of cell-specific transcription by a serine phosphatase at the site of asymmetric division. *Science* **270,** 641–644.
21. Kennelly, P. J., Oxenrider, K. A., Leng, J., Cantwell, J. S., and N. Zhao, N. (1993) Identification of a serine/threonine-specific protein phosphatase from the archaebacterium *Sulfolobus solfataricus. J. Biol. Chem.* **268,** 6505–6510.
22. Leng, J., Cameron, A. J., Buckel, S., and Kennelly, P. J. (1995) Isolation and cloning of a protein-serine/threonine phosphatase from an archaeon. *J. Bacteriol.* **177,** 2763–2769.
23. Oxenrider, K. A. and Kennelly, P. J. (1993) A protein-serine phosphatase from the halophilic archaeon *Haloferax volcanii. Biochem. Biophys. Res. Commun.* **194,** 1330–1335.
24. Oxenrider, K. A., Rasche, M. E., Thorsteinsson, M. V., and Kennelly, P. J. (1993) An okadaic acid-sensitive protein phosphatase from the archaeon *Methanosarcina thermophila* TM–1. *FEBS Lett.* **331,** 291–295.
25. Ingebritsen, T. S. and Cohen, P. (1983) The protein phosphatases involved in cellular regulation 1. Classification and substrate specificities. *Eur. J. Biochem.* **132,** 255–261.
26. Cohen, P. (1991) Classification of protein-serine/threonine phosphatases: Identification and quantification in cell extracts. *Meth. Enzymol.* **201,** 389–398.
27. McGowan, C. H. and Cohen, P. (1988) Protein phosphatase-2C from rabbit skeletal muscle and liver: an Mg^{2+}-dependent enzyme. *Meth. Enzymol.* **159,** 416–426.
28. Chu, Y., Lee, E. Y. C., and Schlender, K. K. (1996) Activation of protein phosphatase 1. Formation of a metalloenzyme. *J. Biol. Chem.* **271,** 2574–2577.
29. Roux, K. H. (1994) Using mismatched primer-template pairs in touchdown PCR. *BioTechniques* **16,** 812–814.
30. Laemmli, U. K. (1970) Cleavage of structural proteins during the assembly of bacteriophage T4. *Nature* **227,** 680–685.
31. Bradford, M. M. (1976) A rapid and sensitive method for the quantitation of microgram quantities of protein utilizing the principle of protein-dye binding. *Anal. Biochem.* **72,** 248–254.

2

Protein Phosphatase Type 1 and Type 2A Assays

S. Derek Killilea, Qi Cheng, and Zhi-Xin Wang

1. Introduction

Protein phosphatases type 1 (PP1) and type 2A (PP2A) are the only activities known in mammalian tissues to dephosphorylate glycogen phosphorylase *a*. Phosphorylase was the first enzyme demonstrated to undergo regulation of catalytic activity via reversible covalent modification involving phosphorylation. It exists in a dephosphorylated, *b* form, which is catalytically active only in the presence of its allosteric activator, AMP, and a phosphorylated, *a* form, which is catalytically active in the absence of AMP. The conversion of phosphorylase *b* to *a* is catalyzed by phosphorylase kinase, a process that results in the phosphorylation of a single serine residue (Ser-14) in each identical subunit of the native dimer. This is one of the advantages of using phosphorylase *a* as a protein phosphatase substrate to study PP1 and PP2A activities in mammalian and nonmammalian eucaryotic cells. Other protein substrates, such as glycogen synthase and phosphorylase kinase, contain multiple phosphorylation sites, the reversible covalent modification of which are not always correlated with changes in enzymic activity. Other advantages include the commercial availability of both phosphorylase and phosphorylase kinase. Phosphorylase *b* can also be conveniently isolated in gram quantities from 1 kg of either fresh or frozen rabbit skeletal muscle by a procedure that can be completed within a week and does not involve any chromatography steps *(1,2)*. Phosphorylase kinase can be isolated within a 24 h period from fresh rabbit skeletal muscle *(3)*.

Several procedures are presented below. Method A describes the preparation of ^{32}P-phosphorylase, and by altering the form of ATP used, unlabeled phosphorylase *a* or thiophosphorylase preparations are also prepared by this procedure. The latter is resistant, but not immune to phosphatase action *(4,5)*.

From: *Methods in Molecular Biology, Vol. 93: Protein Phosphatase Protocols*
Edited by: J. W. Ludlow © Humana Press Inc., Totowa, NJ

Several forms of PP1 and PP2A exist and some require either activation and/or Mn^{2+} for activity. One of these is $PP1_i$, which is inactive and can be activated by glycogen synthase kinase-3, a process that is ionic strength sensitive *(6)*. A more convenient procedure to activate $PP1_i$ involves pretreatment with Mn^{2+} and trypsin (Mn/trypsin), a process that is also ionic strength sensitive *(7)*. This procedure is presented in Method B. Native forms of PP2A are activated by protamine, a process that is also ionic strength dependent and yields maximal activity at a 1:1 molar ratio of phosphorylase *a* monomer:protamine *(8)*. Catalytic subunits of PP1, recombinant PP1, and PP2A are partial or completely dependent on Mn^{2+} for activity *(9)*. References to assay conditions for these different activities are given in Methods C, D, and E.

Methods C and D detail two fixed time point assays for PP1 and PP2A activities. Under the standard conditions of these assays (10 μM phosphorylase) the reactions are linear up to the utilization of 30% of the substrate.

In Method C, protein phosphatase (PPase) activity is determined from the loss in AMP-independent catalytic activity when phosphorylase *a* is converted to phosphorylase *b* *(10)*.

$$\text{Phosphorylase } a + H_2O \xrightarrow{\text{PPase}} \text{Phosphorylase } b + P_i$$

The assay involves two steps. First the phosphatase is incubated with phosphorylase *a* for a fixed period of time. Then the phosphatase activity is terminated by dilution of the assay mixture with a buffer that contains NaF, a phosphatase inhibitor. Samples are then analyzed for the phosphorylase *a* activity *(11)* remaining after the action of the phosphatase. By comparing this activity with that of the starting phosphorylase *a* activity, the fraction of the phosphorylase *a* converted to the *b* form can be determined.

In Method D, phosphatase activity is determined from the release of ^{32}P-inorganic phosphate from ^{32}P-phosphorylase.

$$^{32}P\text{-Phosphorylase} + H_2O \xrightarrow{\text{PPase}} \text{Phosphorylase } b + ^{32}P_i$$

This is the most commonly used and most sensitive of the PP1 and PP2A assays. This assay involves incubation of the phosphatase with ^{32}P-phosphorylase for a fixed period of time. The reaction is terminated by the addition of trichloroacetic acid (TCA) and the precipitated protein is separated from the released ^{32}Pi by centrifugation. The ^{32}Pi is determined by scintillation counting.

Method E is a recently introduced assay system *(9)* that allows for the continuous determination of phosphatase activity. As such it is very useful for kinetic studies. The assay incorporates a coupled assay system in which purine nucleoside phosphorylase uses the inorganic phosphate, released by phosphatase action on phosphorylase a, to convert 7-methyl-6-thioguanosine (MTGuo) to 7-methyl-6-thioguanine (MTGua), a process that results in the increase in absorbance at 360 nm *(12)*.

$$\text{Phosphorylase } a + H_2O \xrightarrow{\text{PPase}} \text{Phosphorylase } b + P_i$$

$$Pi + MTGuo \xrightarrow{\text{PNPase}} MTGua$$

The assay requires the presence of at least 0. 1 M NaCl to maintain phosphorylase a in solution particularly when high levels of this substrate are used. This assay was validated at pH 7. 0 only. This assay system can also be used to study the action of phosphatases on thiophosphorylated substrates *(5)*.

Method F details an assay procedure that can be used to study the action of phosphatases on TCA-soluble substrates such as phosphorylated peptides *(13)*. However, this procedure is also very useful to confirm, when using Method D, that the TCA-soluble material is ^{32}Pi and not acid-soluble ^{32}P-peptide material released from ^{32}P-phosphorylase by proteolytic activity. This is an important control to carry out when studying a new putative protein phosphatase activity. In this procedure, released inorganic phosphate, but not the phosphorylated peptide, is extracted as the phosphomolybdate complex into an organic solvent mixture and quantified by scintillation counting.

2. Materials

2.1. Method A

1. Buffer A:50 mM β-glycerophosphate, pH 7.0, 1 mM dithiothreitol.
2. Buffer B: 0.125 M Trizma base/0.125 M β-glycerophosphate, pH 8.6, 1 mM CaCl$_2$, 0.1 M NaF.
3. Buffer C: 10 mM Tris-HCl, pH 7.0, 1 mM dithiothreitol.
4. Buffer D: 50 mM Bis-Tris, pH 7.0, 5 mM caffeine, 1 mM dithiothreitol, 50% glycerol.
5. Buffer E: 50 mM Tris-HCl, pH 7.0.
6. 1 M Mg acetate.
7. 100 mM [γ-^{32}P]ATP, pH 7.0 (200–500 cpm/pmol) (*see* **Note 1**).
8. Phosphorylase b:100 mg in 3 mL of buffer A (*see* **Note 2**).
9. Phosphorylase kinase: 200 U (Sigma, St. Louis, MO) in 100 µL of buffer A (*see* **Note 2**).
10. Saturated ammonium sulfate (neutralized by the addition of 0.6 mL NH$_4$OH/L).
11. Acid washed active carbon: suspend 0.2 g of Norite A in 2 mL of Buffer A in a small test tube. Centrifuge in a clinical centrifuge for 2 min at low speed to sediment the carbon. Gently add Buffer A to the contents of the tube to allow carbon, floating on the surface of the buffer, to be washed out of the tube.
12. Acrylic or other safety shield, gloves, polyethylene transfer pipets,

2.2. Method B

1. Buffer A: 50 mM Bis-Tris, pH 7.0, 1 mM dithiothreitol.
2. 1 mM MnCl$_2$, 0.5 M NaCl in Buffer A.
3. Trypsin (TPCK-treated): 0.5 mg/mL in 1 mM HCl.
4. Trypsin inhibitor (soybean): 2 mg/mL in buffer A.
5. PP1$_i$ diluted into buffer A.

2.3. Method C

1. Assay buffer: 50 mM Bis-Tris, pH 7.0, 5 mM caffeine, 2 mM dithiothreitol, 1 mg/mL of bovine serum albumin. In assays for Mn/trypsin-activated PP1$_i$ and the catalytic subunits of PP1 and PP2A, 0.5 mM MnCl$_2$ is also included. 0.2 M NaCl is included for PP2A assays (*see* **Note 3**).
2. Protamine: 25 μM in assay buffer containing 0.2 M NaCl. Protamine chloride is used and the molecular weight of protamine is taken as 4,245 *(14)*, which is based on amino acid sequence data (*see* **Note 4**).
3. Phosphorylase *a*: 12.5 μM (1.25 mg/mL) in assay buffer (diluted from stock phosphorylase *a* stored at −20°C and prepared as by Method A using unlabeled ATP. For PP2A assays, dilute the phosphorylase *a* stock to 25 μM (2.5 mg/mL) and add an equal volume of the 25 μM protamine preparation (*see* **Note 5**).
4. Protein phosphatase.
5. Stop buffer: 50 mM imidazole chloride, pH 6.5, 0. 1 M NaF, 0.5 mM dithiothreitol, 0.5 mM EDTA, 1 mg/mL of bovine serum albumin.
6. Phosphorylase *a* substrate: 50 mM imidazole chloride, pH 6.5, 0.15 M glucose 1-phosphate, 2% glycogen.
7. 0.072 M H$_2$SO$_4$.
8. Color Reagent: 1% ammonium molybdate/4% ferrous sulfate in 1 N H$_2$SO$_4$. (This solution is prepared fresh daily using a stock 1% ammonium molybdate in 1 N H$_2$SO$_4$ solution.)

2.4. Method D

1. Assay buffer: 50 mM Bis-Tris, pH 7.0, 5 mM caffeine, 2 mM dithiothreitol, 1 mg/mL of bovine serum albumin. In assays for Mn/trypsin-activated PP1$_i$ and the catalytic subunits of PP1 and PP2A, 0.5 mM MnCl$_2$ is also included. 0.2 M NaCl is included for PP2A assays (*see* **Note 3**).
2. Protamine: 25 μM in assay buffer containing 0.2 M NaCl. Protamine chloride is used and the molecular weight of protamine is taken as 4,245 *(14)*, which is based on amino acid sequence data (*see* **Note 4**).
3. Phosphorylase *a*: 12.5 μM (1.25 mg/mL) in assay buffer (diluted from the stock ^{32}P-phosphorylase preparation stored at −20°C—*see* **Method A**). For PP2A assays, dilute the ^{32}P-phosphorylase stock to 25 μM (2.5 mg/mL) and add an equal volume of the 25 μM protamine preparation (*see* **Note 5**).
4. Protein phosphatase.
5. 10% TCA.

2.5. For Method E

1. Assay buffer: 50 mM Bis-Tris, pH 7.0, 5 mM caffeine, 2 mM dithiothreitol, 1 mg/mL of bovine serum albumin. In assays for Mn/trypsin-activated PP1$_i$ and the catalytic subunits of PP1 and PP2A, 0.5 mM MnCl$_2$ and 0.1 M NaCl are also included. 0.2 M NaCl is included for PP2A assays (*see* **Note 3**).
2. Phosphorylase a:100 µM (10 mg/mL) in assay buffer (diluted from the stock phosphorylase a solution stored −20°C and prepared as by Method A using unlabeled ATP).
3. Purine nucleoside phosphorylase (bacterial, Sigma): 1 mg/mL in assay buffer.
4. Protein phosphatase.
5. 7-Methyl-6-thioguanosine (MTGuo): 7.2 mM in dimethyl sulfoxide (DMSO) prepared daily. The concentration is determined on a 3 µL aliquot, diluted to 1 mL in 50 mM Tris-HCl, pH 7.0, at 331 nm using the molar extinction coefficient of 32,000 m^{-1} cm^{-1} *(12)*. MTGuo is not available commercially, but is readily synthesized by a modification of the procedure of ref. *12* as follows. Dissolve 2-amino-6-chloro-purine ribonucleoside (2 g) in dry dimethylformamide (10 mL). Add methyl iodide (4 mL) and place the mixture in the dark for 14 h at room temperature. Add benzene (5 mL) and pass nitrogen gas over the surface of the solution while warming at about 40°C to remove excess methyl iodide. The precipitate that occurs on addition of the benzene goes back into solution as the benzene and methyl iodide are removed. Add thiourea (2 g) and gently stir for 30 min at room temperature. Then add methanolic ammonia to bring to pH 7.0–7.5 (determined with pH indicator paper) and pour the mixture into stirred acetone (125 mL). After 5 min, the resulting light yellow precipitate is collected on a filter by suction, washed with acetone, dried in a vacuum desiccator at room temperature, and stored desiccated at −70°C. The MTGuo, thus prepared, is judged at least 60% pure by silica gel TLC using the solvent ethyl acetate/1-propanol/water (5:2:1) and can be used without further purification (*see* **Note 6**).
6. Protamine: 100 µM in assay buffer containing 0.2 M NaCl. Protamine chloride is used and the molecular weight of protamine is taken as 4,245 *(14)*, which is based on amino acid sequence data (*see* **Note 4**).

2.6. For Method F

1. **Items 1–4** used in Method D.
2. Reagent A:20 mM silicotungstic acid/1 mM sodium phosphate/1 N H$_2$SO$_4$.
3. Reagent B: 7.5% ammonium molybdate.
4. Isobutanol-benzene (1:1 [v/v]).

3. Methods

3.1. Method A

3.1.1. Preparation of ^{32}P-Phosphorylase

1. In a 50-mL centrifuge tube add together buffer B (820 µL); Mg acetate (40 µL); phosphorylase kinase (100 µL); phosphorylase (3 mL), and [γ-^{32}P]ATP (40 µL).

2. Incubate at 30°C for 1 h.
3. Add 4 mL of saturated ammonium sulfate and incubate at room temperature for 20 min to allow protein to precipitate.
4. Centrifuge at 14,000g for 20 min at 4°C.
5. Redissolve the protein pellet in 8 mL of Buffer A.
6. Add 0.2 g of Norite suspended in 2 mL of buffer A and mix.
7. Centrifuge at 14,000g for 5 min.
8. Carefully transfer the supernatant to a fresh 50 mL centrifuge tube.
9. Add 10 mL of saturated ammonium sulfate and incubate at room temperature as before prior to centrifugation at 14,000g for 20 min.
10. Redissolve the protein precipitate in 5 mL of buffer C and dialyze against 4 L of Buffer C overnight at 4°C to crystallize the ^{32}P-phosphorylase.
11. Carefully transfer the contents of the dialysis bag to a 50 mL centrifuge tube placed in ice.
12. Centrifuge at 14,000g for 10 min at 4°C.
13. Dissolve the ^{32}P-phosphorylase crystals in 4 mL of Buffer D using a glass rod. This may require incubation at 30°C.
14. Centrifuge at 14,000g for 20 min at 25°C to remove denatured protein and remaining Norite A.
15. Dilute 10 μL of the redissolved protein into 990 μL of Buffer E and prepare a blank by diluting 10 μL of Buffer D into 990 μL of Buffer E. Determine the absorbance at 280 nm and calculate the phosphorylase concentration. The A_{280} for 1 mg/mL solution of phosphorylase is 1.31.
16. Carefully transfer the ^{32}P-phosphorylase to a vial that is placed in a lead container and store at –20°C (*see* **Note 2**).
17. When required, remove the lead container from the freezer. Transfer the vial to a second container at room temperature and allow the contents time to equilibrate to room temperature (30 min) before aliquots are removed for appropriate dilution for use in assays.

3.2. Method B

3.2.1. Activation of PP1$_i$ by Mn/Trypsin

1. Pipet 20 μL samples of PP1$_i$ into assay tubes.
2. Add 20 μL of the MnCl$_2$/NaCl solution, vortex mix, and place in a water bath at 30°C (*see* **Note 7**).
3. Add 10 μL of trypsin to each tube at 10 s intervals, mix, and incubate at 30°C for 10 min (*see* **Note 8**).
4. Terminate trypsin activity by the addition of 10 μL of trypsin inhibitor to each tube at 10 s intervals and vortex mix.
5. Remove 10 μL samples for protein phosphatase assays.
6. Scale up the Mn/trypsin treatment if necessary for the continuous spectrophotometric assay (Method E).

3.3 Method C

3.3.1. Assay for Protein Phosphatase
by Changes in Phosphorylase a Activity

1. Carry out the assays in duplicate in labeled tubes. Two additional tubes, labeled T for total, are set up to determine the initial (total) phosphorylase *a* activity.
2. Dilute the protein phosphatase sample in the appropriate assay buffer.
3. Pipet 10 μL of the diluted phosphatase into the assay tubes and 10 μL of assay buffer into the total tubes. Place the tubes in a water bath at 30°C.
4. Initiate the phosphatase assays by adding 40 μL of phosphorylase *a*, pre-equilibrated at 30°C for 5 min, at 10 or 15 s intervals to each tube. Vortex mix and incubate at 30°C for 5 min (*see* **Note 8**).
5. Terminate the phosphatase reactions by the addition of 950 μL of stop buffer to each tube at 10 or 15 s intervals and vortex mix (*see* **Note 9**).
6. Transfer 50 μL from each total and sample tube to a second set of tubes. Include two additional tubes for blanks containing 50 μL from any of the sample or total tubes. Place the tubes in a water bath at 30°C, keeping the blank tubes separate.
7. Initiate the phosphorylase *a* reaction by the addition of 50 μL of phosphorylase *a* substrate mixture to all the tubes, except the blanks, at 10 or 15 s intervals. Vortex mix and incubate at 30°C for 5 min.
8. Add 2 mL of 0.072 M H_2SO_4 to the blanks and then add 50 μL of the phosphorylase *a* substrate and vortex mix (*see* **Note 9**).
9. Terminate the phosphorylase *a* reactions by the addition of 2 mL of 0.072 M H_2SO_4 to each total and sample tube. Vortex mix and place at room temperature (*see* **Note 9**).
10. Add 2 mL of Color Reagent to each of the total, sample, and blank tubes (*see* **Note 9**).
11. Mix and after 2 min determine the absorbance at 600 nm (*see* **Note 10**).
12. Correct the A_{600} for the total and samples by subtraction of the A_{600} of the blank.
13. Calculation: one unit of protein phosphatase converts 1 nmole of phosphorylase *a* to *b* (equivalent to the release of 1 nmole of phosphate from phosphorylase *a*).

$$U/mL = \frac{(\text{Total } A_{600} - \text{Samples } A_{600})}{\text{Total } A_{600}} \times 10 \times 1/_T \times 5$$

Where 10 = phosphorylase concentration in nmoles/mL; T = time of the protein phosphatase assay (5 min in example); 5 = the fold that the protein phosphatase was diluted into the phosphatase assay.

3.4. Method D

3.4.1. Assay for Protein Phosphatase Using ^{32}P-Phosphorylase

1. Carry out assays in duplicate in labeled tubes. Also include two tubes for total counts per min (cpm) and two tubes for blank cpm.
2. Dilute the protein phosphatase into the appropriate assay buffer.

3. Pipet 10 μL of phosphatase samples into the sample tubes. Pipet 10 μL of assay buffer into each of the total and blank tubes. Place the tubes in a water bath at 30°C.
4. Initiate the protein phosphatase reaction by the addition of 40 μL of ^{32}P-phosphorylase, pre-equilibrated at 30°C for 5 min, to each tube at 10 or 15 s intervals. Vortex mix, and incubate at 30°C for 5 min (see Note 8).
5. Terminate the reaction by the addition of 50 μL of 10% TCA to all but the total tubes at 10 or 15 s intervals, and vortex mix. Add 50 μL of water to the total tubes.
6. Centrifuge to pellet the denatured protein using a clinical centrifuge at high speed for 10 min at room temperature.
7. Pipet 50 μL of each supernatant solution and totals into vials containing 3 mL of aqueous-compatible scintillation fluid. Alternatively, spot the 50 μL of each supernatant onto 1 inch squares of Whatman 31 ET paper, dry under a heat lamp, and place in a vial containing 10 mL of an aqueous-incompatible scintillation fluid. The advantage of the latter is that after removal of the paper squares, the vial can be reused after being checked for contamination by ^{32}P$_i$ released on paper fibers.
8. Calculation: 1 U of protein phosphatase releases 1 nmole of phosphate from phosphorylase *a* per minute at 30°C.

$$\text{Units/mL} = \frac{(\text{Sample cpm} - \text{Blank cpm})}{(\text{Total cpm} - \text{Blank cpm})} \times 10 \times {}^1\!/_T \times 5$$

Where 10 = phosphorylase concentration in nmoles/mL; T = time of the protein phosphatase assay (5 min in the example); 5 = the fold that protein phosphatase was diluted into the phosphatase assay.

3.5. Method E

3.5.1. Continuous Spectrophotometric Protein Phosphatase Assay

1. Carry out the continuous spectrophotometric assay for protein phosphatases at 25°C.
2. Add the following components to the cuvet: Buffer A (1410 μL); phosphorylase *a* (180 μL); purine nucleoside phosphorylase (90 μL); MTGuo (20 μL); [For PP2A assays add protamine (180 μL) and adjust the volume of assay buffer added to 1230 μL].
3. Mix and incubate in the spectrophotometer for 4–5 min to allow temperature equilibration and monitor absorbance at 360 nm to establish a blank rate, if any (*see* **Note 11**).
4. Initiate the reaction by the addition of phosphatase (100 μL). If a different volume is to be added adjust the amount of Buffer A added to the cuvet.
5. Record the increase in absorbance at 360 nm, which is a result of the conversion of MTGuo to MTGua in the presence of inorganic phosphate released from phosphorylase *a* by the phosphatase. Quantification of the phosphate release is made using the extinction coefficient of 11,200 m^{-1}cm^{-1} for the phosphate dependent reaction at 360 nm *(15)*.
6. Calculation: 1 U of protein phosphatase releases 1 nmole of phosphate from phosphorylase a per minute at 25°C.

$$\text{Activity (U/mL)} = \frac{\Delta A_{360}/\text{min}}{0.0112 \times X}$$

where X = volume of phosphatase added/mL of assay mixture (in this example X = 0. 1/1.8 mL, where the total volume in the cuvet is 1.8 mL).

3.6. Method F

3.6.1. Organic Extracts of Inorganic Phosphate as the Phosphomolybdate Complex

1. Perform the assays at 30° according to Method D with the exceptions that the final assay volume is 100 µL (i.e., initiate the reaction by the addition of 80 µL of ^{32}P-phosphorylase to 20 µL of phosphatase) and terminate assays by the addition of 100 µL of Reagent A.
2. Remove precipitated protein by centrifugation in a clinical centrifuge at room temperature.
3. Transfer 150 µL of each supernatant solution to a new tube. Add 25 µL of Reagent B and mix.
4. Add 250 µL of the isobutanol-benzene mixture to each tube and vortex the mixture vigorously to extract the phosphomolybdate complex into the organic layer (carry out this step in a fume hood).
5. Separate the aqueous and organic layer by centrifugation in a clinical centrifuge for 2 min.
6. Transfer 125 µL of the organic layer to a scintillation vial containing 5 mL of aqueous-compatible scintillation fluid.
7. To determine the total counts used in the assays, add 80 µL of the 12.5 µM ^{32}P-phosphorylase substrate to 120 µL of assay buffer and mix. Transfer 75 µL to a scintillation vial containing 5 mL of aqueous-compatible scintillation fluid.
8. Calculations are the same as that given in Method D.

4. Notes

1. Unlabeled phosphorylase a and thiophosphorylated phosphorylase can be prepared by substituting ATP or Adenosine 5'-O-(γ-thio)triphosphate.
2. Commercial sources for these proteins include Sigma, Boehringer-Mannheim, and Life Technologies. The latter also provides a kit for the phosphorylation of phosphorylase *b*. Isolated preparations of phosphorylase *(1,2)* and phosphorylase kinase *(3)* can be stored at −20°C for up to at least 3 yr in buffer containing 50% glycerol. These proteins denature when frozen at −20°C in the absence of glycerol. When isolating phosphorylase kinase, care must be exercised in the removal of the rabbit muscle tissues. *See* **ref. 3** for details.
3. Caffeine (or theophylline) is included in phosphorylase phosphatase assays to prevent AMP inhibition (if present) and to stimulate the phosphatase activity by stabilizing the T-state of phosphorylase *a* in which the serine-14 phosphate is exposed *(16)*. Caffeine also prevents the crystallization of the phosphorylase *a*:protamine complex present in PP2A assays *(8)*. 0.2 *M* NaCl in protamine-stimu-

lated PP2A assays prevents precipitation of phosphorylase a by protamine and allows for the maximal activation of PP2A$_1$ *(8)*.

4. In our hands, protamine chloride (Sigma) was more effective as a PP2A activator than the free base and sulfate forms that are also available. The chloride content in protamine chloride preparations are high (15–18%) and can be obtained from the vendor.

5. PP2A preparations should not be preincubated with protamine as this leads to a time-dependent decrease in protamine-stimulated activity in the assays *(8)*. Thus, the protamine is added together with the phosphorylase substrate.

6. It is necessary to remove the excess methyl iodide before the addition of thiourea.

7. Addition of Mn^{2+} and NaCl to the enzyme prior to the addition of trypsin usually results in more activation than if the three components are added at the same time.

8. Initiating reactions at 10 s intervals is possible if the same pipet tip is used. Care must be exercised to prevent the contamination of the tip with phosphatase samples.

9. This solution can be conveniently delivered using a bottle top type dispenser.

10. The remaining glucose 1-phosphate slowly hydrolyzes under the strong acid conditions of the Color Reagent. Absorbance reading should be made as soon as possible after the 2 min color development time.

11. For the continuous spectrophotometric assay it is convenient if the spectrophotometer is equipped with a magnetic stirrer in the cuvet holder. This allows rapid homogeneous mixing of the reaction components and allows acquisition of data within 5 s of the addition of the phosphatase. It is also convenient when the initial rates are determined from the slopes of progress curves acquired using spectrophotometer compatible software.

References

1. Fischer, E. H. and Krebs, E. G. (1962) Muscle phosphorylase b. *Method Enzymol.* **5**, 369–373.

2. Krebs, E. G. and Fischer, E. H. (1962) Phosphorylase b kinase from rabbit skeletal muscle. *Methods Enzymol.* **5**, 373–376.

3. Cohen, P. (1983)Phosphorylase kinase from rabbit skeletal muscle. *Methods Enzymol.* **99**, 243–250.

4. Li, H. -C., Simonelli, P. F., and Huan, L. -J. (1988) Preparation of protein phosphatase-resistant substrate using adenosine 5'-O-(γ-Thio)triphosphate. *Methods Enzymol.* **159**, 346–356.

5. Wang, Z. -X., Cheng, Q., and Killilea, S. D. (1995) A continuous spectrophotometric assay for phosphorylase kinase. *Anal. Biochem.* **230**, 55–61.

6. Henry, S. P. and Killilea, S. D. (1993) Hierarchical regulation by casein kinase I and II of the activation of protein phosphatase-1$_i$ by glycogen synthase kinase-3 is ionic strength dependent. *Arch. Biochm. Biophys.* **301**, 53–57.

7. Schuchard, M. D. and Killilea, S. D. (1989)Salt stimulation of the activation of latent protein phosphatase, Fc. M, by Mn^{2+} and Mn/trypsin. *Biochem. Int.* **18**, 845–849.

8. Cheng, Q., Erickson, A. K., Wang, Z. -X., and Killilea, S. D. (1996) Stimulation of phosphorylase phosphatase activity of protein phosphatase $2A_1$ by protamine is ionic strength dependent and involves interaction of protamine with both substrate and enzyme. *Biochemistry* **35**, 15,593–15,600.
9. Cheng, Q. , Wang, Z. -X. , and Killilea, S. D. (1995). A continuous spectrophotometric assay for protein phosphatases. *Anal. Biochem.* **226**, 68–73.
10. Brandt, H. , Capulong, Z. L., and Lee, E. Y. C. (1975) Purification and properties of rabbit liver phosphorylase phosphatase. *J. Biol. Chem.* **250**, 8038–8044.
11. Hedrick, J. L. and Fischer, E. H. (1965) On the role of pyridoxal 5'-phosphate in phosphorylase. Absence of classical vitamin B_6-dependent enzymic activities in muscle glycogen phosphorylase. *Biochemistry* **4**, 1337–1343.
12. Webb, M. R. (1992) A continuous spectrophotometric assay for inorganic phosphate and for measuring phosphate release kinetics in biological systems. *Proc. Natl. Acad. Sci. USA* **89**, 4884–4887.
13. Killilea, S. D., Mellgren, R. L., Aylward, J. H., and Lee, E. Y. C. (1978) Inhibition of phosphorylase phosphatase by polyamines. *Biochem. Biophys. Res. Commun.* **81**, 1040–1046.
14. Ando, T. and Watanabe, S. (1969) A new method for fractionation of protamines and the amino acid sequences of salmine and three components of iridine. *Int. J. Protein Res.* **1**, 221–224.
15. Sergienko, E. A. and Srivastava, D. K. (1994) A continuous spectrophotometric method for the determination of glycogen phosphorylated-catalyzed reaction in the direction of glycogen synthesis. *Anal. Biochem.* **221**, 348–355.
16. Madsen, N. B. (1986) Glycogen Phosphorylase, in *The Enzymes* (Boyer, P. D. and Krebs, E. G., eds.), 17, Academic, London, New York, pp. 365–394.

3

Analyzing Gene Expression with the Use of Serine/Threonine Phosphatase Inhibitors

Axel H. Schönthal

1. Introduction

This chapter describes the use of phosphatase-inhibitory compounds to study gene regulation. Serine/threonine protein phosphatase inhibitors can be divided into two groups: The first group (*see* **Table 1**) comprises environmental toxins and other natural products that are mostly produced by micro-organisms, such as blue-green algae or soil bacteria. They are structurally very diverse molecules that share the ability to inhibit phosphatase activity *(1–3)*. The best known and most widely used compound of this group is okadaic acid, a polyether fatty acid derivative that is produced by marine dinoflagellates *(4)*. Its use in research became quite widespread during the last few years, and the number of publications describing its effects have increased enormously. In addition, a number of natural and synthetic compounds have been discovered recently, which are useful as inhibitors of protein phosphatase type 2B (also called calcineurin) *(5,6)* (*see* **Table 1**).

The second group of protein phosphatase-inhibitory products comprises endogenous heat-stable proteins that are synthesized by the cell itself. Examples for members of this group are Inhibitor-1 (I-1), I-2, DARPP, and several others. The potential use of this second group of inhibitory compounds for the study of gene regulation will not be discussed here; a recent review has summarized the latest findings *(7)*.

For use in the analysis of gene regulation, a selected phosphatase-inhibitory compound (*see* **Note 2**) is dissolved in the appropriate solvent and added into the cell culture medium. After the desired incubation period, the cells are harvested and analyzed for various parameters, such as overall mRNA or protein

From: *Methods in Molecular Biology, Vol. 93: Protein Phosphatase Protocols*
Edited by: J. W. Ludlow © Humana Press Inc., Totowa, NJ

Table 1
Protein Phosphatase Inhibitors (see Note 1)

Inhibitory compound	Inhibitor activity	Cell-membrane permeable
Okadaic acid	PP2A > PP1*	yes
Cantharidin	PP2A > PP1*	yes
Calyculin	PP2A = PP1*	yes
Microcystin LR	PP2A = PP1*	no**
Nodularin	PP2A = PP1*	no**
Tautomycin	PP2A < PP1*	yes
Cycolosporin A	PP2B	yes
Cypermethrin	PP2B	yes
Deltamethrin	PP2B	yes
Fenvalerate	PP2B	yes

* At very high concentrations (>1 μM) there is also some inhibitory activity on PP2B.
** Liver cells appear to have an uptake system capable of transporting this compound.
PP2A = PP1: Inhibitory concentration (IC_{50}) is similar for PP2A and PP1.
PP2A > PP1: Inhibitory concentration is larger for PP2A than for PP1.
PP2A < PP1: Inhibitory concentration is smaller for PP2A than for PP1.

levels (e.g., by Northern blot, Western blot), mRNA stability, or transcriptional activity (nuclear run-off analysis). Similarly, after transfection of suitable reporter plasmid constructs the cell lysate can be analyzed for e.g., luciferase or chloramphenicol acetyltransferase (CAT) activity, which is indicative of the respective promoter activity. Moreover, signal transduction pathways can be studied by analyzing the activity of their various components. This can be accomplished by immunoprecipitating the respective component, for example, one of the MAP kinases, and subsequent in vitro kinase assays with suitable substrates. This latter approach may be quite informative, because inhibited phosphatase activity may result in increased kinase activity, which in turn may impinge on transcription factor activity, and hence, on gene expression (8).

2. Materials

1. Okadaic acid is commercially available in three salt forms (sodium, potassium, and ammonium). The use of the salt forms is preferred over the free acid form because of their increased stability during storage, both in the unopened vial and after dissolution in solvent. Because in cell culture medium okadaic acid is ionized to its salt form, the biological activity of either form is the same (see Note 3). Okadaic acid salts are soluble in DMSO, ethanol, or water. DMSO is the solvent of choice when okadaic acid analogs are to be included as negative controls (see below). The stock solution typically is made to 0.1–1.0 mM. It should be stored at –20°C protected from light. Okadaic acid is offered by several suppliers, for

example, Alexis (San Diego, CA), Calbiochem (La Jolla, CA), Sigma (St. Louis, MO), and Boehringer Mannheim (Indianapolis, IN).

2. Several okadaic acid analogs with similar physical and chemical properties, but reduced or no phosphatase inhibitory activity, are commercially available from Sigma, Calbiochem, and Alexis. These compounds are suitable as negative controls for okadaic acid. 1-norokadaone appears to be most useful as a negative control because its chemical structure closely matches that of okadaic acid. Methyl-okadaate, which has been used as a negative control in the past, is not recommended for whole-cell and tissue assays, as it is suspected to be metabolized back to the active okadaic acid *(9)*. The above mentioned okadaic acid analogs are soluble in DMSO and should be stored at −20°C protected from light.

3. Calyculin A, microcystin-LR, nodularin, tautomycin, and cantharidin are available through Sigma, Alexis, and Calbiochem. They are soluble in DMSO or ethanol. They should be stored at −20°C. Tautomycin and calyculin A need to be protected from light (*see* **Note 4**).

4. Cyclosporin A is soluble in ethanol and methanol. It is available through Sigma, Calbiochem, and other suppliers.

5. Cypermethrin, deltamethrin, and fenvalerate are synthetic type II pyrethroids (*see* **Note 5**). They are available through Calbiochem and Alexis. They are soluble in common organic solvents such as DMSO, acetone, or ethanol, and exhibit good chemical stability. However, plastic tubes should be avoided. Glass glass tubes should be treated with 1% PEG (polyethyleneglycoll) in ethanol and heated to 250°C for 30 min before use.

6. Resmethrin is a synthetic type I pyrethroid and useful as a weakly active negative control for the highly active PP2B-inhibitory type II pyrethroids. It is soluble in common organic solvents and needs to be protected from light. Resmethrin is available through Calbiochem or Alexis. The latter supplier provides further negative control compounds such as allethrin and permethrin.

Most of the phosphatase-inhibitory compounds are potent toxins and tumor promoters *(1)*. Okadaic acid, through its accumulation in filter feeding marine organisms, is able to enter the human food chain and is the causative agent of diarrhetic shellfish poisoning. In animal experiments okadaic acid and calyculin A promote tumorigenesis in tissues like skin and gut *(10)*. Microcystin and nodularin have been shown to accumulate in the liver and cause hepatic tumors *(11)*. It is, therefore, imperative to take appropriate care during handling, use, and disposal of these compounds.

3. Methods

1. One day before the addition of inhibitor, seed the cells into the culture dish so that they will be 50–70% confluent the next day.

2. The next day, add the desired inhibitor (*see* **Notes 5,6**). This can be done by either adding it directly into the medium in the culture dish, or by first adding it to fresh medium in the bottle and then distributing it into the culture dish (*see* **Note 5**).

3. After the appropriate incubation time, harvest the cells for the desired analysis (*see* **Note** 7), e.g., mRNA for Northern blots; proteins for Western blots; individual, immunoprecipitated kinases for in vitro kinase reactions; nuclei for mRNA run-off analysis; and so on according to standard procedures.

4. Notes

1. For use in cell culture, the inhibitor used needs to be able to enter the cell. As listed in **Table 1**, microcystin and nodularin do not penetrate the cell membrane and, therefore, cannot be used with most cells; their use is restricted to cells with an appropriate uptake system, e.g., liver cells *(11)*.

2. Different inhibitors target the various phosphatases with different efficiencies *(1,3,4)*. This can be beneficial, as a combination of inhibitors with different IC_{50} values (the concentration of drug that inhibits 50% of the enzymatic activity) may help to preliminary characterize which type of serine/threonine protein phosphatase may be involved in a certain biological process.

3. The IC_{50} values that have been published are an approximation; they depend somewhat on the cell type and on the respective substrate that is being used. Importantly, these values are valid for in vitro dephosphorylation reactions only; for use in cell culture the efficient concentration generally is higher. For example, the published IC_{50} for okadaic acid is 0.2–1.0 n*M* with respect to PP2A activity in vitro (when added to diluted cellular lysate) *(4)*. However, when used in cell culture with the murine fibroblast cell line NIH3T3, the IC_{50} is around 30 n*M* *(12)*.

4. The phosphatase inhibitory compounds are rather cytotoxic, especially at higher concentrations and upon longer incubation periods; it is, therefore, advisable to determine whether an observed effect (e.g., the down-regulation of a gene) is caused by a general shut-down of cellular functions. In the same vein, different cell types vary greatly in their sensitivity to the cytotoxic effects of these inhibitors. Certain cells cannot be treated long enough to study a desired process. Usually, the murine fibroblast cell line NIH3T3 is relatively "sturdy" and tolerates elevated levels of inhibitors, e.g., 50 n*M* okadaic acid for a few days, or up to 500 n*M* for a few hours. Calyculin A is extremely cytotoxic, presumably because of its highly efficient simultaneous inhibition of PP1 and PP2A; its use in cell culture usually is restricted to 0.5–5.0 n*M*.

5. Special care must be taken when using the calcineurin inhibitors cypermethrin, deltamethrin, or fenvalerate. These inhibitors should be added directly into the medium of the cell culture dish, rather than into the bottled medium.

6. At higher concentrations, okadaic acid inhibits protein synthesis *(13)*. It is likely that other phosphatase-inhibitory compounds have similar effects. This needs to be taken into consideration when using elevated concentrations of a drug. This may be especially critical when analyzing gene expression via transfection of reporter plasmids and subsequent measurement of luciferase or CAT activity.

7. In addition to the well-established protein phosphatases (PP1, PP2A, PP2B, PP2C), new members of this expanding family have been identified and cloned, such as PP3, PP4, and PP5 *(14–16)*. These novel phosphatases, and likely some

others that are still to be discovered, are sensitive to inhibition by okadaic acid-type compounds as well. Thus, after treatment of cells with a phosphatase-inhibitory drug, it is difficult to ascribe an observed effect directly to one specific phosphatase. To establish the involvement of a particular phosphatase, additional experiments are necessary, such as, for example, the transfection of expression vectors for an individual phosphatase *(12)*.

References

1. Fujiki, H. and Suganuma, M. (1993) Tumor promotion by inhibitors of protein phosphatase 1 and 2A: the okadaic acid class of compounds. *Adv. Cancer Res.* **61**, 143–194.
2. Schönthal, A. (1992) Okadaic acid—a valuable new tool for the study of signal transduction and cell cycle regulation? *New Biologist* **4**, 16–21.
3. Li, Y. -M., and Casida, J. E. (1992) Cantharidin-binding protein: identification as protein phosphatase 2A. *Proc. Natl. Acad. Sci. USA* **89**, 11,867–11,870.
4. Bialojan, C. and Takai, A. (1988) Inhibitory effect of a marine-sponge toxin, okadaic acid, on protein phosphatases. *Biochem. J.* **256**, 283–290.
5. Enan, E. and Matsumura, F. (1992) Specific inhibition of calcineurin by type II synthetic pyrethroid insecticides. *Biochem. Pharm.* **43**, 1777–1784.
6. Liu, J., Farmer, Jr. J. D., Lane, W. S., Friedman, I., and Schreiber, S. L. (1991) Calcineurin is a common target of cyclophilin-cyclosporin A and FKBP-FK506 complexes. *Cell* **66**, 807–815.
7. Shenolikar, S. (1995) Protein phosphatase regulation by endogenous inhibitors. *Sem. Cancer Biol.* **6**, 219–227.
8. Schönthal, A. H. (1995) Regulation of gene expression by serine/threonine protein phosphatases. *Sem. Cancer Biol.* **6**, 239–248.
9. Nishiwaki, S., Fujiki, H., Suganuma, M., Furuya-Suguri, H., Matsushima, R., Iida, Y., Ojika, M., Yamada, K., Uemura, D., Yasumoto, T., Schmitz, F. J., and Sugimura, T. (1990) Structure-activity relationship within a series of okadaic acid derivatives. *Carcinogenesis* **11**, 1837–1841.
10. Suganuma, M., Fujiki, H., Furuya-Suguri, H., Yoshizawa, S., Yasumoto, S., Kato, Y., Fusetani, N., and Sugimura, T. (1990) Calyculin A, an inhibitor of protein phosphatases, a potent tumor promoter on CD-1 mouse skin. *Cancer Res.* **50**, 3521–3525.
11. Yoshizawa, S., Matsushima, R., Watanabe, M.F., Harada, K. -I, Ichihara, A., Carmichael, W. W., and Fujiki, H. (1990) Inhibition of protein phosphatases by microcystins and nodularin associated with hepatotoxicity. *J. Cancer Res. Clin. Oncol.* **116**, 609–614.
12. Jaramillo-Babb, V., Sugarman, J. L., Scavetta, R., Wang, S. -J., Berndt, N., Born, T. L., Glass, C. K., and Schönthal, A. H. (1996) Positive regulation of cdc2 gene activity by protein phosphatase type 2A. *J. Biol. Chem.* **271**, 5988–5992.
13. Redpath, N. T. and Proud, C. G. (1989) The tumor promoter okadaic acid inhibits reticulocyte lysate protein synthesis by increasing the net phosphorylation of elongation factor 2. *Biochem. J.* **262**, 69–75.

14. Honkanen, R. E., Zwiller, J., Daily, S. L., Khatra, B. S., Dukelow, M., and Boynton, A. L. (1991) Identification, purification, and characterization of a novel serine/threonine protein phosphatase from bovine brain. *J. Biol. Chem.* **266,** 6614–6619.
15. Brewis, N. D., Street, A. J., Prescott, A. R., and Cohen, P. T. W. (1993) PPX, a novel protein serine/threonine phosphatase localized to centrosomes. *EMBO J.* **12,** 987–996.
16. Chen, M. X., McPartlin, A. E., Brown, L., Chen, Y. H., Barker, H. M., and Cohen, P. T. W. (1994) A novel human protein serine/threonine phosphatase, which possesses four tetratricopeptide repeat motifs and localizes to the nucleus. *EMBO J.* **13,** 4278–4290.

4

Inhibitor-1, a Regulator of Protein Phosphatase 1 Function

John H. Connor, Hai Quan, Carey Oliver, and Shirish Shenolikar

1. Introduction

With the discovery of numerous eukaryotic protein serine/threonine phosphatases, a classification scheme was developed that exploited some of the biochemical properties of these enzymes to expedite their identification *(1)*. These properties included the preferential dephosphorylation of either α or β-subunits of phosphorylase kinase, the metal ion dependency displayed by some phosphatases for enzyme activity and their inhibition by two mammalian thermostable proteins, inhibitor-1 (I-1) and inhibitor-2 (I-2).

The susceptibility of phosphorylase kinase to proteolysis, particularly the β-subunits, and the comigration of α with the proteolyzed β on SDS-PAGE makes it difficult to accurately monitor the selective dephosphorylation of these subunits. Metal ion dependency is also not the clearest discriminator of protein phosphatases. For instance, PP2B or calcineurin is defined as a calcium-calmodulin-activated protein phosphatase. However, when purified to homogeneity, PP2B shows little activity in the presence of calcium and calmodulin *(2,3)* but it is fully reactivated by millimolar concentrations of Mn^{2+}. In addition, phosphatases like PP1 and PP2A spontaneously inactivate and are also reactivated by Mn^{2+}. A growing number of phosphatases identified by molecular cloning, when expressed in bacteria or insect cells, also show a Mn^{2+} requirement for enzyme activity *(4)*. In light of this, inhibition by I-1 and I-2 becomes a more dependable criterion for distinguishing type-1 and type-2 protein serine/threonine phosphatases. Type-1 protein phosphatases (PP1) from many species are inhibited by nanomolar concentrations of these inhibitors. Type-2 phosphatases are universally unaffected by low concentrations of I-1 and I-2. I-2 is often favored over I-1 for characterizing PP1 as it is a constitu-

From: *Methods in Molecular Biology, Vol. 93: Protein Phosphatase Protocols*
Edited by: J. W. Ludlow © Humana Press Inc., Totowa, NJ

tive inhibitor of this enzyme. In contrast, I-1 requires prior phosphorylation by cAMP-dependent protein kinase (PKA) to inhibit PP1. The need for stoichiometric phosphorylation of I-1 and its possible dephosphorylation by type-2 phosphatases in tissue extracts also encouraged the widespread use of I-2. However, PP1 inhibition by nanomolar concentrations of I-2 requires the prior incubation of the two proteins. Lack of adequate preincubation reduces the extent of PP1 inhibition and provides a false measure of PP1 activity. In addition, prolonged incubation of I-2 with crude extracts encourages proteolysis of this largely unstructured protein and reduces its efficacy as a PP1 inhibitor. In these situations, higher concentrations of I-2 are often used to ensure PP1 inhibition. This may be undesirable as low micromolar concentrations of I-2 inhibit PP2A, the major type-2 phosphatase *(5)*. By comparison, I-1 is an instantaneous inhibitor of PP1, which retains its selectivity over a much wider range of concentrations. Moreover, there appears to be little or no dephosphorylation and inactivation of I-1 (present in nanomolar concentrations) in the presence of a phosphoprotein substrate (usually at micromolar concentrations). In any case, the potential dephosphorylation of I-1 can be circumvented by thiophosphorylation of I-1. This produces a potent PP1 inhibitor that is not significantly dephosphorylated by cellular phosphatases. Thus, thiophosphorylated I-1 can be injected into living cells to inhibit PP1 activity and define its physiological role *(6,7)*.

   I-1 has drawn considerable attention as a prototype mechanism for kinase-phosphatase crosstalk. Hormones that elevate intracellular cAMP promote I-1 phosphorylation in skeletal muscle *(8–10)*, cardiac muscle *(11,12)*, adipocytes *(13)*, neurons *(14)*, and liver *(15)*. PP1 inhibition by the hormone-activated I-1 parallels PKA activation and provides tremendous amplification to cAMP signaling cascades. PP1 inhibition by I-1 may also broaden the spectrum of hormone-induced phosphorylations to include substrates of many cAMP-independent kinases that are essential components of the physiological response. The high evolutionary conservation of the I-1 structure *(16,17)* also argues for its importance in signal transduction in higher eukaryotes. This chapter will discuss procedures utilized by this laboratory to isolate I-1 from mammalian tissue as well as two expression systems that produce recombinant I-1 proteins and peptides. Methods for phosphorylating and activating I-1 and its subsequent use as an inhibitor of PP1 in vitro and in vivo will also be described.

2. Materials

   1. Sodium pentobarbitol (Euthatol).
   2. 1% (w/v) trichloroacetic acid.
   3. 100% (w/v) trichloroacetic acid
   4. 0.5 M Tris-HCl, pH 8.0 (at 25°C).
   5. 5 mM Tris-HCl, pH 8.0 (at 25°C).

6. 10 M NH$_4$OH.
7. (NH$_4$)$_2$SO$_4$.
8. 5 mM Tris-HCl, pH 8.0 containing 1 mM EDTA and 0.005% Brij 35.
9. Sephadex G100 superfine (Pharmacia, Uppsala, Sweden).
10. ^{32}P-γ-ATP (Amersham, Arlington heights, IL, PB10168).
11. PKA Catalytic Subunit (Promega, Madison, WI).
12. Terrific Broth containing ampicillin (50 μg/mL).
13. Isopropyl 1-thio-β-D-galactopyranoside (IPTG).
14. 50 mM Tris-HCl, pH 7.5 containing 1% (v/v) nonidet P-40, 5 mM EDTA, 5 mM EGTA, 5 mM benzamidine, and 1 mM PMSF.
15. Glutathione-Sepharose (Pharmacia).
16. 20 mM Tris-HCl, pH 7.5 containing 0.15 M NaCl (TBS).
17. 20 mM Tris-HCl, pH 7.5 containing 0.15 M NaCl and 1.0% (v/v) nonidet P-40.
18. 5 mM Tris-HCl, pH 7.5 containing 0.005% Brij 35.
19. 50 mM Tris-HCl, pH8.5 containing 10 mM glutathione.
20. 50 mM Tris-HCl, pH 8.5 containing 5 mM CaCl$_2$.
21. Thrombin.
22. Phenyl methyl sufonyl fluoride (PMSF).
23. Luria Bertani medium containing ampicillin (50 μg/mL).
24. 500 mM Tris-HCl, pH 7.5.
25. 50 mM Tris-HCl, pH 7.5 containing 100 μM ATP and 1 mM MgCl$_2$
26. Sephadex G25 (Pharmacia).
27. Centricon 30 filter (Amicon, Danvers, MA).
28. 50 mM Sodium glycerophosphate, pH 6.8 containing 50 μM ATP, 1 mM MgCl$_2$, and 200 μCi ^{32}P-γ-ATP (Specific activity > 4000 Ci/mmole—Amersham).
29. 25% (w/v) Trichloroacetic acid.
30. Bovine serum albumin (10 mg/mL)
31. 50 mM sodium glycerophosphate, pH 6.8 containing 0.2 mM ATP-γ-S and 2 mM MgCl$_2$
32. ^{32}P-labeled Phosphorylase a.
33. Protein phosphatase-1 (PP1) catalytic subunit (UBI).
34. 50 mM Tris HCl, pH 7.5 containing 1 mM EDTA, 1% (w/v) BSA, and 0.3% (v/v) β-mercaptoethanol
35. Scintillation fluid (Safety-Solve, RPI).
36. Bovine brain protein phosphatase-2B (PP2B or calcineurin, UBI).
37. 50 mM Tris HCl, pH 7.5 containing 0.1 mM EGTA, 5 mM MgCl$_2$, 0.1 mM MnCl$_2$, 0.2 mM CaCl$_2$, 0.1 mM dithiothreitol (DTT), 10 μg/mL calmodulin, and 0.1 mg/mL BSA.

All reagents were obtained from Sigma unless otherwise stated.

3. Methods

3.1. Isolation of I-1 from Rabbit Skeletal Muscle

The purification of I-1 is commonly undertaken using rabbit skeletal muscle, an excellent source for this protein. The isolation procedure for I-1 is a modification of the one first developed by Nimmo and Cohen *(18)*. Several modifications have been made to the original procedure that shorten the time taken to purify I-1 to homogeneity without decreasing yield *(19,20)*. Most earlier procedures isolated I-1 in its phosphorylated state. In contrast, the procedure described below (**steps 1–16**) is equally effective in isolating phosphorylated and dephosphorylated I-1 and primarily yields dephosphorylated I-1, as might be expected from unstimulated tissues. In this respect, the purification protocol also offers us the unique ability to estimate hormone-induced changes in I-1 phosphorylation and activity.

1. Place female New Zealand White rabbit (approx 2 kg body weight) in an appropriate restraining harness and euthanize with the slow injection of sodium pentobarbital or "Euthatol" (approx 1.5 mL/kg) through the ear vein. Exsanguinate the animal by an incision in the jugular vein. Rapidly excise the skeletal muscle from the two hind limbs and back, and place on ice. Typically, three to four animals are processed together. All subsequent steps are carried out in the coldroom or 4°C, unless otherwise stated.
2. Mince the tissue in a meat grinder, weigh and then homogenize in 5 vol of ice cold 1% (w/v) trichloroacetic acid in a commercial Waring blender (5 L capacity) at medium speed for 45 s.
3. Stir the homogenate for 1 h at 4°C before subjecting to centrifugation at 6000*g* for 45 min.
4. Discard the pellet. Adjust the supernatant to 15% (w/v) trichloroacetic acid using slow addition of a stock 100% (w/v) tricholoroacetic acid solution. Stir this suspension continuously for at least 3 h and then centrifuge at 20,000*g* for 15 min. Discard the supernatant.
5. Homogenize the pellet in 0.5 *M* Tris-HCl, pH 8.0 (1/20 of the original extract volume) using a Waring blender. Adjust pH of the suspension to 7.5 by the addition of 10 *M* ammonium hydroxide and dialyze overnight against 10–20 vol of 5 m*M* Tris-HCl, pH 8.0 with at least one change of the dialysis buffer after 4–5 h.
6. A heavy precipitate will be seen at this stage, representing proteins that failed to renature after dialysis. Remove the precipitated proteins by centrifugation at 20,000*g* for 15 min. Retain the supernatant. Homogenize the precipitated proteins twice more in a Waring blender, each time using an equal volume of 5 m*M* Tris-HCl, pH 8.0, to extract all residual I-1.
7. Pool all supernatants from the previous step and heat in a conical flask (maximum 500 mL supernatant per 2 L flask) placed in a boiling water bath (*see* **Note 1**). Stir the solution constantly. Once the temperature of the solution has reached 90°C, incubate for 10–15 min. Remove the flask and cool on ice. Remove all denatured proteins by centrifugation at 20,000*g* for 30 min.
8. Add solid ammonium sulphate to the supernatant to achieve 60% saturation and adjust the suspension to pH 7.0 using ammonium hydroxide. Stir for 2–4 h (*see* **Note 2**).
9. Centrifuge the suspension at 20,000*g* for 30–40 min and discard the supernatant. Resuspend the precipitate in approx 10 mL of 5 m*M* Tris-HCl, pH 7.5 containing 1.0 m*M* EDTA and 0.005% (w/v) Brij 35 using a Dounce glass homogenizer.

10. Dialyze the suspension for 1–2 h against 5 mM Tris-HCl, pH 7.5 containing 1.0 mM EDTA and 0.005% (w/v) Brij 35 and load on Sephadex G100 Superfine (5 × 70 cm) equilibrated in the same buffer. Collect 10–15 mL fractions at a flow rate of 60 mL/h (*see* **Notes 3** and **4**).

11. Monitor the elution of I-1, the primary PKA substrate at this stage, by phosphorylating aliquots (20–50 μL) of the fractions using ^{32}P-γ-ATP and PKA catalytic subunit (*see* the phosphorylation protocol described in **Subheading 3.4.**). Rabbit skeletal muscle I-1 is identified as a ^{32}P-labeled protein of apparent molecular weight 26 kDa on SDS-PAGE, followed by autoradiography (*see* **Note 5**).

12. Pool all fractions containing ^{32}P-labeled 26 kDa protein and lyophilize.

13. Dissolve the lyophilized sample in 2 mL of SDS-PAGE sample buffer and subject to preparative gel electrophoresis (Biorad Model 491, Prep Cell Column-37 mm ID) using a 9% (w/v) acrylamide gel containing 0.1% (w/v) SDS. The 60 mL gel should be previously calibrated with known proteins as molecular weight standards to facilitate the subsequent collection of all fractions (5–6 mL) containing I-1.

14. Identify all fractions containing I-1 by phosphorylation of aliquots (10–20 μL) of every fourth or fifth fraction and SDS-PAGE on Biorad mini-gels.

15. Pool fractions containing I-1 and dialyze extensively against 5 mM Tris-HCl, pH 7.5 containing 1.0 mM EDTA and 0.005% Brij 35 (5 changes of 5 L each) to remove most of the SDS.

16. Store the purified I-1 as frozen aliquots at –80°C or lyophilize and store in a desiccator at –20°C.

3.2. Expression and Purification of Recombinant I-1—Method 1

Problems of obtaining sufficient tissue from species such as human means that one must resort to molecular cloning and the expression of cDNAs encoding I-1, to obtain the recombinant proteins. Two strategies that successfully yield pure recombinant human I-1 that has been expressed in *E. coli* are described (*see* **Note 6**).

The I-1 cDNA, cloned from a human brain library *(17)*, was inserted into the pGEX-2T (Pharmacia) vector and expressed in *E. coli* as a fusion protein with glutathione-S-transferase under the control of the inducible promoter.

1. Grow a seed culture (5 mL) of *E. coli* BL21 transformed with pGEX-hI-1 overnight at 37°C in Terrific Broth *(21)* containing ampicillin (50 μg/mL).

2. Add the seed culture to 250 mL of Terrific Broth at 37°C and continue bacterial growth in an environmental orbital shaker at 37°C. Monitor absorbance of the bacterial suspension periodically (every 30 min) at 600 nm until A_{600} reaches 0.6.

3. Cool the bacterial culture to 25°C and continue growth in a shaking incubator at 25°C until A_{600} of the culture reaches 0.8. Add the inducer, IPTG, to the culture medium to a final concentration of 1 mM. This will initiate the induction of GST-I-1 protein. Continue incubation of the bacterial culture in the shaking incubator

at 25°C for a further 3 h to obtain maximal levels of the recombinant protein (*see* **Note 7**).

4. Centrifuge bacteria at 3,000*g* for 15 min. Resuspend bacteria in 20 mL of lysis buffer—50 m*M* Tris-HCl, pH 7.5 containing 1% Nonidet P-40, 5 m*M* EDTA, 5 m*M* EGTA, 5 m*M* benzamidine, 1 m*M* PMSF—at 4°C and lyse by sonication (Branson Sonifier, Output Control 0.5, Duty Cycle 50%, 5 × 10 s pulses) (*see* **Note 8**).

5. Shake bacterial lysate with the affinity gel, glutathione-Sepharose (10 mL), equilibrated with 20 m*M* Tris-HCl, pH 7.5 containing 0.15 *M* NaCl (TBS) for 10 min at 4°C using a Thermolyne Varimix rocker set at low speed.

6. Centrifuge the affinity matrix at 1,000*g*. Discard the supernatant. Wash the affinity gel twice with lysis buffer (10–20 mL) using gentle shaking and centrifugation, followed by three washes (20 mL each) with TBS containing 1% Nonidet P-40.

7. Elute GST-I-1 using three washes (10 mL each) of 50 m*M* Tris-HCl, pH 8.5 containing 10 m*M* glutathione.

8. Pool the eluates and dialyze against 50–100 vol of 5 m*M* Tris-HCl, pH 7.5, containing 0.005% (w/v) Brij 35 overnight at 4°C. Lyophilize and store the GST-I-1 protein at –20°C.

9. To obtain highly purified GST-I-1, use preparative SDS-PAGE as described in **Subheading 3.1., step 13** (*see* **Fig. 1**). Dialyze all fractions containing GST-I-1 (apparent molecular weight 47,000) against 5 m*M* Tris-HCl pH 7.5, containing 0.005% Brij 35 at 4°C, lyophilize, and store at –80°C.

10. For proteolytic cleavage of GST-I-1 with thrombin at a unique site constructed in the linker region to yield I-1 free of GST, digest 5 mg of GST-I-1 in 10 mL of 50 m*M* Tris-HCl, pH 8.5, containing 5 m*M* CaCl$_2$ and thrombin (0.5–1.0 NIH U/mL) at 30°C for 30 min. Terminate the reaction by adding PMSF to a final concentration of 1 m*M*, followed by heating the sample in a boiling water bath for 5 min (*see* **Note 9**).

11. Centrifuge the reaction mixture at 40,000*g* for 20 min and subject the supernatant to preparative SDS-PAGE. Pool fractions containing a 30 kDa polypeptide that represents human I-1, dialyze and lyophilize as described in **step 9**.

3.3. Expression and Purification of Recombinant I-1—Method 2

Initial attempts to express I-1 as an unfused protein in bacteria, fungi, and insect cells were unsuccessful, in part because of rapid and extensive proteolysis of I-1 in these systems. Hence, systems are employed such as that described above (**Subheading 3.2.**) in which I-1 could be readily expressed, albeit fused to GST *(17)*. However, concerns about the interference from GST in structure-function analysis of I-1 and the considerable loss of protein during cleavage of GST-I-1 with thrombin, encouraged us to reexamine this issue. An expression system using the pT7-7 bacterial expression vector developed by Tabor and colleagues *(22)* is described that finally yielded recombinant human I-1 as an unfused protein (*see* **Note 10**).

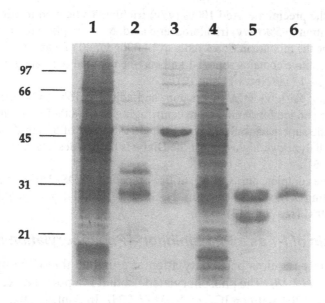

Fig. 1. Purification of Recombinant Human I-1. GST-I-1 was expressed in *E. coli* BL21 using the pGEX-2T vector and purified as described in **Subheading 3.2.** Lane 1 shows bacterial extract; lane 2 is the eluate from glutathione affinity chromatography and lane 3 is GST-I-1 purified by preparative SDS-PAGE. Recombinant unfused I-1 was expressed using the pT7-7 vector and purified as described in **Subheading 3.3.** Lane 4 shows the bacterial extract, lane 5 is I-1 obtained by fractionation with 5–15% (w/v) TCA and lane 6 shows I-1 following preparative SDS-PAGE.

1. Inoculate a seed culture of LB medium (5 mL) containing ampicillin (50 µg/mL) and grow overnight at 37°C in an orbital shaking incubator.
2. Transfer the seed culture to a 1 L conical flask containing 250 mL LB medium containing ampicillin and continue bacterial growth at 37°C until absorbance of the culture at 600 nm reaches 0.6.
3. Induce I-1 production by addition of IPTG to a final concentration of 1 mM and continue bacterial growth at 37°C for 3 h.
4. Centrifuge the bacteria at 3,000g for 15 min at 4°C and discard the supernatant.
5. Resuspend the bacterial pellet in 20 mL ice cold lysis buffer (*see* **Subheading 3.2., step 4**) and sonicate for 4 × 10 s using a Branson Sonifier (Duty Cycle 30% and Output power 0.5).
6. Centrifuge the sonicated extract at 3000g for 5 min and discard the pellet representing bacterial cell debris.
7. Slowly add 100% (w/v) trichloroacetic acid to the supernatant to a final concentration of 1% (w/v) and mix. Place mixture on ice for 10 min and centrifuge in Corex tubes at 12,000g for 20 min at 4°C.

8. Discard the precipitate. Add 100% (w/v) trichloroacetic acid to adjust the final concentration to 5% (w/v) trichloroacetic acid. Mix and place on ice for 10 min. Centrifuge the mixture in Corex tubes at 12,000g for 20 min at 4°C. The 5% (w/v) TCA precipitate contains some I-1 and may be retained separately to extract any remaining I-1 (*see* **Notes 11** and **12**).

9. Retain the 5% (w/v) TCA supernatant and adjust to 15% (w/v) trichloroacetic acid. Place on ice for 5 min before centrifuging at 12,000g for 20 min.

10. Discard the supernatant. Resuspend the pellet in 3 mL of 500 mM Tris-HCl, pH 7.5, and dialyze overnight against the Dialysis buffer (described in **Subheading 3.1., step 5**).

11. To achieve complete purity of I-1, utilize preparative SDS-PAGE as described in **Subheading 3.1., step 13** (*see* **Fig. 1**). Store purified I-1 as a lyophilized powder at −80°C (*see* **Note 13**).

3.4. Analysis of I-1 as a PP1 Inhibitor—Phosphorylation by PKA

I-1 absolutely requires phosphorylation at a unique threonine by PKA for PP1 inhibition *(17,18)*. The phosphorylated protein is a potent mixed competitive inhibitor of PP1 with an IC$_{50}$ of 1 nM *(17,23)*. In contrast, the phosphorylated I-1 functions as an excellent substrate for PP2A and PP2B and will, therefore, competitively inhibit these enzymes. However, inhibition of PP2A and PP2B by I-1 shows a much higher IC$_{50}$ of 32 and 6 μM, respectively.

1. Incubate I-1 (0.25 mg/mL) in 50 mM Tris-HCl, pH 7.5 containing 100 μM ATP, 1 mM MgCl$_2$ and the catalytic subunit of PKA (0.2 μg/mL), isolated from bovine heart *(24)* or obtained commercially from Promega, at 30°C for 1–4 h. Monitor the time-dependent phosphorylation of I-1 by including a trace amount of ^{32}P-γ-ATP in the reaction. Follow incorporation of ^{32}P-phosphate into I-1 by SDS-PAGE and autoradiography. Concentration of I-1 can also be verified from the incorporation of 1 mol of ^{32}P-phosphate per mole of I-1 in this reaction.

2. Once maximal incorporation of phosphate into I-1 has been observed, inactivate PKA by heating the reaction mixture at 95°C for 10 min. Remove excess ATP, a potential phosphatase inhibitor, by desalting the reaction mixture on a Sephadex G25 column or centrifugation through a Centricon-30 filter (Amicon) (*see* **Notes 14–16**).

3.5. Analysis of I-1 as a Substrate for PP2B and PP2A—Phosphorylation by PKA

I-1 is dephosphorylated in vitro and in vivo by PP2B and/or PP2A *(25–28)*. Indeed, I-1 is one of the best in vitro substrates of PP2B, also known as calcineurin *(1,19)*. In contrast to the nanomolar concentrations of phosphorylated I-1 that are used to selectively inhibit PP1, higher concentrations of I-1 (5 μM or higher) are required for the analysis of I-1 as a PP2A/PP2B substrate (*see* **Note 17**).

1. Phosphorylate I-1 or GST-I-1 (4-6 mg/mL) with PKA catalytic subunit (1000 U or 0.1–0.3 mg/mL) in 50 mM sodium glycerophosphate buffer, pH 6.8 (or 50 mM Tris-HCl pH 7.5) containing 50 μM ATP, 1 mM MgCl$_2$ and 200 μCi of ^{32}P-γ-ATP (specific activity 4000–6500 Ci/mmole).
2. After 2 h incubation at 37°C, remove 2 μL of the reaction mixture and precipitate the phosphorylated protein by adding 1 mL of 25% (w/v) TCA and 100 μL of BSA (5–10 mg/mL). Wash twice with 1 mL each of 25% (w/v) TCA and count the pellets (Cerenkov) to determine ^{32}P incorporation into I-1.
3. Repeat **step 2** at 30 min intervals until stoichiometric phosphorylation of I-1 has been achieved (usually 3–4 h).
4. Remove excess labeled ATP using Centricon 30 as described in **Subheading 3.4., step 2**. Repeat the dilution and concentration steps at least twice to remove most of the labeled ATP. Finally, dilute the radiolabeled substrate to a total volume of 1–2 mL with 50 mM sodium glycerophosphate pH 6.8. Count 2 μL aliquots of this solution before and after TCA precipitation to ensure that the background (i.e., non-TCA-precipitable counts) is below 1–2% of the total counts.

3.6. Thiophosphorylation of I-1 with PKA (for affinity purification of PP1 and PP2B and for use as a PP1 inhibitor in crude extracts and intact cells)

Thiophosphorylated proteins are very poorly dephosphorylated by protein phosphatases *(29)* and have been successfully used for the affinity purification of protein phosphatases *(30,31,32)*. Thiophosphorylated I-1, in particular, is essentially resistant to dephosphorylation by cellular phosphatases either in vitro or in vivo *(6,7)* (*see* **Note 18**). This has promoted the use of thiophosphorylated I-1 to inhibit PP1 activity in crude extracts or in intact cells where there are serious concerns about its potential dephosphorylation and inactivation by one or more type-2 phosphatases.

1. Incubate I-1 (4-6 mg/mL), purified by preparative SDS-PAGE, with PKA catalytic subunit (1000 U or 0.1–0.3 mg/mL) in 50 mM sodium glycerophosphate buffer, pH 6.8 (or 50 mM Tris-HCl, pH 7.5) containing 0.2 mM ATP-γ-S and 2 mM MgCl$_2$. The incubation should be carried out at 30°C for at least 24 h. In some cases, it may take up to 72 h to complete thiophosphorylation of I-1. Incubate a parallel or control reaction containing all components except PKA (*see* **Note 19**).
2. Filter the reaction mixtures through Centricon-30 to remove excess ATP-γ-S as described in **Subheading 3.4., step 2**.
3. Analyze an aliquot of the two mixtures by SDS-PAGE to ensure that the I-1 has not been proteolyzed during the lengthy incubation.
4. Incubate aliquots of the control and thiophosphorylated I-1 with ^{32}P-γ-ATP as described in the radiolabeling protocol (**Subheading 3.4.**) that is used to prepare I-1 for PP1 inhibition. The back-phosphorylation determines the extent of the prior thiophosphorylation of I-1. In general, depending on the specific activity of

PKA catalytic subunit (preferably greater than 300 U/mg), more than 90% of I-1 becomes thiophosphorylated in 24 h. In other words, the thiophosphorylated I-1 is radiolabeled to a maximum of 5–10% of control unphosphorylated I-1.

3.7. PP1 Inhibitor Assays

One unit of PP1 activity is defined as that which hydrolyses 1 nmol of phosphorylase a in 1 min. I-1 activity, on the other hand, can be defined either in terms of units where 1 unit inhibits 0.02 U of PP1 by 50% in this assay or an IC_{50} of 1 nM for PP1 inhibition by purified I-1. However, PP1 is also present in the phosphatase assays at low nanomolar concentrations, so that effective PP1 inhibition requires knowing both the purity of the I-1 preparation and the concentration of PP1 to be used in this assay (*see* **Note 20**).

1. Incubate 20 μL of ^{32}P-labeled phosphorylase a (6 mg/mL; 100,000–200,000 cpm) with 20 μL phosphorylated I-1 in 50 mM Tris-HCl, pH 7.5 containing 1 mM EDTA and 0.005% Brij 35 (Buffer A) in a 1.5 mL microfuge tube at 37°C for 2 min (*see* **Note 21**).
2. Add 20 μL of PP1 in 50 mM Tris-HCl, pH 7.5 containing 1 mM EDTA, 1% (w/v) BSA, 0.3% (v/v) β-mercaptoethanol (Buffer B) and continue the incubation at 37°C for 10 min.
3. Terminate the assay by adding 100 μL of ice cold 25% (w/v) TCA and 100 μL of BSA (5 mg/mL). Vortex and place on ice for 5 min.
4. Centrifuge in a microcentrifuge at 15,000g for 5 min.
5. Pipet 200 μL of supernatant into a minivial. Add 2 mL of scintillation fluid (Safety-Solve, RPI) and count ^{32}P-phosphate released (*see* **Notes 22–24**).

3.8. Inhibitor-1 as a Substrate for PP2B

I-1 phosphorylated by PKA is primarily a substrate for PP2A and PP2B with PP2B being significantly more efficient (two- to fivefold) as an I-1 phosphatase (*see* **Note 25**).

1. Incubate ^{32}P-labeled I-1 (5 μM) with bovine brain PP2B (20–30 ng; UBI) diluted in 50 mM Tris-HCl, pH 7.5 containing 0.1 mM EGTA (or EDTA), 5 mM MgCl$_2$, 0.1 mM MnCl$_2$, 0.2 mM CaCl$_2$, 0.1 mM DTT, calmodulin (10 μg/mL) and BSA (0.1 mg/mL) in a total volume of 100 μL at 30°C.
2. After incubating for 10 min at 30°C, terminate the reaction by adding 0.1 mL 25% (w/v) TCA and 0.1 mL of BSA (5–10 mg/mL) (*see* **Note 26**).
3. Stand on ice for 5 min and centrifuge in a microcentrifuge at 15,000g for 5 min.
4. Analyze ^{32}P-release in 200 μL of supernatant as described in **Subheading 3.7., step 5**).

4. Notes

1. Many earlier purification procedures *(18–20)* exploited I-1's thermostability for the first step of its isolation. In these procedures, the tissue extracts were heated

in a boiling water bath to denature the skeletal muscle proteins. Given the volumes involved in large scale preparations, it often took 15–20 min to raise the extract temperature from 4°C to above 90°C. Combined with the further 15 min heating that was used to ensure extensive denaturation of the muscle proteins, this resulted in extensive proteolysis (20–60%) of I-1. This was determined by including the 30 kDa recombinant human I-1 as an internal marker in the rabbit skeletal muscle extracts and immunoblotting with a polyclonal antibody generated against the human I-1 (Endo,S. and Shenolikar,S., unpublished data) before and after the heating step. Using ^{32}P-labeled human I-1, it was also found that heating promoted the complete dephosphorylation of I-1 in the muscle extracts. An alternate method *(33)*, which denatured the muscle proteins more rapidly, involved the addition of the extract to 5 vol of heated buffer (50 mM Tris-HCl, pH 7.5) at 100°C with constant stirring. Slow addition of the ice cold tissue extract and the continued heating prevented the temperature of the mixture from falling below 90°C. This significantly reduced the extent of I-1 proteolysis (< 10%), but did not protect against its dephosphorylation. This procedure also increased in the extract volume, making it impractical for large scale preparations of I-1. Thus, heat treatment as a first step in purification is probably better suited for small scale preparations (100–200 g tissue). By homogenizing in 1% (w/v) trichloroacetic acid, complete protection was achieved against both proteolysis and dephosphorylation of I-1. At the same time, the sample is concentrated, thereby facilitating later steps of the purification.

2. The fractionation procedure which used 45–80% (v/v) ethanol in 50 mM Tris-HCl, pH 8.5 containing 1.0 mM EDTA and 200 mM NaCl was eliminated as an added step in the purification of I-1 from tissues *(19,20)*. Variable yields of I-1 were also observed using this procedure and we have not yet resolved the reasons behind this variability.

3. Earlier purification procedures *(18–20)* also utilized ion-exchange chromatography on DEAE-Cellulose, DEAE-Sephadex or Mono-Q FPLC. This was an excellent step for separating I-1 from I-2. Ion-exchange chromatography also partially separates I-1 from PKI, which can potentially inhibit its in vitro activation by PKA. A later step of the procedure repeated the ion-exchange chromatography taking full advantage of the fact that phosphorylated I-1 eluted at a higher salt concentrations than the unphosphorylated protein. However, unless earlier steps involved heat treatment of cell extracts or other procedures that promoted the dephosphorylation of I-1, ion exchange chromatography as an early purification step would be predicted to separate phospho- and dephospho-I-1 and thereby reduce the overall yield of I-1.

4. Gel filtration on Sephadex G100 readily separates I-1, apparent molecular weight 65,000, from PKI, which elutes with an apparent molecular weight of 25,000. This permits efficient phosphorylation of I-1 by the PKA subunit isolated from bovine heart *(24)* and, thus, provides a convenient assay for I-1. However, gel filtration does not completely separate I-1 from I-2, so that PP1 inhibitor assays cannot be utilized to monitor I-1 elution. However, the preparative SDS-PAGE

completely separates rabbit I-1 (apparent M_r 26,000) and I-2 (apparent molecular weight 32,000) and the sequential use of gel filtration and preparative SDS-PAGE yields homogeneously pure I-1.

5. Rabbit, rat, and human I-1 have no tryptophans or tyrosines and only a single phenylalanine *(17)*. Hence, it has a very low absorbance at 280 nm and UV absorbance cannot be used to follow its purification.

6. GST-I-1 is highly susceptible to proteolysis when expressed in most *E.coli* strains. In strains such as DH5α and JM109, particularly when protein induction is carried out at 30–37°C, no full-length 47 kDa GST-I-1 fusion protein is seen. Instead, a fusion protein of apparent molecular weight 37,000 by SDS-PAGE is obtained. This polypeptide results from a specific cleavage at residue 61 of human I-1 *(17)*. However, immunoblotting with a polyclonal antibody against human I-1 failed to visualize any peptides representing the C-terminus of human I-1. This suggests that the C-terminal 110 amino acids of I-1 are very rapidly hydrolyzed by bacterial proteases. In contrast, the N-terminal region, either because of its intrinsic structure or because of its fusion with GST, is resistant to further proteolysis. Degradation of I-1 is significantly reduced when GST-I-1 is expressed in a low-protease ompT- strain such as *E. coli* BL21 or SF100. Additional steps that reduce I-1 proteolysis include the expression of the GST-I-1 fusion protein in bacteria grown at 25°C rather than 30 or 37°C. Bacterial growth at 25°C in the presence of inducer, IPTG, for 3 h, is slightly lower (20–30%) than that seen when the bacteria are grown at 37°C. This reduces the total GST-I-1 protein obtained. However, the protein purified by affinity chromatography is largely intact (i.e., apparent M_r 47 kDa by SDS-PAGE). Yields of GST-I-1 from BL21 cells grown at 25°C are often as high as 40 mg/L.

7. Time-dependent induction of GST-I-1 by IPTG in the *E. coli* BL21 cultures was maximal after 2–3 h growth at 25°C. Further growth of the bacteria in the induction medium was not utilized as this increased the levels of proteolytic products of GST-I-1 without increasing the overall yield of fusion protein.

8. The bacterial pellet following IPTG induction of a 250 mL culture can be resuspended in 20 mL ice cold lysis buffer and stored at −80°C for periods of up to 4 mo without significant reduction in the recovery of GST-I-1.

9. The fusion of I-1 to GST slowed its phosphorylation by PKA by three- to fivefold. Moreover, the phosphorylated GST-I-1 was a less effective PP1 inhibitor than rabbit skeletal muscle I-1 (IC_{50} 32 n*M* compared with 1 n*M*). However, the phosphorylated GST-I-1 remained a highly specific PP1 inhibitor with an IC_{50} that was 10,000-fold lower than that for PP2A. Following thrombin cleavage, recombinant human I-1 was equipotent to rabbit muscle I-1 as a PP1 inhibitor. The phosphorylated GST-I-1 was also dephosphorylated by PP2A and PP2B at rates comparable to PKA phosphorylated I-1 purified from tissues.

10. Expression of I-1 as an unfused protein using the pT7-7 expression vector was surprisingly achieved in BL21 cells at 37°C without significant proteolysis. Moreover, the levels of I-1 expressed were very similar to those of GST-I-1 expressed in the BL21 cells at 25°C. The time-course of I-1 expression using the pT7-7

vector was also very similar, reaching a maximum after approximately 3 h. Bio-chemical analyses show that recombinant protein obtained by this procedure shares all the properties previously described for I-1 purified from rabbit skeletal muscle.

11. TCA fractionation takes advantage of the remarkable acid stability of I-1 and allows for the rapid and quantitative purification of this protein. Previous proce-dures have established that I-1 was not precipitated by 1% (w/v) TCA. In con-trast, I-1 is quantitatively precipitated by 15% (w/v) TCA. An intermediate fractionation step was introduced, namely 5% (w/v) TCA, that more effectively removes residual bacterial proteins, yielding essentially pure I-1 following pre-cipitation with 15% (w/v) TCA.

12. A viscous yellow sludge distinct from the white pellet that represents denatured bacterial proteins in the 5% (w/v) TCA precipitate has been noted. Analysis of this fraction shows that it contains largely I-1 (that accounts for 15–20% of total I-1 in the bacterial extract). This fraction can be carefully separated from the white pellet and redissolved in 0.5 *M* Tris-HCl, pH 8.5. Following dialysis against 50 m*M* Tris-HCl, pH 7.5, the fraction can be directly precipitated with 15% (w/v) TCA to yield essentially pure I-1. The basis for obtaining two different fractions of I-1 by TCA fractionation remains unclear.

13. As with the bacterially expressed GST-I-1, there is some proteolysis of I-1 even with the pT7-7 expression system. The major proteolytic products are visualized by loading high concentrations of the purified I-1 protein on SDS-PAGE and stain-ing with silver stain or by phosphorylation of I-1 with ^{32}P-γ-ATP and PKA prior to electrophoretic separation and autoradiography. As described in **Subheading 3.2.**, complete purification of I-1 requires the use of preparative SDS-PAGE.

14. Centricon-30 is generally used to remove excess ATP from the phosphorylation reaction. In this method, the reaction mixture (100–500 μL) is combined with 1.5 mL of 50 m*M* sodium glycerophosphate buffer and centrifuged at 1500g (Beckman JA20 rotor) for 1 h to reduce the total volume to 100 μL. The sodium glycerophos-phate buffer (1.9 mL) is once again added and the centrifugation is repeated to "wash" out most of the residual ATP. The Centricon-30 tube is then inverted into a microfuge tube and spun at 500g for 10 min to collect the phosphorylated I-1.

15. GST-I-1 (0.5 mg/mL) can also be phosphorylated using essentially the same proto-col. However, duration of the phosphorylation reaction must be increased as PKA phos-phorylates GST-I-1 three-to fivefold slower than I-1. Moreover, heat treatment or TCA precipitation cannot be used to terminate the phosphorylation reaction as these treat-ments irreversibly denature GST-I-1. Hence, desalting on Sephadex G25 or Centricon-30 is routinely used to remove excess ATP. The phosphorylated GST-I-1 is generally used immediately in the PP1 inhibitor assay (described in **Subheading 3.7.**). The dilu-tion (over 1000-fold) of phosphorylated GST-I-1 required for PP1 inhibition reduces PKA to a level where there is little or no interference in the phosphatase assay. In any case, control incubations lacking GST-I-1 are highly recommended.

16. No specific steps (e.g., heat or acid treatment) can be taken to inactivate contami-nating proteases that do not also result in loss of GST-I-1. Both phosphorylated and unphosphorylated GST-I-1 slowly degrade upon storage at 4°C. This yields

two or three closely migrating bands ranging in size from 30–37 kDa on SDS-PAGE. Only one or two of these fragments of GST-I-1 are effectively phosphorylated by PKA (White, J. E. and Shenolikar, S., unpublished observation). Moreover, once phosphorylated, these proteins poorly inhibit PP1 activity unless first cleaved from GST with thrombin *(17)*.

17. Phosphorylated I-1 is a significantly better substrate than the chromogenic substrate, p-nitrophenyl phosphate, which is most often used to monitor PP2B activity in vitro. The improved substrate properties of radiolabeled I-1 provides an extremely sensitive assay for PP2B, not only reducing the amount of enzyme required by nearly 1000-fold, but also allowing a more accurate analysis of PP2B regulation by activators like calcium/calmodulin and the inhibitors, cyclosporin, and FK506 (George, S. and Shenolikar, S., unpublished observations).

18. The thioester linkage in the thiophosphorylated I-1 is hydrolyzed 500–1000-fold slower than the phosphoester bond by PP2A or PP2B. These properties contribute to I-1's enhanced efficacy as a PP1 inhibitor in the presence of these and other cellular phosphatases.

19. PKA, like many protein kinases, utilizes thio-ATP much less efficiently than ATP. For example, PKA hydrolyses thio-ATP 10–50-fold more slowly than ATP under the conditions described in **Subheading 3.4.** Hence, higher concentrations of thio-ATP (overcoming the lower Km for this substrate) and prolonged incubations are required to ensure stoichiometric thiophosphorylation.

20. I-1 has the unusual property that it is dephosphorylated by PP1 only in the absence of other phosphoproteins and then only in assays containing Mn^{2+}. Manganese ions also increase PP1 activity against a few selected substrates including histone H1 and p-nitrophenyl phosphate. More commonly, the divalent cations inhibit PP1 activity against substrates like phosphorylase *a*. However, I-1's potency as a PP1 inhibitor is independent of the substrate used. In other words, there is no measurable dephosphorylation and inactivation by PP1 in the presence of other substrates.

21. I-2, although a constitutive inhibitor, requires preincubation with PP1 to effectively inhibit its activity. In contrast, phosphorylated I-1 is a virtually instantaneous inhibitor.

22. The phosphatase assay monitors release of ^{32}P-counts from phosphoproteins that can be readily precipitated by TCA. The ^{32}P release seen as TCA-soluble counts could arise either from the dephosphorylation of ^{32}P-labeled substrate or by cleavage by contaminating proteases to yield small phosphopeptides. In this regard, phosphorylation site sequences are located near the surface of phosphoproteins and are equally accessible to phosphatases and proteases. These two activities, however, can be readily distinguished by a procedure that specifically quantitates ^{32}P-phosphate *(19)*. Briefly, 0.1 mL of the TCA supernatant (PP1 inhibitor assay, **Subheading 3.7., step 5**) is mixed with 0.2 mL of 1.25 m*M* KH_2PO_4 in 1 N H_2SO_4 in a microfuge tube. Then, 0.5 mL of isobutanol-toluene (1:1) is added, followed by 0.1 mL of 5% (w/v) ammonium molybdate. The solution is mixed thoroughly and centrifuged at 15,000*g* for 2 min in a microcentrifuge. Then 0.3 mL of the upper (organic) phase that contains the phosphomolybdate complex is

counted in a scintillation vial containing 2.0 mL of scintillation fluid. Although somewhat more laborious, this variation of the phosphatase assay is specific for phosphate release and does not measure phosphopeptides. This assay should always be carried out when assaying tissue extracts to ensure that you are measuring phosphatases and not contaminating proteases.

23. Phosphorylase phosphatases, like PP1 and PP2A, are also distinguished from proteases by assays that include toxin inhibitors like microcystin-LR or okadaic acid (1 μM). These compounds are available from several commercial vendors and potently inhibit the major cellular serine/threonine phosphatases *(34)*.

24. With the advent of cloning, a variety of cDNAs encoding PP1 catalytic subunits have been identified. The recombinant PP1 catalytic subunits obtained by expression of these cDNAs share similar activity to the tissue-purified PP1 and can associate with a variety of regulators. However, the recombinant PP1 catalytic subunits were severely compromised in their ability to be inhibited by I-1, with IC_{50} ranging from 100–600 nM *(4)*.

25. I-1 is an unusual protein with little secondary structure *(17,35)*. This accounts for its remarkable resistance to heat, acid, detergents, and organic solvents. The lack of ordered structure also accounts for I-1's anomalous separation on gel exclusion chromatography. Whereas PKA phosphorylation enhances I-1's activity as a PP1 inhibitor by more than 50,000-fold, it does not result in a discernable change in the I-1 structure. I-1 is structurally and functionally related to the neuronal phosphoprotein known as DARPP-32 *(25–27)*. The structural homology between the two proteins is restricted to an N-terminal region defined by I-1 and DARPP-32 peptides as the functional domain of these PP1 inhibitors *(36–38)*. On the other hand, I-1 shows little or no structural homology to two other PP1 inhibitors, I-2 and NIPP-1. The molecular basis for PP1 inhibition by I-1 and other endogenous regulators is not fully understood.

   I-1 phosphorylation and activity is modulated by a variety of hormones in the heart *(11,12)*, liver *(13)*, skeletal muscle *(8–10)*, adipose tissue *(13)*, and brain *(14)*. While I-1 has been implicated in control of such diverse processes as glycogen metabolism *(27,39,40)*, muscle contraction *(41)* and learning and memory *(6)*, its physiological role remains largely unexplored.

26. PP2B activity against several substrates including I-1 is abolished by the presence of the immunosuppressive drug, FK506 (1 nM) and its intracellular receptor, FKBP-12 (100 nM). The enzyme is also inhibited by cyclosporin (10–100 nM) in the presence of cyclophilin A (1 μM). On the other hand, the dephosphorylation of I-1 by PP2A is unaffected by the immunosuppressive drugs, but is potently inhibited by okadaic acid (3 nM), microcystin LR (10 nM) and other toxins that target this phosphatase. Thus, the I-1 phosphatase may be characterized by the use of these toxins.

References

1. Ingebritsen, T. S. and Cohen, P. (1983) The protein phosphatases involved in cellular regulation. 1. Classification and substrate specificities. *Eur. J. Biochem.* **132,** 244–261.

2. Stewart, A. A., Ingebritsen, T. S., and Cohen, P. (1983) The protein phosphatases involved in cellular regulation. 5. Purification and properties of a Ca^{2+}/calmodulin-dependent protein phosphatase (2B) from rabbit skeletal muscle. *Eur. J. Biochem.* **132,** 289–295.

3. Klee, C. B., Krinks, M. H., Manalan, A. S., Cohen, P., and Stewart, A. A. (1983) Isolation and characterization of bovine brain calcineurin: a calmodulin-stimulated protein phosphatase. *Methods Enzymol.* **102,** 227-244.

4. Alessi, D., Street, A. J., Cohen, P., and Cohen, P. T. W. (1993) Inhibitor-2 functions like a chaperone to fold three expressed isoforms of mammalian protein phosphatase-1 into a conformation with the specificity and regulatory properties of the native enzyme. *Eur. J. Biochem.* **213,** 1055–1066.

5. Brautigan, D. L., Gruppuso, P. A., and Mumby, M. (1986) Protein phosphatase type-1 and type-2 catalytic subunits both bind inhibitor-2 and monoclonal antibodies. *J. Biol. Chem.* **261,** 14,924–14,928.

6. Mulkey, R. M., Endo, S., Shenolikar, S., and Malenka, R. C. (1994) Involvement of a calcineurin/inhibitor-1 phosphatase cascade in hippocampal long-term depression. *Nature* **369,** 486–488.

7. Endo, S., Critz, S, Byrne, J. H., and Shenolikar, S. (1995) Protein phosphatase-1 regulates outward K+ currents in sensory neurons in *Aplysia californica. J. Neurochem.* **64,** 1833–1840.

8. Foulkes, J. G. and Cohen, P. (1981) The hormonal control of glycogen metabolism. Phosphorylation of protein phosphatase inhibitor-1 *in vivo* in response to adrenaline. *Eur. J. Biochem.* **97,** 251–256.

9. Foulkes, J. G., Cohen, P., Strada, S. J., Everson, W. V., and Jefferson, L. S. (1982) Antagonistic effects of insulin and beta-adrenergic agonists on the activity of protein phosphatase inhibitor-1 in skeletal muscle of the perfused rat hemicorpus. *J. Biol. Chem.* **257,** 12,493–12,496.

10. Khatra, B. S., Chiasson, J. L., Shikama, H., Exton, J. H., and Soderling, T. R. (1980) Effect of epinephrine and insulin on the dephosphorylation of protein phosphatase inhibitor-1 in perfused rat skeletal muscle. *FEBS Lett.* **114,** 253–256.

11. Ahmad, Z., Green, F. J., Subuhi, H. S., and Watanabe, A. M. (1989) Autonomic regulation of type 1 protein phosphatase in cardiac muscle. *J. Biol. Chem.* **264,** 3859–3866.

12. Neumann, J., Gupta, R. C., Schmitz,W., Scholz, H., Nairn. A. C., and Watanabe, A. M. (1991) Evidence for isoproterenol-induced phosphorylation of phosphatase inhibitor-1 in intact heart. *Circulation Res.* **69,** 1450–1457.

13. Nemenoff, R. A., Blackshear, P. J., and Avruch, J. (1983) Hormonal regulation of protein dephosphorylation. Identification and hormonal regulation of protein phosphatase inhibitor-1 in rat adipose tissue. *J. Biol. Chem.* **258,** 9437–9443.

14. Snyder, G. L., Girault, J. A., Chen, J. Y., Czernik, A. J., Kebabian, J. W., Nathanson, J. A., and Greengard, P. (1992) Phosphorylation of DARPP-32 and protein phosphatase inhibitor-1 in rat choroid plexus : regulation by factors other than dopamine. *J. Neurosci.* **12,** 3071–3083.

15. MacDougall, L. K., Campbell, D. G., Hubbard, M. J., and Cohen, P. (1989) Partial structure and hormonal regulation of rabbit liver inhibitor-1; distribution of inhibitor-1 in rabbit and rat tissues. *Biochim. Biophys. Acta.* **1010,** 218–226.

16. Elbrecht, A., DiRenzo, J., Smith, R. G., and Shenolikar, S. (1990) Molecular cloning of the cDNA for protein phosphatase inhibitor-1 and expression in rat and rabbit tissues. *J. Biol. Chem.* **265,** 13,415–13,418.
17. Endo, S., Zhou, X. Z., Connor, J., Wang, B., and Shenolikar, S. (1996) Multiple structural elements define the specificity of recombinant human inhibitor-1 as a protein phosphatase-1 inhibitor. *Biochemistry* **35,** 5220–5228.
18. Nimmo, G. A. and Cohen, P. (1978) The regulation of glycogen metabolism. Purification and characterization of protein phosphatase inhibitor-1 from rabbit skeletal muscle. *Eur. J. Biochem.* **87,** 341-351.
19. Shenolikar, S. and Ingebritsen, T. S. (1984) Protein (serine,threonine) phosphatases. *Methods Enzymol.* **107,** 102–129.
20. Cohen, P., Foulkes, J. G., Holmes, C. F., Nimmo, G. A., and Tonks, N. K. (1988) Protein phosphatase inhibitor-1 and inhibitor-2 from rabbit skeletal muscle. *Methods Enzymol.* **159,** 427–437.
21. Tartof, K. D. and Hobbs, C. A. (1988) New cloning vectors and techniques for easy and rapid restriction mapping. *Gene* **67,** 169–182.
22. Tabor, S. and Richardson, C. C. (1992) A bacteriophage T7 RNA polymerase/promoter system for controlled exclusive expression of specific genes. *Biotechnology* **24,** 280–284.
23. Foulkes, J. G., Strada, S. J., Henderson, P. J., and Cohen, P. (1983) A kinetic analysis of the effects of inhibitor-1 and inhibitor-2 on the activity of protein phosphatase-1. *Eur. J. Biochem.* **132,** 309–313.
24. Beavo, J. A., Bechtel, P. J., and Krebs, E. G. (1974) Preparation of homogenous cyclic AMP-dependent protein kinase(s) and its subunits from rabbit skeletal muscle. *Methods Enzymol.* **38,** 299–308.
25. Cohen, P. (1989) The structure and regulation of protein phosphatases. *Annu. Rev. Biochem.* **58,** 453–508.
26. Shenolikar, S. and Nairn, A. (1991) Protein phosphatases: recent progress. *Adv. Second Messenger Phosphoprotein Res.* **23,** 1–121.
27. Bollen, M. and Stalmans, W. (1992) The structure, role and regulation of type-1 protein phosphatases. *Crit. Rev. Biochem. Mol. Biol.* **27,** 227–281.
28. Shenolikar, S. (1994) Protein serine/threonine phosphatases: new avenues for cell regulation. *Annu. Rev. Cell Biol.* **10,** 55–86.
29. Gratecos, D. and Fischer, E. H (1974) Adenosine 5'-o(3-thiotriphosphate) in the control of phosphorylase activity. *Biochem. Biophys. Res. Commun.* **58,** 960–967.
30. McGowen, C. H. and Cohen, P. (1988) Protein phosphatase-2C from rabbit skeletal muscle and liver: an Mg^{2+}-dependent enzyme. *Methods Enzymol.* **159,** 416–426.
31. Pato, M. D. and Kerc, E. (1988) Purification of smooth muscle myosin phosphatase from turkey gizzard. *Methods Enzymol.* **159,** 446–453.
32. Tonks, N. K., Diltz, C. D., and Fischer, E. H. (1991) Purification of protein-tyrosine phosphatases from human placenta. *Methods Enzymol.* **201,** 427–442.
33. Yang, S. D., Vandenheede, J. R., and Merlevede, W. (1981) A simplified procedure for the purification of the protein phosphatase modulator (inhibitor-2) from rabbit skeletal muscle. *FEBS Lett.* **132,** 293–295.

34. Holmes, C. F. B. and Boland, M. (1993) Inhibitors of protein phosphatase-1 and -2A; two major serine-threonine phosphatases involved in cellular regulation. *Curr. Biol.* **3,** 934–943.
35. Cohen, P., Nimmo, G. A., Shenolikar, S., and Foulkes, J. G. (1979) The role of inhibitor-1 in the cyclic AMP mediated control of glycogen metabolism in skeletal muscle. "Cyclic Nucleotides and Protein Phosphorylation in Cell Regulation" (Krause, E.G., ed.) Pergamon, Oxford, *FEBS Symp.* **54,** 161–169.
36. Shenolikar, S., Foulkes,J. G., and Cohen, P. (1978) Isolation of an active fragment of protein phosphatase inhibitor-1 from rabbit skeletal muscle. *Biochem. Soc. Trans.* **6,** 111–113.
37. Aitken, A. and Cohen, P. (1984) Isolation and characterization of active fragments of protein phosphatase inhibitor-1 from rabbit skeletal muscle. *FEBS Lett.* **147,** 54–58.
38. Hemmings, H. C. Jr., Nairn, A. C., Elliott, J. I., and Greengard, P. (1990) Synthetic peptide analogs of DARPP-32 (M_r 32,000 dopamine- and cAMP-regulated phosphoprotein), an inhibitor of protein phosphatase-1. *J. Biol. Chem.* **265,** 20,369–20,376.
39. Cohen, P. (1988) Protein phosphorylation and hormone action. *Proc. Royal Soc. Lond. Biol. Sci.* **234,** 115–144.
40. Nakielny, S., Campbell, D. G., and Cohen, P. (1991) The molecular mechanism by which adrenalin inhibits glycogen synthesis. *Eur. J. Biochem.* **199,** 713–722.
41. Gupta, R. C., Neumann, J., and Watanabe, A. M. (1993) Comparison of adenosine and muscarinic receptor-mediated effects on protein phosphatase inhibitor-1 activity in the heart. *J. Pharmacol. Exp. Ther.* **266,** 16–22.

5

I_1^{PP2A} and I_2^{PP2A}

Two Potent Protein Phosphatase 2A-Specific Inhibitor Proteins

Mei Li and Zahi Damuni

1. Introduction

Protein phosphatase 2A (PP2A) is a protein serine/threonine phosphatase that regulates the activities of key rate-limiting enzymes and proteins involved in diverse cellular processes *(1–4)*. However, although much is known about the subunit composition and sequence of different forms of the enzyme *(1–7)*, little information is available on its regulation. Nonetheless, evidence has emerged that the catalytic subunit of PP2A is subject to posttranslational modifications including phosphorylation on threonines *(8–10)* and tyrosines *(11)*, and methylation at Leu[309] *(12–14)*. Phosphorylation at either threonine or tyrosine residues inactivates PP2A *(8–10)*. By contrast, methylation appears to activate the phosphatase *(14)*.

More recently, the authors have purified to apparent homogeneity and identified by molecular cloning methods two potent heat-stable PP2A-specific inhibitor proteins which were tentatively designated I_1^{PP2A} and I_2^{PP2A} *(15–17)*. Both proteins inhibit PP2A with several substrates including ^{32}P-labeled MBP, histone H1, pyruvate dehydrogenase complex, protamine kinase, and phosphorylase *(15)*. In contrast, I_1^{PP2A} and I_2^{PP2A} exhibit little or no effect on PP2A activity with ^{32}P-labeled casein as substrate *(15)*. By contrast to PP2A, protein phosphatase-1, protein phosphatase 2B, and protein phosphatase 2C, the other three major mammalian protein serine/threonine phosphatases *(1–4)*, appear to be unaffected by I_1^{PP2A} and I_2^{PP2A} *(15–17)*. These proteins inhibit PP2A in a noncompetitive and potent manner, exhibiting apparent K_i values of about 30 nM and 25 nM, respectively, as determined with ^{32}P-labeled MBP as substrate *(15)*. Because I_1^{PP2A} and I_2^{PP2A} inhibit purified preparations of the catalytic subunit as well as native forms of PP2A, viz. PP2A$_1$ and PP2A$_2$, the inhibitor proteins appear to act, at least in part, by binding to the catalytic subunit of the phosphatase *(15–17)*. In this

From: *Methods in Molecular Biology, Vol. 93: Protein Phosphatase Protocols*
Edited by: J. W. Ludlow © Humana Press Inc., Totowa, NJ

chapter, instructions are provided on the procedure the authors developed and use routinely to purify I_1^{PP2A} and I_2^{PP2A} from extracts of bovine kidney. They also describe how to express and purify from bacteria recombinant human I_1^{PP2A} and I_2^{PP2A}.

2. Materials

2.1. Preparation of ^{32}P-labeled MBP

1. To purify MBP from bovine brain, use the procedure described by Deibler et al. *(18)*. Alternatively, purified MBP may be obtained from Sigma (St. Louis, MO).
2. Cytosolic Protamine Kinase (cPK) should be highly purified as described by Damuni et al. *(19)*.

2.2. Assay of PP2A, I_1^{PP2A} and I_2^{PP2A}

1. Assay buffer: 50 mM Tris-HCl, pH 7.0, 10% glycerol, 1 mg/mL bovine serum albumin, 1 mM benzamidine, 0.1 mM phenylmethanesulfonyl fluoride (PMSF), 0.03% Brij35, and 14 mM β-mercaptoethanol.

2.3. Purification of I_1^{PP2A} and I_2^{PP2A}

1. Buffer A: 25 mM Tris-HCl, pH 7.4, 0.25 M sucrose, 1 mM ethylenediaminetetraacetic acid (EDTA), 1 mM benzamidine, 0.1 mM PMSF, and 14 mM β-mercaptoethanol.
2. Buffer B: 25 mM Tris-HCl, pH 7.4, 10% glycerol, 1 mM EDTA, 1 mM benzamidine, 0.1 mM PMSF, and 14 mM β-mercaptoethanol.
3. Buffer C: 25 mM Tris-HCl, pH 7.4, 10% glycerol, 0.1 mM EDTA, 1 mM benzamidine, 0.1 mM PMSF, and 14 mM β-mercaptoethanol.
4. Buffer D: 25 mM Tris-HCl, pH 7.4, 10% glycerol, 4 mM EDTA, 1 mM benzamidine, 0.1 mM PMSF, and 14 mM β-mercaptoethanol.

2.4. Cloning and Expression of I_1^{PP2A} and I_2^{PP2A}

1. Reverse transcription (RT) buffer: 10 mM Tris-HCl, pH 8.3, 50 mM KCl and 5 mM MgCl$_2$.
2. Polymerase chain reaction (PCR) buffer: 20 mM Tris-HCl, pH 8.3, 10 mM KCl, 10 mM $(NH_4)_2SO_4$, 0.1% Triton X-100, 2 mM MgCl$_2$, and 0.1 mg/mL bovine serum albumin.
3. Ligation reaction (LR) buffer: 20 mM Tris-HCl, pH 7.6, 1 mM ATP, 5 mM MgCl$_2$, 5 mM dithiothreitol, and 50 mg/mL bovine serum albumin.

3. Methods

3.1. Assay Methods

3.1.1. Preparation of ^{32}P-Labeled MBP (see **Note 1**)

1. At 30°C, incubate 0.5 mg MBP in 50 mM Tris-HCl, pH 7.0, 10% glycerol, 1 mM benzamidine, 0.1 mM PMSF, 14 mM β-mercaptoethanol, 10 mM MgCl$_2$, 0.2 mM [γ-^{32}P]ATP (2000 cpm/pmol), and 50 U cPK in a final volume of 0.5 mL.

2. After 60 min, add 0.5 mL of 20% (w/v) trichloroacetic acid to terminate the reaction and centrifuge the mixture at 14,000g for 2 min in a Marathon K/M microcentrifuge (Fisher Scientific, Pittsburgh, PA).
3. Discard the supernatant fluid and wash the pellet 10 times with 1-mL portions of 10% trichloroacetic acid followed by three times with 1-mL portions of 99% ethanol.
4. Resuspend the pellet in 5 mL of 50 mM Tris-HCl, pH 7.0, containing 10% glycerol and 1 mM benzamidine.
5. Aliquot and store the solution at −70°C.

3.1.2. Assay of PP2A (see *Note 2*)

1. Dilute PP2A with assay buffer, and mix 0.04 mL of the diluted enzyme with 0.005 mL of 20 mM Mn^{2+}.
2. Start the reactions with 0.005 mL of ^{32}P-labeled MBP.
3. After 10 min at 30°C, add 0.1 mL of 10% (w/v) trichloroacetic acid to terminate the reactions and centrifuge the mixture at 14,000g for 2 min in a microcentrifuge.
4. Mix a 0.12 mL aliquot of the supernatant with 1 mL of scintillant (ScintiSafe™ Econo 1, Fisher Scientific).
5. Determine the radioactivity using a liquid scintillation counter.

3.1.3. Assay of I_1^{PP2A} and I_2^{PP2A} (see *Note 3*)

1. Measure PP2A activity as described above in the absence and presence of the inhibitor protein preparations.

3.2. Purification of I_1^{PP2A} and I_2^{PP2A} from Extracts of Bovine Kidney (see *Note 4*)

1. Perform all operations at 4°C unless otherwise indicated.
2. Transport bovine kidney from a local slaughterhouse to the laboratory on ice. Remove the kidney cortex and homogenize in a Waring blender at high setting for 1 min with 2 vol of Buffer A.
3. Centrifuge the homogenate for 30 min at 14,000g in a Beckman JA-10 rotor.
4. Discard the pellets and adjust the pH of the supernatant (extract) to 7.4 by dropwise addition, with stirring, of a 1 M NaOH solution.
5. To the extract, add, with stirring, 0.08 vol of 50% (w/v) poly(ethylene)glycol 8000 (J. T. Baker, Philipsburg, PA), followed by NaCl to a final concentration of 0.5 M.
6. Readjust the pH of the mixture to 7.4 with 1 M NaOH. After stirring for 30 min, centrifuge the mixture and discard the pellets.
7. Heat the supernatant, with constant stirring at 100°C for 10 min, and then cool to 4°C on ice.
8. Centrifuge the mixture, discard the pellets, and to the supernatant, add, with stirring, 20 g/L of trichloroacetic acid.
9. After 10 min, centrifuge the solution and discard the supernatant.
10. Resuspend the pellets in 500 mL of 70% ethanol (v/v in water), centrifuge the mixture at 30,000g in a Beckman JA-14 rotor, and discard the supernatant. Repeat this four times.

11. Resuspend the final pellets in 300 mL of buffer B, and adjust the pH to 7.4 by dropwise addition, with stirring, of 1 M Tris base.

12. Clarify the solution by centrifugation for 30 min at 30,000g in a Beckman JA-14 rotor.

13. Apply the clarified solution onto a Büchner funnel with fritted disk containing DEAE- cellulose (10 × 15 cm) equilibrated with Buffer B. Wash the column under mild suction (4 L/h) with 8 L of buffer B containing 0.2 M NaCl (see Note 5) followed by 200 mL of Buffer B containing 0.5 M NaCl. Elute inhibitor protein activity with an additional 700 mL of Buffer B containing 0.5 M NaCl.

14. Dilute this active fraction with 4 vol of Buffer B, and apply the mixture onto a Büchner funnel with fritted disk containing Q-Sepharose (5 × 4 cm) equilibrated with Buffer B. Wash under mild suction (4 L/h) with 8 L of Buffer B containing 0.2 M NaCl, followed by 100 mL Buffer B containing 0.5 M NaCl. Elute inhibitor protein activity with an additional 350 mL of Buffer B containing 0.5 M NaCl.

15. Dilute the active fraction with 4 vol of Buffer B, and apply the mixture onto a column (2.5 × 5 cm) of poly(L-lysine)agarose equilibrated with Buffer B. Wash the column with 1 L of Buffer B followed by 1 L of Buffer B containing 0.2 M NaCl. Then, apply Buffer B containing 0.8 M NaCl and collect sixty 6-mL fractions.

16. Pool about 100 mL of the active fractions and mix these, with stirring for 5 min at room temperature, with 2 vol of 99% ethanol. Centrifuge the mixture for 15 min at 39,000g in a Beckman JA-20 rotor. Discard the supernatant, and resuspend the pellets in 2 mL of Buffer B. Centrifuge the resupended pellets to remove insoluble material.

17. Apply the clarified solution onto a calibrated column (2.5 × 90 cm) of Sephacryl S-200 equilibrated and developed with Buffer B containing 0.1 M NaCl and 0.03% Brij35. Collect 1.8-mL fractions after discarding the first 150 mL of the eluate.

18. Pool the active fractions (up to 20 mL each) containing I_1^{PP2A} and I_2^{PP2A} separately (see Note 6), and mix each pool with 4 vol of Buffer B.

19. Apply each pool onto a column (1.5 × 1.5 cm) of Q-Sepharose equilibrated with Buffer B. Wash each column with 50 mL of Buffer B, and elute I_1^{PP2A} and I_2^{PP2A} with Buffer B containing 1 M NaCl. Collect 20 0.2-mL fractions.

20. Pool the active fractions containing I_1^{PP2A} and I_2^{PP2A} separately and dialyze each pool, with three changes in 16 h, against 20 vol of Buffer C.

21. Aliquot the purified I_1^{PP2A} and I_2^{PP2A} preparations (see Note 7), and store at −20°C (see Note 8).

3.3. Cloning and Expression of Recombinant Human I_1^{PP2A} and I_2^{PP2A}

3.3.1. Generation of Human I_1^{PP2A} and I_2^{PP2A} cDNA

1. To generate first strand I_1^{PP2A} cDNA, incubate 0.5 μg of human kidney polyadenylated RNA (Clontech Laboratories, Palo Alto, CA), 50 pmol of the 3'-I_1^{PP2A}-based oligonucleotide (GGATTCCACTTAGTCATCATCTT), 2.5 U of MuLV reverse transcriptase (Perkin-Elmer, Branchburg, NJ) and 1 mM of each dNTP at 42°C for 30 min in 0.02-mL RT buffer containing 20 U of RNase inhibitor.

2. Heat the mixture at 99°C for 5 min, and then cool to 4°C.
3. Add a 0.005-mL aliquot of the mixture to 0.095 mL of PCR buffer containing 100 pmol of the 3'- and a 5'- I_1^{PP2A}-based (GCGAGCCATGGAGATGGGCAGAC) oligonucleotides, 0.2 mM of each dNTP, and 5 U of recombinant *pfu* DNA polymerase (Stratagene Cloning Systems, La Jolla, CA). After heating for 5 min at 94°C, perform PCR amplification for 30 cycles in a Temptronic Series 669 thermocycler (Thermolyne, Dubuque, IA) as follows: 30 s at 94°C, 30 s at 55°C, and 3 min at 72°C, except that, in the last cycle, carry out the incubation at 72°C for 7 min.
4. Ligate the amplified cDNA (766 bp) into the *Sma*I site of pUC18 in LR buffer, and then transform *Escherichia coli* ONE SHOT™ by incubating the cells with the ligation mixture on ice for 30 min, followed by heating at 42°C for 1.5 min as described by Sambrook et al. *(20)*.
5. After growth at 37°C in Luria-Bertani medium *(20)* containing 100 μg/mL ampicillin, purify the pUC18 containing I_1^{PP2A} cDNA from the transformed cells by alkaline lysis *(20)*.
6. To generate I_2^{PP2A} cDNA, perform the RT-PCR exactly as described in **Subheading 3.3.1.**, **steps 1–3** for I_1^{PP2A}, except that, instead of the I_1^{PP2A}-based oligonucleotides, use the 3' (GGATCCTAGTCATCTTCTCCTTC) and 5' (CCG CGGAGCAGCCATATGTCGGC) I_2^{PP2A}-based oligonucleotides. Also, perform PCR amplification as described above in **step 3** (this section), except that, during the first 29 cycles, incubate at 72°C for 2.5 min. To ligate, transform, and purify I_2^{PP2A} cDNA (854 bp), follow the instructions outlined above for I_1^{PP2A} cDNA in **steps 4** and **5**.

3.3.2. Expression of I₁^{PP2A} and I₂^{PP2A} cDNA

1. Excise human I_1^{PP2A} cDNA from pUC18 with *Bam*H I and *Nco*I, and purify the cDNA from low melting agarose with GLASSMILK® (Gene Clean II, Bio 101, Vista, CA) following gel electrophoresis.
2. Linearize the bacterial expression vector pET-3d DNA (Novagen , Madison, WI) with *Bam*H I and *Nco* I, and dephosphorylate it with calf intestinal alkaline phosphatase.
3. Incubate 50 ng of the linearized pET-3d cDNA with 10 ng of the purified I_1^{PP2A} cDNA in 0.01 mL LR buffer in the presence of 4 U of T4 DNA ligase (Invitrogen, San Diego, CA) overnight at 16°C, and transform BL21(DE3)pLysS *E. coli* (Novagen) with this mixture as described above in **Subheading 3.3.1.**, **step 4**.
4. Inoculate a 5-mL seed culture of the transformed cells into 500 mL of Terrific Broth *(20)* containing 100 μg/mL ampicillin. After growth at 37°C to log phase ($A_{600\ nm} \sim 0.4$), add isopropyl-1-thio-β-D-galactopyranoside (IPTG) to a final concentration of 0.5 mM, and grow the cells for another 2 h.
5. Express human I_2^{PP2A} cDNA exactly as described above for I_1^{PP2A} cDNA, except that, instead of pET-3d and BL21(DE3)pLysS, use pET-21a DNA as the expression vector and *E. coli* BL21(DE3) as host cells. Also, instead of *Bam*H I and *Nco*I, linearize the I_2^{PP2A} cDNA and pET-21a vector with *Bam*H I and *Nde*I.

3.3.3. Purification of Recombinant Human I_1^{PP2A} and I_2^{PP2A}

Use the procedure described below to purify recombinant human I_1^{PP2A} or I_2^{PP2A}.

1. Perform all operations at 4°C unless otherwise indicated.
2. Centrifuge the IPTG-treated bacterial cells prepared as described in **Subheaidng 3.3.2., step 5**, in a Beckman JA-10 rotor for 15 min at 2000g. Discard the supernatant and wash the pellet once with 100 mL of Buffer D. After a second centrifugation, resuspend the pellet in 30 mL of Buffer D and subject the mixture to 1200 psi using a French press. Centrifuge the mixture at 39,000g for 30 min in a Beckman JA-20 rotor and discard the pellet (*see* **Note 9**).
3. Apply the supernatant onto a column (2.5 × 8.5 cm) of poly(L-lysine)agarose equilibrated with Buffer B. Wash the column with 1 L of Buffer B, and then develop it with a 600-mL linear gradient from 0 to 1.0 M NaCl. Collect 160 4-mL fractions.
4. Pool up to 120 mL of the active fractions (*see* **Note 10**) and mix with 2 vol of a 99% ethanol solution at room temperature. Centrifuge the solution at 30,000g for 10 min in a Beckman JA-14 rotor, and resuspend the pellet with 2 mL of buffer B. Centrifuge the resuspended pellet to remove insoluble material.
5. Apply the clarified solution onto a calibrated column (2.5 × 90 cm) of Sephacryl S-200 equilibrated and developed with Buffer B containing 0.1 M NaCl and 0.03% Brij 35. Discard the first 150 mL of the eluate, and then collect 1.8-mL fractions (*see* **Note 11**).
6. Pool the active fractions (up to 20 mL), and add 20 g/L trichloroacetic acid with stirring. Centrifuge the mixture in a Beckman JA-20 rotor for 10 min at 39,000g, and discard the supernatant.
7. Resuspend the pellet in a solution of 70% ethanol (v/v in water). Centrifuge the mixture and discard the supernatant. Repeat this procedure three times.
8. Resuspend the final pellet with 1 mL of Buffer C and dialyze against 20 vol of Buffer C with three changes in 16 h.
9. Aliquot the purified recombinant human I_1^{PP2A} and I_2^{PP2A} preparations separately (*see* **Notes 12** and **13**), and store at −20°C (*see* **Note 8**).

4. Notes

1. ^{32}P-labeled MBP (or histone H1) is used as substrate to assay I_1^{PP2A} and I_2^{PP2A} activity because PP2A is most sensitive to the inhibitor proteins with this substrate.
2. Assay of PP2A activity is based on measurement of the initial rate of release of ^{32}P-labeled phosphoryl groups from MBP which has been phosphorylated with $[\gamma\text{-}^{32}P]ATP$. To ensure linearity with respect to time and the amount of phosphatase, the extent of dephosphorylation of substrate is limited to < 10%. One unit of PP2A activity is defined as the amount of phosphatase that releases 1 nmol of phosphoryl groups per minute from ^{32}P-labeled MBP.
3. Inhibition of PP2A by I_1^{PP2A} and I_2^{PP2A} is linear up to 50%. One unit of inhibitor protein activity is defined as the amount that inhibits 1 U of PP2A by 50% in the standard assay.

4. This procedure is modified from Li et al. *(15)*, and can be completed within 3 d. It is based on the heat-, acid-, and ethanol-stabilities of I_1^{PP2A} and I_2^{PP2A}, and the relatively tight binding of these inhibitor proteins to poly(L-lysine)agarose. I_1^{PP2A} and I_2^{PP2A} are separated at the very last step of the procedure by gel permeation chromatography on Sephacryl S-200.

5. About 30% of the inhibitor protein activity is detected in this wash.

6. I_1^{PP2A} elutes from Sephacryl S-200 in the void vol. In contrast, I_2^{PP2A} elutes with an apparent Mr of about 80,000.

7. The purified I_1^{PP2A} and I_2^{PP2A} preparations from bovine kidney consist of single polypeptides of apparent Mr ~ 30,000 and 20,000, respectively, as determined by sodium dodecyl sulfate-polyacrylamide gel electrophoresis (SDS-PAGE).

8. These preparations are stable for at least one year.

9. To reduce proteolysis, 4 mM EDTA is included in Buffer D and heating of the preparations at 100°C is omitted.

10. Recombinant I_1^{PP2A} and I_2^{PP2A} are recovered from poly(L-lysine)agarose at about 0.65 M NaCl.

11. Recombinant I_1^{PP2A} and I_2^{PP2A} elute from Sephacryl S-200 in the void vol.

12. The purified preparations of recombinant human I_1^{PP2A} consist of a single polypeptide of apparent M_r ~ 30,000 as estimated by SDS-PAGE, similar to the apparent M_r of the bovine kidney I_1^{PP2A}. By contrast, the purified preparations of recombinant human I_2^{PP2A} consist of a single polypeptide of apparent M_r ~ 40,000 as estimated by SDS-PAGE. This value is different from the apparent M_r (~ 20,000) of purified preparations of bovine kidney I_2^{PP2A}. The reason for this discrepancy appears to be that the bovine kidney preparations are proteolyzed during the purification procedure.

13. The purified preparations of recombinant human I_1^{PP2A} and I_2^{PP2A} inhibit PP2A potently and specifically similar to the bovine kidney inhibitor protein preparations.

Acknowledgments

    This work was supported by grants BE-247 from the American Cancer Society and MCB-9513672 from the National Science Foundation.

References

1. Cohen, P. (1989) The structure and regulation of protein phosphatases. *Ann. Rev. Biochem.* **58,** 453–508.

2. Shenolikar, S. and Nairn, A. C. (1991) Protein phosphatases: recent progress. *Adv. Second Messenger Phosphoprotein Res.* **23,** 1–121.

3. Mumby, M. C. and Walter, G. (1993) Protein serine/threonine phosphatases: structure, regulation and function in cell growth. *Physiol. Rev.* **73,** 673–699.

4. Wera, S., and Hemmings, B. A. (1994) Serine/threonine protein phosphatases. *Biochem. J.* **311,** 17–29.

5. McCright, B. and Virshup, D. M. (1995) Identification of a new family of protein phosphatase 2A regulatory subunits. *J. Biol. Chem.* **270,** 26,123–26,128.

6. Csortos, C., Zolnierowicz, S., Bako, E., Durbin, S. D., and DePaoli-Roach, A. A. (1996) High complexity in the expression of the B' subunit of protein phosphatase

2A0. Evidence for the existence of at least seven novel isoforms. *J. Biol. Chem.* **271**, 2578–2588.

7. Zolnierowicz, S., Van Hoof, C., Andjelkovicm M., Cron, P., Stevens, I., Merlevede, W., Goris, J., and Hemmings, B. A. (1993) The variable subunit associated with protein phosphatase $2A_0$ defines a novel multimember family of regulatory subunits. *Biochem. J.* **268**, 15,267–15,276.

8. Guo, H., Reddy, S. A. G., and Damuni, Z. (1993) Purification and characterization of an autophosphorylation-activated protein serine threonine kinase that phosphorylates and inactivates protein phosphatase 2A. *J. Biol. Chem.* **268**, 11,193–11,198.

9. Guo, H. and Damuni, Z. (1993) Autophosphorylation-activated protein kinase phosphorylates and inactivates protein phosphatase 2A. *Proc. Natl. Acad. Sci. USA* **90**, 2500–2504.

10. Damuni, Z., Xiong, H., and Li, M. (1994) Autophosphorylation-activated protein kinase inactivates the protein tyrosine phosphatase activity of protein phosphatase 2A. *FEBS Lett.* **352**, 311–314.

11. Chen, J., Martin, B. L., and Brautigan D. L. (1992) Regulation of protein serine threonine phosphatase type-2A by tyrosine phosphorylation. *Science* **257**, 1261–1264.

12. Lee, J. and Stock, J. (1993) Protein phosphatase 2A catalytic subunit is methylesterified at its carboxyl terminus by a novel methyltransferase. *J. Biol. Chem.* **268**, 19,192–19,195.

13. Xie, H. and Clarke, S. (1994) Protein phosphatase 2A is reversibly modified by methyl esterification at its C-terminal leucine residue in bovine brain. *J. Biol. Chem.* **269**, 1981–1984.

14. Favre, B., Zolnierowicz, S., Turowski, P., and Hemmings, B. A. (1994) The catalytic subunit of protein phosphatase 2A is carboxyl-methylated *in vivo*. *J. Biol. Chem.* **269**, 16,311–16,317.

15. Li, M., Guo, H., and Damuni, Z. (1995) Purification and characterization of two potent heat-stable protein inhibitors of protein phosphatase 2A from bovine kidney. *Biochemistry* **34**, 1988–1996.

16. Li, M., Makkinje, A., and Damuni, Z. (1996) Molecular identification of I_1^{PP2A}, a novel potent heat-stable inhibitor protein of protein phosphatase 2A. *Biochemistry* **35**, 6998–7002.

17. Li, M., Makkinje, A., and Damuni, Z. (1996) The myeloid leukemia-associated protein SET is a potent inhibitor of protein phosphatase 2A. *J. Biol. Chem.* **271**, 11,059–11,062.

18. Deibler, G. E., Boyd, L. F., and Kies, M. W. (1984) Proteolytic activity associated with purified myelin basic protein. *Prog. Clin. Biol. Res.* **146**, 249–256.

19. Damuni, Z., Amick, G. D., and Sneed, T. R. (1989) Purification and properties of a distinct protamine kinase from the cytosol of bovine kidney cortex. *J. Biol. Chem.* **264**, 6412–6416.

20. Sambrook, J., Fritsch, E. F., and Maniatis, T. (1989) *Molecular Cloning: A Laboratory Manual,* 2nd ed., Cold Spring Harbor Laboratory, Cold Spring Harbor, NY.

6

Control of PP1 Activity Through Phosphorylation by Cyclin-Dependent Kinases

Norbert Berndt

1. Introduction

If it was not for the protein kinases, protein phosphatases would probably not have a right to exist. Naturally, protein kinases most often respond to an extracellular signal and catalyze a reaction that bears a positive signal—the incorporation of phosphate into a protein—a process that can be studied in a straightforward manner. It is only after the protein kinase has completed its task, that a protein phosphatase might or might not be called to action. In fact, protein phosphatases were considered to be housekeeping, if not boring, enzymes. Thus, protein dephosphorylation was—for decades—treated as a mere afterthought. This view had to be revised during the last few years, since the use of cell-permeable protein phosphatase inhibitors in many biological systems has suggested that phosphatases play an active role in most cellular processes.

We hypothesized that a serine/threonine-specific phosphatase, PP1 in our case, may be a substrate for a protein kinase, after it became clear that almost all PP1 isozymes bear a motif near the C-terminus that corresponds to a site preferentially phosphorylated by cyclin-dependent kinases (CDKs) *(1)*. Since PP1 is involved in cell-cycle control in fission yeast and fungi *(2,3)* and the dephosphorylation of the Rb protein *(4–7)*, this hypothesis became all the more attractive, because inhibitory phosphorylation would provide an elegant way of downregulating PP1 activity when CDKs phosphorylate PP1 substrates. This chapter will describe detailed methods to assay for in vitro phosphorylation of PP1, including incubation conditions, and various analyses, such as effect of phosphorylation on PP1 activity, and phosphorylation stoichiometry. Our strat-

From: *Methods in Molecular Biology, Vol. 93: Protein Phosphatase Protocols*
Edited by: J. W. Ludlow © Humana Press Inc., Totowa, NJ

egy to determine the phosphorylation site *(1)* will be briefly described. Approaches currently used in my lab to examine in vivo phosphorylation of PP1 will be discussed. Until recently, it used to be difficult to trap phosphatases in their phosphorylated form in vivo *(8)*; this may explain why phosphorylation of a serine/threonine-specific phosphatase in cells has been demonstrated in a few cases to date: The PP1-like *dis2* protein is phosphorylated by *cdc2* kinase in *Schizo-saccharomyces pombe* *(9)*, and PP2A is phosphorylated on tyrosine in fibroblasts *(10)*. This year, cell-cycle-dependent phosphorylation of PP1α was demonstrated with phosphorylation site-specific antibodies *(11)*.

2. Materials

2.1. Nonstandard Laboratory Equipment and Materials

1. Electrophoresis and blotting apparatus: X-Cell module from Novex, San Diego, CA.
2. Scanning densitometer GS300 and software GS370 v3.0, Hoefer Scientific Instruments, San Francisco, CA.
3. Several flexible exposure cassettes, 8 × 10 in., Sigma, St. Louis, MO.
4. X-ray film: Kodak X-Omat AR, 8 × 10 in.
5. SpeedVac System SC110 plus vacuum pump VP100, Savant Instruments, Farmingdale, NY.
6. Merck precoated TLC plates, 20 × 20 cm, thickness 0.1 mm. EM Science, Gibbstown, NJ.
7. Horizontal electrophoresis system to run microcellulose TLC plates. We are using the Hunter Thin Layer Peptide Mapping System HTLE-7000, which is attached to an electrophoresis power supply EPS-2000 Series II, and a Large Chromatography Tank LCT-100. All these instruments are from CBS Scientific Company, Del Mar, CA.
8. A scintillation counter to determine the PP1 activity.

2.2. Reagents and Solutions

All commercially available chemicals are from Sigma, if not otherwise indicated. Apart from the items listed below, reagents for SDS gel electrophoresis are also required.

1. Purified-protein phosphatase 1α (*see* Chapter 15).
2. A source of cyclin-dependent kinase: purified CDK1/cyclin A *(1)*, or recombinant CDK/cyclin complexes expressed in the baculovirus system, kindly provided to us by David Morgan, University of California San Francisco *(12)*, and Charles J. Sherr, St. Jude Children's Research Hospital, Memphis, TN *(13)*.
3. 5X concentrated reaction buffer: 250 m*M* Tris-HCl, pH 7.5, 5 m*M* dithiothreitol, 0.5 m*M* EDTA, 100 m*M* NaCl. Store at 4°C for up to 6 mo.
4. 100 m*M* $(CH_3COO)_2Mg$. Store at 4°C for up to 6 mo.
5. Stock solution of 100 m*M* ATP (Boehringer-Mannheim, Indianapolis, IN). Dissolve 600 mg in 8 mL H_2O. Adjust to pH 7.0 with 0.1 *M* NaOH. Adjust volume to 10 mL, and store in small aliquots at –70°C.

6. [γ-^{32}P]ATP, ICN, Irvine, CA. Store frozen at –20°C.
7. Stock solution of 100 μ*M* okadaic acid (LC Laboratories, Woburn, MA) in DMSO. Store in small aliquots at –20°C for at least 1 yr. For use in phosphatase assays, dilute to 6 n*M* okadaic acid in solution B: 50 m*M* Tris-HCl, pH 7.0, 0.1% (v/v) 2-mercaptoethanol.
8. 2X concentrated stop buffer: 100 m*M* Tris-HCl, pH 7.5, 8 m*M* EDTA, 200 m*M* NaF, 2 mg/mL bovine serum albumin (BSA), and freshly added 0.2% (v/v) 2-mercaptoethanol.
9. GF/C filter disks of 2.5 cm diameter (Whatman, Hillsboro, OR).
10. Stock solutions of 30, 10, and 5% (w/v) trichloroacetic acid (TCA). Store at 4°C indefinitely.
11. Ethanol, ether, and a 1:1 (v/v) mixture of ethanol/ether.
12. pH 1.9 buffer for thin-layer electrophoresis: 50 mL 88% formic acid, 156 mL acetic acid, and 1794 mL H$_2$O.
13. Phosphochromatography buffer for ascending chromatography: 750 mL *n*-butanol, 500 mL pyridine, 150 mL acetic acid, and 600 mL H$_2$O.
14. RIPA buffer for lysing ^{32}P-labeled cells: 50 m*M* Tris-HCl, pH 7.5, 100 m*M* NaCl, 1% (v/v) Nonidet P-40, 0.5% (v/v) Na deoxycholate, 0.1% (w/v) sodium dodecyl sulfate (SDS), 2 m*M* EDTA, and 5 μ*M* okadaic acid.

3. Methods

3.1. In Vitro Phosphorylation of PP1

3.1.1. Kinase Reaction

Depending on the nature of the planned analysis (PP1 activity assays, phosphorylation stoichiometry, SDS gel electrophoresis, phosphopeptide mapping), the incubation conditions, the amounts to be used should be carefully chosen and appropriate preparations should be taken. In a study carried out in our lab, it became apparent that different CDK/cyclin complexes—which promote distinct phases of the cell-cycle—vastly differ in their ability to phosphorylate PP1α (manuscript in preparation). The different analyses are described in subsequent sections.

1. Choose a specific radioactivity of the ATP, e.g., 2000 cpm/pmol. Prepare a working solution of ATP by diluting [γ-^{32}P]ATP with "cold" 100 m*M* ATP and H$_2$O to a final concentration of 2 m*M* (*see* **Note 1**).
2. Adjust the concentration/activity of both proteins involved in such a way that you can comfortably pipet the required amounts. For PP1, this means a concentration of approx 200 μg/mL. Especially in the case of comparing different kinases with one another, it may be advisable to normalize the kinase activity with the aid of a classical CDK substrate, such as histone or retinoblastoma protein.
3. For a typical incubation of 50 μL final volume, transfer into microcentrifuge tubes: 10 μL 5X reaction buffer, 5 μL magnesium acetate, 5 μL 2 m*M* [γ-^{32}P]ATP or "cold" ATP, 10 μL 0.1–0.5 mg/mL PP1α (1–5 μg), and H$_2$O to a final volume

of 50 µL. Start the reaction by adding 5 µL of appropriately diluted CDK and place the tube into a water bath of 30°C for up to 1 h (*see* **Note 2**).

Variations: The aforementioned volumes leave enough leeway to include either negative or positive effectors in the reaction, e.g., CDK or phosphatase inhibitors. For phosphatase activity measurements, the radioactive ATP has to be omitted, since this would interfere with the assay (*see* Chapter 15 and **Note 3**).

Controls: To control for the "autophosphorylation" of either CDK catalytic subunit or the cyclin, include a sample without added PP1. To control for possible contaminations in the PP1 preparation, include a sample without added kinase. If activity measurements are to be done, include a sample without added ATP to rule out the possibility that any components in the system inhibit PP1 activity. The volumes of these control reactions can most likely be reduced.

3.1.2. Effect of Phosphorylation on PP1 Activity

1. Before starting the reaction described above, prepare a set of microcentrifuge tubes containing 1 mL of solution B + 1 mg/mL BSA and place them on ice.
2. At specified time intervals ranging from 2 to 60 min of incubation, remove an aliquot from the reaction mixture, and transfer it into solution B containing BSA. The dilution should be chosen such that the subsequent PP1 activity assay yields results that fall within the linear range (*see* **Note 4**).
3. Immediately carry out the PP1 assay as described in the accompanying chapter (Chapter 15).

3.1.3. Detection of Phosphate Incorporation into PP1 by Electrophoresis and Autoradiography

1. Stop the phosphorylation reaction by mixing equal volumes of reaction mixture and 2X concentrated SDS sample buffer, and boil the samples for 5 min (*see* **Note 5**).
2. Separate the proteins on a 12% SDS-polyacrylamide gel. Load at least 0.25 µg PP1/lane. This would allow for comfortable detection of PP1 protein in the gel after Coomassie blue R staining.
3. Stain the gel with 0.1% (w/v) Coomassie blue R in destaining solution (43% [v/v] methanol, 7% [v/v] acetic acid, and 2% [v/v] glycerol) for 1 h at room temperature while gently shaking. Place the gel in destaining solution and continue shaking. Replace with fresh destaining solution every hour or so until the background is clear.
4. Place the wet gel onto a clean glass plate, and cover both with Saran wrap so that no liquid can escape. Place this assembly into a flexible film cassette, and expose to Kodak X-Omat AR film for 1–18 h. Start with a short exposure, and depending on the result, re-expose the gel for a longer time (*see* **Note 6**).

3.1.4. Phosphorylation Stoichiometry

The following protocol is based on a method used years ago to assess the phosphorylation of glycogen phosphorylase *(14)* that can be adopted virtually unaltered for PP1 or any other protein. The phosphorylation stoichiometry is defined as moles of phosphate incorporated into the protein per mole of protein. To determine this, three parameters are needed: the specific radioactivity of the ATP used (in cpm/mol), the specific radioactivity of the phosphoprotein to be analyzed (in cpm/mol), and the purity of the phosphoprotein (in fractions of 1).

The concentration of PP1 in the kinase reaction (**Subheading 3.1.1.**) should be at least 5 μg/50 μL or 0.1 mg/mL. The volume of the kinase reaction depends on the number of aliquots to be analyzed.

1. Incubate PP1 with CDK and [γ-^{32}P]ATP as described in **Subheading 3.1.1.** For a "blank," set up one tube without PP1 to account for the autophosphorylation of proteins present in the kinase preparation.
2. For different conditions, label GF/C filters with a permanent marker pen. At the appropriate times, dilute an aliquot of the mixture (10–50 μL) with an equal volume of 2X concentrated stop buffer. Spot 90% of this mix onto a GF/C filter, and immediately place into 10% TCA. Do the same with the "blank" sample.
3. To determine the specific radioactivity of the ATP, spot a small aliquot (5 or 10% of the reaction mixture) onto a GF/C filter and proceed to **step 5**. Do not place into TCA.
4. Wash the filters 2 × 15 min in 10% TCA, 1 × 15 min in 5% TCA, 1 × 15 min in ethanol, 1 × 15 min in diethyl ether/ethanol, and finally 1 × 15 min in ethanol.
5. Dry the filters thoroughly at 60–80°C.
6. Place the filters in scintillation fluid, and count the radioactivity.
7. Evaluate the phosphorylation stoichiometry as follows. First compute the amount of PP1, a (in pmol) that is bound per filter:

$$a = \frac{c \times v \times p}{M_r} \quad (1)$$

where c is PP1 concentration during the assay (in μg/mL), v is volume of aliquot spotted onto the filter (in mL), p is purity of PP1 (in fractions of 1) (*see* **Note 7**), and M_r is molecular mass of PP1 = 0.0374 μg/pmol.

Then compute the incorporation of phosphate into PP1, i, expressed in moles P$_i$/mol protein:

$$i = \frac{\text{Net cpm}}{SRA} \times a^{-1} \quad (2)$$

where SRA is specific radioactivity of the ATP (in cpm/pmol) and a is amount of PP1 bound to the filter (in pmol).

A typical experiment would yield the following results: Assume that c = 100 μg/mL, v = 0.025 mL, and p = 0.8, and hence a = 53.5 pmol. The specific radioactivity of the ATP in the reaction was adjusted to 1000 cpm/pmol. This has to be verified by checking the radioactivity on the "total" filter. If you count 25,000 cpm, then i = 0.47 moles of P$_i$/mol of PP1 (*see* **Note 8**).

3.1.5. Determination of the Site Phosphorylated by CDKs

The techniques involved are mentioned here, since some of them are also being used to study the in vivo phosphorylation of PP1 (*see* **Subheading 3.2.**).

Briefly, to determine the phosphorylation site, our approach was as follows: One-dimensional phosphopeptide mapping and phosphoamino acid analysis established that CDKs phosphorylate a threonine residing in a C-terminal fragment of 12 kDa. Since this fragment contains four threonines, only one of which, Thr-320, corresponds to a CDK phosphorylation site, we replaced this Thr with an alanine by site-directed mutagenesis. This mutant PP1αT320A could not be phosphorylated or inhibited by incubation with CDK1/cyclin A and ATP (1). Second, when PP1α phosphorylated by CDK was subjected to two-dimensional phosphopeptide mapping, only one major phosphopeptide was observed *(7)*. In combination, these two findings strongly suggest that Thr-320 is the only site in PP1α phosphorylated by CDKs.

To obtain a 2D phosphopeptide map requires multiple manipulations of the protein and approx 3–4 d of work. An excellent step-by-step protocol for this experimental strategy, including a discussion of many potential pitfalls, can be found in refs. *(15)* and *(16)*. What follows is a description of the major steps as well as a listing of conditions that are unique for PP1:

1. Isolate CDK-phosphorylated PP1, either by immunoprecipitation and/or SDS-PAGE.
2. Identify the radioactive PP1 band and excise from the gel, grind the gel slice with a disposable Kontes micro tissue grinder, elute the protein from the gel, and precipitate with TCA.
3. Incubate with 10 µg of protease for up to 4 h at 37°C, add another 10 µg of protease, and continue to incubate overnight. Which protease to use depends very much on the expected peptide fragments. For CDK-phosphorylated PP1a, trypsin is a good choice.
4. Repeatedly dry the arising peptides in the SpeedVac.
5. Apply the peptides to a TLC microcellulose plate, and separate them by electrophoresis and then ascending chromatography. The running conditions depend on the phosphoprotein and the nature of the arising phosphopeptides. For PP1α, we found that the following conditions produced satisfactory results:
   Electrophoresis conditions—solvent: pH 1.9 buffer, time: 50 min, constant voltage: 1000 V.
   Chromatography conditions—solvent: phosphochromatography buffer, time: 8–9 h.
6. Expose the TLC plates to X-ray film. The exposure time can reasonably be up to 1 wk. This depends on the amount of radioactivity that is being applied to the TLC plates and how many radioactive phosphopeptides have been generated. As a guideline, for 1000 cpm, an overnight exposure will normally suffice.

3.2. In Vivo Phosphorylation of PP1

Inhibition of PP1 activity by catalytic subunit phosphorylation in certain phases of the cell-cycle may be a simple mechanism, whereby a cell ensures maximal phosphorylation of PP1 substrates. However, evidence for cell-cycle-dependent phosphorylation of PP1 has been difficult to obtain. The odds for a successful isolation of phosphorylated PP1 from mammalian cells may be slim for the following reasons: First, PP1 is not one enzyme, but a family of multitask enzymes. Therefore, only a fraction of all PP1 present in the cell may be involved in cell-cycle control. In turn, only a fraction of PP1 may ever be phosphorylated. Second, given these circumstances, the method of isolating phospho-PP1 becomes all the more important. The PP1α-specific antibody generated in the author's laboratory is directed toward the C-terminal region, i.e., amino acids 316–330 *(17)*, and is incapable of immunoprecipitating PP1α phosphorylated in Thr-320, most likely because the presence of a phosphate group prevents efficient antibody binding. Using a phosphorylation site-specific antibody or an antibody directed toward residues 294–309 that were kindly provided by Angus Nairn, Rockefeller University, New York, NY *(11)* or Emma Villa-Moruzzi, University of Pisa, Italy *(18)*, we have recently been able to demonstrate in vivo phosphorylation of PP1α at two points in the cell-cycle: from late G1 to early S-phase and again in M-phase (manuscript in preparation).

The methodology to demonstrate in vivo phosphorylation of PP1 is very similar to the one that has successfully been used in so many other cases and is briefly outlined as follows:

1. Synchronize the cells in different phases of the cell-cycle. Typically, serum starvation yields cells arrested in G0/G1. Mimosine, aphidicolin, and nocodazole arrests cells in late G1, at the G1/S boundary, and in M-phase, respectively. Cells enriched in S-phase can be obtained by releasing cells from the aphidicolin blockade (*see* **Note 9**).
2. Stimulate the cells to re-enter the cell-cycle by incubating them in phosphate-free medium containing 10% serum that has been dialyzed against phosphate-free medium. After 30 min, add radioactive inorganic phosphate for 2–3 h. My laboratory uses 1–2 mCi of ^{32}P/10^6 cells. It is crucial the cells are covered with the radioactive medium during the incubation time. Nonetheless, the labeling should be done in the smallest volume possible to maximize the concentration, and, therefore, the uptake of inorganic phosphate into the cells. Therefore, the flasks with the cells should either be leveled properly or placed on a slowly moving rocking platform. During the labeling of the cells, phosphorylate purified PP1 with CDK in vitro (*see* **Subheading 3.1.1.**).
3. Lyse the cells with RIPA buffer containing 2 mM EDTA (to inhibit protein kinases), 5 μM okadaic acid (to inhibit protein phosphatases), and protease inhibitors (Chapter 15).
4. Immunoprecipitate PP1 with the appropriate antibody.
5. Separate the in vitro phosphorylated PP1 and the immunoprecipitated PP1 by SDS-PAGE and expose the gel to X-ray film.
6. Prepare the samples for 2D phosphopeptide mapping (**Subheading 3.1.5.**). For each experimental condition that produced in vivo phosphorylated PP1, prepare at least three samples before digesting the protein with trypsin: in vivo phospho-

rylated PP1, in vitro phosphorylated PP1, and a mixture of the two. Provided the same sites are phosphorylated in vitro and in vivo, the first two TLC plates should yield very similar maps, whereas the third (the mixture) should yield a map that contains only comigrating peptides (*see* **Note 10**).

Meanwhile, we have adopted a different strategy to probe for the potential significance of cell-cycle-dependent PP1 phosphorylation. The questions we asked were: Do wild-type PP1α and a constitutively active mutant, PP1αT320A, have an effect on cell-cycle progression and, more importantly, do the effects differ from each other? Therefore, we have developed a technique to introduce PP1α protein into mammalian cells by electroporation. The major conclusion of this study was that unless it is phosphorylated by CDKs, PP1α has the potential to cause G1 arrest, which depends on the presence of functional retinoblastoma protein *(19)*.

4. Notes

1. My laboratory has successfully worked with specific radioactivities ranging from 1000–10,000 cpm/pmol.
2. The ATP concentration in the reaction depends on two factors. Ideally, the ATP concentration should be way above the K_M value for the protein kinase used. This, however, may not be known, or there may not sufficient kinase available to determine it. ATP concentrations between 0.1 and 0.2 mM have worked quite well in our lab. In any case, the ATP concentration should be one or two orders of magnitude higher than the concentration of the phosphoprotein substrate, PP1. This is the case, because the 5 μg of PP1 used equals $5 \times 26.7 = 133.5$ pmol. In a 50-μL reaction, this would amount to a concentration of 2.67 μM.
3. The first phosphorylation of PP1 catalytic subunit to be discovered was catalyzed by pp60[src] and involved a tyrosine residue near the C-terminus *(20)*. This was confirmed in our laboratory. At this point, it is not yet clear whether this reaction has any physiological relevance. However, this finding demonstrates that there possibly are multiple sites in PP1 that can be phosphorylated. Small mole-wht inhibitors of CDKs, such as olomoucine *(21,22)*, can quickly establish whether or not the phosphorylation of PP1 is caused by CDK activity.

    Second, in our experiments the phosphorylation stoichiometry was never better than 0.5 mol/mol *(1)*, which raises the issue whether PP1 dephosphorylates itself or the kinase catalytic subunit during the incubation with the CDK, thereby decreasing the incorporation of phosphate. This can be addressed by either heat-inactivating PP1 prior to the phosphorylation assay or by including 5–10 μM okadaic acid in the phosphorylation assay. We found that both of these steps did not significantly increase the phosphorylation of PP1 by CDK1/cyclin A *(7)*.
4. The specific activity of PP1 catalytic subunit is on the order of 20,000 U/mg = 20 U/μg. Assuming you utilize 2 μg = 40 U of PP1/50 μL of reaction mixture, the volume activity amounts to 800 U/mL. Since the activity in the phosphatase assay has to be in the range of 1 U/mL sample, dilute the aliquot approx 800-fold.

5. The radioactive background can be reduced by precipitating the proteins with TCA. In this case, terminate the reaction by mixing 1 part reaction mixture, 1 part 0.3% (w/v) BSA, and 1 part 30% (w/v) TCA. Place on ice for 30 min. Centrifuge for 5 min at 10,000g, and wash the protein pellet three times with a 1:1 mixture of ethanol/ether to remove residual acid. Resuspend the protein pellet in 2X concentrated SDS sample buffer containing 0.05% (w/v) bromophenol blue, and place in a boiling water bath. If the sample turns yellow, add NaOH until it regains its blue color. TCA-precipitated proteins typically require longer boiling until they dissolve.

6. Only dry the gel, if the experiment is complete at this point. Although autoradiography films yield better resolutions when exposed to dried gels, it is not necessary and perhaps not even desirable to do so, if further analyses of the gel are required. The manipulation of a wet gel is certainly much easier than that of a dried gel.

7. The purity of PP1 can easily be assessed as follows. Run the PP1 preparation to be used on a 12% SDS polyacrylamide gel, and stain the gel with Coomassie blue R. After complete destaining, scan the gel, and determine the relative intensity of each protein band. A detailed description and discussion of this method can be found in an earlier volume of this series *(23)*.

8. It makes sense to go over the numbers involved before carrying out the actual experiment. Regarding our typical experiment (50 µL reaction volume, 5 µg or 133.5 pmol of PP1, 0.2 mM ATP at 1000 cpm/pmol). This means: The reaction mix contains 10 nmol of ATP with 10 ξ 10^6 cpm. Stoichiometric, that is, complete phosphorylation of PP1 (assuming there is only one site phosphorylated by the kinase) would result in 133.5 ξ 1000 = 133,500 cpm. Thus, the method described (**Subheading 3.1.4.**) is sensitive enough to detect comfortably even a low level phosphorylation of < 0.1 mol P_i/mol protein.

   If the proteins to be used are scarce, there is an alternative approach to obtaining a good estimate for phosphate incorporation into the protein of interest: Load increasing amounts (cpm) of a ^{32}P-labeled phosphoprotein with a known phosphate content (e.g., phosphorylase-*a* with ideally 1 mol P/mol protein) onto adjacent lanes of an SDS-polyacrylamide gel. Expose the dried gel to X-ray film, scan the autoradiograph in a densitometer, and plot the intensity (arbitrary units) against the cpm loaded onto the gel. This will result in a linear curve over quite a wide range. Comparing your actual result with this standard curve will give you a fairly good estimate of the phosphorylation stoichiometry, provided of course that the amount of phosphatase (pmoles) loaded onto the gel and the specific radioactivity of the ATP used are known.

9. The synchronization by low-serum or drug-induced cell-cycle arrest may lead to nearly perfect synchrony at distinct stages, but to detect PP1 that is being phosphorylated in other phases, it may be necessary to release cells from the various cell-cycle blocks and then stop them after defined intervals.

10. Provided antibodies that are specific for the CDK-phosphorylated form of PP1 are being used, lysates from unlabeled, synchronized cells could simply be ana-

lyzed by Western blots. However, since this would only yield a snapshot of the momentarily phosphorylated PP1, my lab still prefers to use ^{32}P-labeled cells followed by immunoprecipitation, since this better accommodates the dynamic nature of protein phosphorylation reactions. At least, the need for the cumbersome 2D phosphopeptide mapping procedure would be eliminated.

Acknowledgments

Work in my laboratory was supported in part by grants from the Tobaco-related Disease Research Program of the State of California (3KT-0193) and the National Institutes of Health (1R01-CA54167). I thank Mariam Dohadwala and Cathy Liu for their contributions to this work.

References

1. Dohadwala, M., Da Cruz e Silva, E. F., Hall, F. L., Williams, R. T., Carbonaro-Hall, D. A., Nairn, A.C., et al. (1994) Phosphorylation and inactivation of protein phosphatase 1 by cyclin-dependent kinases. *Proc. Natl. Acad. Sci. USA* **91**, 6408–6412.
2. Ohkura, H., Kinoshita, N., Miyatani, S., Toda, T., and Yanagida, M. (1989) The fission yeast dis^{2+} gene required for chromosome disjoining encodes one of two putative type 1 protein phosphatases. *Cell* **57**, 997–1007.
3. Booher, R. N. and Beach, D. (1989) Involvement of a type 1 protein phosphatase encoded by bws^{1+} in fission yeast mitotic control. *Cell* **57**, 1009–1016.
4. Ludlow, J. W., Glendening, C. L., Livingston, D. M., and DeCaprio, J. A. (1993) Specific enzymatic dephosphorylation of the retinoblastoma protein. *Mol. Cell. Biol.* **13**, 367–372.
5. Alberts, A. S., Thorburn, A. M., Shenolikar, S., Mumby, M. C., and Feramisco, J. R. (1993) Regulation of cell-cycle progression and nuclear affinity of the retinoblastoma protein by protein phosphatases. *Proc. Natl. Acad. Sci. USA* **90**, 388–392.
6. Durfee, T., Becherer, K., Chen, P.-L., Yeh, S.-H., Yang, Y., Kilburn, A.E., et al. (1993) The retinoblastoma protein associates with the protein phosphatase type 1 catalytic subunit. *Genes Dev.* **7**, 555–569.
7. Berndt, N. (1995) Phosphorylation of protein phosphatase 1 by cyclin-dependent kinases: A novel mechanism for cell-cycle control? *Adv. Protein Phosphatases* **9**, 63–86.
8. Brautigan, D. L. (1995) Flicking the switches: Phosphorylation of serine/threonine protein phosphatases. *Semin.Cancer Biol.* **6**, 211–217.
9. Yamano, H., Ishii, K., and Yanagida, M. (1994) Phosphorylation of dis2 protein phosphatase at the C-terminal cdc2 consensus and its potential role in cell-cycle regulation. *EMBO J.* **13**, 5310–5318.
10. Chen, J., Parsons, S., and Brautigan, D. L. (1994) Tyrosine phosphorylation of protein phosphatase 2A in response to growth stimulation and v-*src* transformation of fibroblasts. *J. Biol. Chem.* **269**, 7957–7962.
11. Kwon, Y.-G., Lee, S.-Y., Choi, Y., Greengard, P., and Nairn, A. C. (1997) Cell-cycle-dependent phosphorylation of mammalian protein phosphatase 1 by cdc2 kinase. *Proc. Natl. Acad. Sci. USA* **94**, 2168–2173.

12. Desai, D., Wessling, H. C., Fisher, R. P., and Morgan, D. O. (1995) Effects of phosphorylation by CAK on cyclin binding by CDC2 and CDK2. *Mol. Cell. Biol.* **15**, 345–350.

13. Kato, J.-Y., Matsuoka, M., Strom, D. K., and Sherr, C. J. (1994) Regulation of cyclin D-dependent kinase 4 (cdk4) by cdk4-activating kinase. *Mol. Cell. Biol.* **14**, 2713–2721.

14. Berndt, N. and Rösen, P. (1984) Isolation and partial characterization of two forms of rat heart glycogen phosphorylase. *Arch. Biochem. Biophys.* **228**, 143–154.

15. Boyle, W. J., Van der Geer, P., and Hunter, T. (1991) Phosphopeptide mapping and phosphoamino acid analysis by two-dimensional separation on thin-layer cellulose plates. *Methods Enzymol.* **201**, 110–149.

16. Van der Geer, P., Luo, K., Sefton, B. M., and Hunter, T. (1993) Phosphopeptide mapping and phosphoamino acid analysis on cellulose thin-layer plates, in *Protein Phosphorylation—A Practical Approach*, 1st ed. (Hardie, D.G., ed.), IRL, Oxford, England, pp. 31–60.

17. Runnegar, M. T., Berndt, N., Kong, S., Lee, E. Y. C., and Zhang, L. (1995) *In vivo* and *in vitro* binding of microcystin to protein phosphatases 1 and 2A. *Biochem. Biophys. Res. Commun.* **216**, 162–169.

18. Villa-Moruzzi, E., Puntoni, F., and Marin, O. (1996) Activation of protein phosphatase-1 isoforms and glycogen synthase kinase-3β in muscle from *mdx* mice. *Int. J. Biochem. Cell Biol.* **28**, 13–22.

19. Berndt, N., Dohadwala, M., and Liu, C. W. Y. (1997) Constitutively active protein phosphatase 1α causes Rb-dependent G1 arrest in human cancer cells. *Curr. Biol.* **7**, 375–386.

20. Johansen, J. W. and Ingebritsen, T. S. (1986) Phosphorylation and inactivation of protein phosphatase 1 by pp60[v-src]. *Proc. Natl. Acad. Sci. USA* **83**, 207–211.

21. Vesely, J., Havlicek, L., Strnad, M., Blow, J. J., Donella-Deana, A., Pinna, L., et al. (1994) Inhibition of cyclin-dependent kinases by purine analogues. *Eur. J. Biochem.* **224**, 771–786.

22. Abraham, R. T., Acquarone, M., Andersen, M., Asensi, A., Bellé, R., Berger, F., et al. (1995) Cellular effects of olomoucine, an inhibitor of cyclin-dependent kinases. *Biol. Cell* **83**, 105–120.

23. Smith, B.J. (1994) Quantification of proteins on polyacrylamide gels (nonradioactive), in *Methods in Molecular Biology*, vol. 32, *Basic Protein and Peptide Protocols* (Walker, J.M., ed.), Humana, Totowa, NJ, pp. 107–111.

7

Regulation of Neuronal PP1 and PP2A During Development

Elizabeth Collins and Alistair T. R. Sim

1. Introduction

A variety of studies in recent years have indicated that modulation of protein phosphatases perturbs brain function. For example, a mutation in PP1 is associated with defective habituation and associative learning in drosophila *(1)*, and the activity of PP1 and/or PP2A is essential for induction and maintenance of long term depression (LTD) in hippocampal neurons *(2)*. Recent studies link reduced PP2A activity with hyperphosphorylation of Tau in Alzheimer's disease *(3)*. A number of studies using inhibitors of PP1 and PP2A have also implicated a role for PP1 and PP2A in regulating neurotransmitter release. Studies from this laboratory suggest a role for these enzymes in priming or modulation rather than triggering of release *(4)*. At the molecular level, regulation of PP1 and PP2A has been implicated in modulating the activity of several receptors and ion channels. For example, inhibition with okadaic acid leads to stimulation of non-NMDA-type glutamate receptors *(5)* and activation of NMDA-type receptors induces specific dephosphorylation and inactivation of DARPP-32, a neuronal homolog of PP1-inhibitor 1 *(6)*. Given the ubiquitous distribution and substrate overlap of the catalytic subunits of PP1 and PP2A, both between cell types and within cells, the current challenge is to delineate and localize the individual mechanisms regulating PP1 and PP2A in the nervous system.

1.1. Unique Features of PP1 and PP2A in Brain

Surprising little is known about the regulation of neuronal PP1 and PP2A. However, it would appear that there are a number of unique features of PP1

From: *Methods in Molecular Biology, Vol. 93: Protein Phosphatase Protocols*
Edited by: J. W. Ludlow © Humana Press Inc., Totowa, NJ

and PP2A in brain. It has been shown that PP1 and PP2A are responsible for the majority of protein dephosphorylation in isolated nerve endings under basal and depolarization conditions *(7)*. However, under basal conditions, the inhibitor okadaic acid only raised the level of phosphorylation by approx 60%. This is in contrast to the severalfold increases observed in other tissues, and suggests that the activities of PP1 and PP2A are normally low in neuronal tissue. It has also been shown that the majority of both PP1 and PP2A are associated with membrane fractions of brain and that these forms are severalfold less active than their cytosolic counterparts or counterparts in other tissues *(8)*. This inactivation is reversible such that there exists considerable potential for control of both PP1 and PP2A in brain.

1.2. Brain Development and Memory

The development of functional synaptic connections can be divided into two broad phases: synapse formation (where there is a net increase in the number of synaptic contacts) and synapse maturation (where there is no net change in the number of synapses, but functional synaptic networks are changed into those with adult properties). Changes occurring during the maturation phase are believed to model the changes in adult brain during learning or memory tasks and, therefore, also represent a model for neuronal plasticity. In most systems, the processes of synapse formation and maturation overlap making it difficult to examine the processes specific to maturation. However, in the developing chicken brain, these two phases are temporally separated, with the maturation phase occurring over a particularly protracted period. Thus, the developing chicken brain provides an excellent system for studying the specific molecular events occurring during learning and memory *(9)*.

1.3. Developmental Changes in PP1 and PP2A in Brain

Studies in the laboratory have shown that PP1 and PP2A are selectively regulated during chicken brain development *(10)*. Whereas the concentration of PP2A in membrane and cytosolic fractions remains relatively constant throughout development, there is a relative decrease in the specific activity of the membrane-associated PP2A during the synapse formation phase. In contrast, the concentration of membrane-associated PP1 increased approx 40% during synapse formation and a further 60% during the maturation phase. During this same maturation period, the activity of membrane-associated PP1 decreased four-fold with little change in cytosolic PP1 activity. These results suggest that different mechanisms for the regulation of PP1 and PP2A occur developmentally, and that PP1 may have a specific role in regulating the maturation of neuronal circuits and therefore memory formation.

1.4. General Methods Used to Measure PP1 and PP2A Activity

1.4.1. PP1 and PP2A Activity In Vitro

PP1 and PP2A activity are routinely measured by quantitating the release of radio label from a substrate that has been prior phosphorylated using ATP-γ ^{32}P. Phosphatase activity is very dependent upon the substrate being examined; under identical conditions, activity toward one substrate may be elevated while activity toward another substrate is diminished. It is, therefore, essential to identify the substrate relevant to the physiological process of interest. Assays utilizing peptide substrates may also be used, although PP1 does not dephosphorylate peptide substrates efficiently *(11)*. Thus, model substrate peptides are available for PP2A, but not PP1. There is considerable overlap in the substrate specificity of PP1 and PP2A such that dephosphorylation of the substrate being used cannot necessarily be taken as an indicator of a particular phosphatase activity. However, PP1 and PP2A may be distinguished by their sensitivity to two heat stable inhibitor proteins (I-1 and I-2) *(12)* or the marine toxin okadaic acid. PP1 is specifically inhibited by I-1 and I-2, whereas under appropriate conditions, 2 nM okadaic acid will specifically inhibit PP2A. A combination of both inhibitors is routinely used.

1.4.2. PP1 and PP2A Activity In Vivo

Because of the considerable overlap in substrate specificity, measurement of activity in intact tissue is difficult unless absolute specific substrates or sites are known. However, by using a number of different cell permeable toxins, it is possible to gain an indication of the phosphatase involved in a particular dephosphorylation event. Okadaic acid has high affinity for PP2A over PP1, whereas Calyculin A is equipotent toward both phosphatases *(13)*. Thus, effects observed at low calyculin A concentrations that are not observed with low concentrations of okadaic acid may be taken to indicate PP1 inhibition effects. Alternatively, effects at low concentrations of both toxins, may indicate effects caused by PP2A inhibition. However, the permeability of cell membranes to these toxins may vary, and the subcellular location of the particular phosphatase may be different such that results obtained from this approach should be interpreted with caution.

Depolarization of isolated synaptosomes results in a time dependent activation of protein phosphatases. In response to potassium depolarization of ^{32}P-labeled synaptosomes, the dephosphorylation of specific proteins can be observed by polyacrylamide gel electophoresis (PAGE) and autoradiography *(14)*. The careful use of specific inhibitors may allow the identification of specific phosphatase substrates.

1.4.3. Nonradioactive Methods

A number of nonradioactive methods have been developed to measure phosphatase activity. These methods differ in that the released phosphate is usually determined colorimetrically. Whereas these methods are satisfactory for some applications, they are less sensitive and are prone to high background. Recently a method based on iron-chelate chromatography has been developed (*see* **Subheading 3.4.**). Using minicolumns of iron charged chelating sepharose, phosphatase activity is measured by separating phospho and dephospho forms of a dye-tagged peptide substrate. This assay is optimized for PP2A, but it is possible to adapt the method to any peptide substrate.

1.4.4. Subcellular Fractionation and Sample Preparation

A number of issues relating to sample preparation must be considered. Phosphatase activity is particularly sensitive to the buffer conditions, which must, therefore, be standardized for all conditions. Furthermore, increasing sample dilution increases phosphatase activity/mg protein, but diminishes the actual dephosphorylation measured, such that the conditions chosen represent a compromise between the maximum possible activity/mg and the maximum possible sample dilution. Under standard conditions phosphatase activity is only linear up to 30% dephosphorylation, and factors including substrate concentration, temperature, proteolysis, and heat denaturation, must be taken into account. PP1 and PP2A activity can be measured after freezing samples, but this does lead to differential changes in activity, such that fresh tissue is preferable.

1.5. Quantitation of PP1 and PP2A

1.5.1. Use of Activity Measurements

Many studies often relate the activity of PP1 and PP2A in a given cell extract to the amount of enzyme present. Thus, it has been published that the lack of PP2A activity in membrane fractions indicates lack of enzyme. In brain this is not the case, with significant amounts of PP2A being shown to be present by Western blotting techniques.

1.5.2. Post-translational Modifications

A number of antibodies are now commercially available for the catalytic subunits of PP1 and PP2A. However, the antigen used to raise these specific antibodies must be known before results obtained with them can be properly interpreted. The majority of antibodies to PP1 are raised against a sequence at the C-terminus of the catalytic subunit. This sequence is contained within a region that is particularly susceptible to proteolysis *(15)*. Thus, lack of signal does not indicate lack of enzyme. Indeed, proteolysis of this region can lead to activation of the enzyme. Similarly, the majority of antibodies to PP2A are raised against a C-terminal peptide. It has now been shown that the C-terminal Leucine of PP2A is carboxymethylated and this posttranslational modification inhibits the binding of antibody *(16)*. Thus, once again, lack of signal does not indicate lack of enzyme. With these PP2A antibodies maximum signal can be obtained by incubating nitrocellulose after transfer with 0.1 *M* NaOH to demethylate the protein prior to blotting.

1.6. Purification of PP1 and PP2A

PP1 and PP2A are both multisubunit enzymes comprising ubiquitous, highly conserved catalytic subunits, and variable regulatory subunits. The regulatory subunits, direct the substrate specificity and subcellular location of the catalytic subunits and are, therefore, tissue and subcellular location-specific. For purification of the catalytic subunits from brain, standard protocols developed using skeletal muscle can be applied. The regulatory subunits of PP1 and PP2A in brain, however, are largely unidentified. Recently, Microcystin-affinity chromatography has been applied *(17)* to the purification of PP1- and PP2A-binding proteins as candidate regulatory subunits (*see* **Subheading 3.8.**). The principle behind this method is the high and specific affinity of the catalytic subunits of PP1 and PP2A for the algal toxin, microcystin-LR. Microcystin-LR is first derivatized to create a free-amino group in a region of the molecule not required for binding the phosphatase. The derivatized microcystin is then linked to CH sepharose for affinity chromatography. Purification by this method can identify binding proteins from a single rat brain. Whereas this method is highly specific for PP1 and PP2A, it is first necessary to separate PP1 and PP2A in order to determine which holoenzyme the binding proteins are derived from. It is found that in brain, phenyl-sepharose provides the best separation of PP1 and PP2A.

2. Materials
2.1. Equipment

Items of equipment generally found in a well-equipped laboratory are not included in this list.

1. Isotope laboratory or isolated area for handling radioactive materials.
2. Freeze drier or vacuum drier suitable for drying 1 mL volumes.
3. 30, 10, and 400 mL-glass homogenizers and Teflon pestle.
4. Peristaltic pump.
6. C18 Sep-paks.
7. HPLC with C18 column (to check derivitization of microcystin for affinity chromatography).
8. 1, 5, 10, and 20-mL Luer-Lock syringes.

9. 18 and 22-gage needles.
10. Tygon microbore tubing (0.3 × 0.09).
11. Densitometer (for quantitation of Western blots).
12. Large and minigel running apparatus.
13. Nitrocellulose membrane.
14. Western blot transfer apparatus for large and minigels.
15. Motorized homogenizer.

2.2. Reagents

2.2.1. Crude Subcellular Fraction

1. 30 mM Tris-HCl, pH 7.4. Add protease inhibitors before use (see **Note 1**).
2. Homogenizing Buffer: 0.32 M sucrose, 1.6 mM HEPES, protease inhibitors, pH 7.4. Store for less than 1 week.

2.2.2. Isolation of Synaptosomes

1. 4X Gradient Buffer: 1.28 M Sucrose, 4 mM EDTA, 8 mM HEPES.
2. Gradient buffer: 0.25 mM DTT, four times Gradient buffer diluted four times, pH 7.4. Protease inhibitors.
3. Standard Krebs Buffer: 118 mM NaCl, 4.7 mM KCl, 1.2 mM CaCl$_2$, 1.2 mM MgCl$_2$, 25 mM NaHCO$_3$. On day of use add 10 mM glucose and make to volume. Aerate with 5% CO$_2$ + 95% O$_2$ for 30 min, which should bring the pH to 7.4.
4. Minimal Ca^{2+} Krebs 118 mM NaCl, 4.7 mM KCl, 0.1 mM CaCl$_2$, 1.2 mM MgCl$_2$, 25 mM NaHCO$_3$. On day of use add 10 mM glucose and make to volume. Aerate with 5% CO$_2$ + 95% O$_2$ for 30 min, which should bring the pH to 7.4.

2.2.3. ^{32}P Labeling of Glycogen Phosphorylase

1. 75% Saturated ammonium sulfate pH 7.5: 5 mM Tris-HCl, 23.75 g ammonium sulfate/50 mL double deionized water (d.d.H$_2$O).
2. 45% saturated ammonium sulfate pH 7.5: 5 mM Tris-HCl, 12.9 g ammonium sulfate/50 mL d.d.H2O.
3. Reaction buffer for 20 mg/mL phosphorylase kinase: 250 mM Tris-HCl, 16.7 mM MgCl$_2$,1.67 mM Adenosine-5'-triphosphate (ATP) disodium salt, 0.83 mM CaCl$_2$, 133 mM Na$^+$ glycerophosphate, pH 8.2 (see **Note 2**).
4. Solubilization buffer: 50 mM Tris-HCl, 0.1 mM EDTA, 15 mM caffeine, 0.1% mercaptoethanol, pH 7.0.
5. 0.5 mCi ^{32}P-ATP:

$$[0.5 \text{ mCi}/10 \text{ mCi/mL}] \times \text{decay factor} \times 1000 = \text{mL needed}$$

Where 10 mCi/mL = radioactive concentration of stock ^{32}P-ATP.

2.2.4. ^{32}P-Phosphorylase Assay for PP1 and PP2A

1. 20% Trichloroacetic acid (TCA).
2. Dilution Buffer: 50 mM Tris-HCl, 0.1 mM EDTA, 1 mg/mL bovine serum albumin (BSA), pH 7.4, 0.1% mercaptoethanol.

3. Assay buffer: 50 mM Tris-HCl, 0.1 mM EDTA, 2 mM EGTA, pH 7.4, 0.1% mercaptoethanol.
4. 6 nM Okadaic acid (OA): Add 304 mL 10% dimethlysulfoxide (DMSO) or ethanol to fresh 100 μg vial of OA (400 μM). Store in 4 μL aliquots in –20°C freezer. On day of use, dilute to 6 nM with assay buffer.

2.2.5. ^{32}P Labeling of PP2A Substrate Peptide

1. 10 mM Ammonium Hydrogen Carbonate.
2. 5 mg peptide substrate, RRATVA (*see* **Note 3**).
3. 39 mM Dithiothreitol (DTT) in 50% glycerol.
4. Protein kinase A catalytic subunit 1000 U, reconstituted in 400 μL of 6 mg/mL DTT, 50% glycerol.
5. Peptide assay buffer: 222 mM Tris-HCl, 89 mM MgCl$_2$, pH 7.4.
6. 30% Glacial acetic acid.
7. AG1X8 anion exchange resin in 2 mL disposable columns.
8. 40 mM ATP. Make less than 24 h before use.
9. 30 μCi ^{32}P-ATP:

$$[30/10 \times \text{decay factor}] = \text{mL to be used in each tube}$$

Where 10 = radioactive stock concentration in mCi/mL

2.2.6. ^{32}P-Peptide PP2A Assay

1. 30% Glacial acetic acid.
2. Dilution buffer: as in **Subheading 2.2.4.**
3. Assay buffer: as in **Subheading 2.2.4.**
4. 6 nM OA: As in **Subheading 2.2.4.**

2.2.7. Nonradioactive Assay for PP2A

1. 2 mg/mL Methyl-Red labeled peptide substrate (*see* **Note 4**)
2. Assay buffer as in **Subheading 2.2.4.**
3. Dilution buffer as in **Subheading 2.2.4.**
4. Chelating sepharose (iminodiacetic acid sepharose).
5. 100 mM Sodium acetate, pH 3.3.
6. Ferric chloride 27mg/mL.
7. 20 mM Sodium dihydrogen phosphate, pH 7.0.
8. Column regeneration buffer: 50 mM Tris-HCl, 0.5 M NaCl, 0.1 mM EDTA, 0.02% Sodium azide, pH 7.5.

2.2.8. Dephosphorylation of Intact Tissue

1. ^{32}Pi radioisotope: 10 mCi/mL, ≥8500 Ci/mmole.
2. Isolated synaptosomes (*see* **Subheading 3.1.2.**).
3. Depolarization buffer: 46 mM NaCl, 77 mM KCl, 25 mM NaHCO$_3$, 1.2 mM MgCl$_2$, 100 μM CaCl$_2$, 10 mM glucose, bubbled with 5% CO$_2$ in O$_2$ for 30 min. Make fresh.

2.2.9. Raising PP1 and PP2A-Specific Antisera

1. PP1 catalytic subunit peptide conjugated to diphtheria toxoid (RPITPPRNSAKAKK).
2. PP2A catalytic subunit peptide conjugated to diphtheria toxoid (PHVTRRTPDYFL).
3. Freund's complete adjuvant (FCA).
4. Freund's incomplete adjuvant.

2.2.10. Aminoethanethiol-Microcystin

1. 1 mM HCl.
2. 50 mM NaHCO$_3$, pH 9.2.
3. Dimethyl sulfoxide (DMSO).
4. 5 N NaOH.
5. Nitrogen gas.
6. 0.1% Trifluoroacetic acid (TFA).
7. 100% Acetonitrile in 0.1% TFA.
8. 10% Acetonitrile in 0.1% TFA.
9. 0.5 M NaCl.
10. Microcystin-LR (0.5 mg).
11. CH-Sepharose.
12. Aminoethanethiol (1 g).

2.2.11. Purification of PP1 and PP2A Binding Proteins by Microcystin Affinity Chromatography

1. Homogenizing buffer: As in **Subheading 2.2.1.**
2. 30 mM Tris-HCl, pH 7.4. Add protease inhibitors to amount of buffer to be used on the day.
3. Salt elution buffer: 75 mM Triethanolamine/HCl, 0.15 mM EGTA, 7.5% glycerol, 1.5 M NaCl, 1.5 mM MnCl$_2$, 0.15 mM DTT, pH 7.5, 0.1% mercaptoethanol, protease inhibitors.
4. Detergent elution buffer: 75 mM Triethanolamine/HCl, 0.15 mM EGTA, 30% glycerol, 0.75% Triton 100, 1.5 mM MnCl$_2$, 0.15 mM DTT, pH 7.5, protease inhibitors, 0.1% mercaptoethanol.
5. 1/10 SM-2 column buffer: 5 mM Tris-HCl, 0.01 mM EDTA, 0.5% glycerol, 0.05 M NaCl, 0.1 mM MnCl$_2$, pH 7.5, 0.1% mercaptoethanol.
6. SM-2 column buffer: 50 mM Tris-HCl, 0.1 mM EDTA, 5% glycerol, 0.5 M NaCl, 1 mM MnCl$_2$, pH 7.5, 0.1% mercaptoethanol.
7. 1/5 Microcystin column buffer: 10 mM triethanolamine, 0.02 mM EGTA, 1% glycerol, 0.1 M NaCl, 0.2 mM MnCl$_2$, 0.02 mM DTT, pH 7.5, 0.1% mercaptoethanol.
8. Microcystin column stock buffer (3x): 150 mM triethanolamine, 0.3 mM EGTA, 15% glycerol, 1.5 M NaCl, 3 mM MnCl$_2$, 0.3 mM DTT, pH 7.5.
9. Microcystin column buffer: Microcystin column buffer diluted three times, 0.1% mercaptoethanol, protease inhibitors.
10. Phosphatase eluting buffer: microcystin column buffer diluted three times, 0.1 M NaCl, 3 M potassium thiocyanate (KSCN), pH 7.5, 0.1% mercaptoethanol, protease inhibitors.

11. Microcystin column storage buffer: 0.02% Na azide in 20-mL microcystin column buffer.
12. SM-2 detergent removal resin (or similar).

2.2.12. Western Blotting

1. Tris buffered saline (TBS): 10 mM Tris-HCl, 0.15 M NaCl, pH 8.0.
2. Tris Buffered Saline Tween (TBST): 10 mM Tris-HCl, 0.15 M NaCl, pH 8.0, 0.05% Tween.
3. Blocking solution 1.: 5% Skim milk in TBS.
4. Blocking solution 2.: 1.5% gelatin in TBS. Ensure gelatin is between 22–30°C.
5. Antibody buffer: 0.75% gelatin in TBST, 0.02% Sodium azide.
6. Alkaline phosphate buffer: 100 mM Tris-HCl, 100 mM NaCl, 5 mM MgCl$_2$.
7. Nitroblue tetrazolium chloride (NBT) solution: 0.05 g NBT/1 mL 70% dimethyl fluoride (DMF). NB. Store in the dark less than 1 mo.
8. BCIP solution: 0.025 g BCIP/mL 100% DMF. Store in the dark less than 1 mo.
9. Alkaline phosphate color development solution: 120 mL of NBT solution and BCIP solution in 20 mL alkaline phosphate buffer. Use immediately.
10. Ponceau S stain: 0.5% Ponceau, 1% glacial acetic acid.

3. Methods

3.1. Sample Preparation

3.1.1. Crude Subcellular Fractionation

1. Weigh 25 mL beaker with 3 mL sucrose. Stun and decapitate animal and dice brain immediately in cold Homogenizing Buffer. Weigh brain plus Homogenizing Buffer and note weight. Homogenize brain with 3 mL Homogenizing Buffer/g wet brain weight on ice with 10 up and down strokes at 700 rpm in a 30 mL glass homogenizing tube and Teflon pestle (*see* **Note 5**).
2. Dilute to 10 mL/g wet brain wt with Homogenizing Buffer. Centrifuge at 500g for 5 min at 4°C.
3. Collect supernatant (S1) into 50 mL centrifuge tubes. Add 2 mL Homogenizing Buffer/g brain wt to pellet (P1). Homogenize pellet with three up and down strokes. Centrifuge pellet 500g for 5 min at 4°C.
4. Collect supernatant from this second spin and pool with the original S1. Centrifuge at 20,000g for 20 min at 4°C.
5. Put supernatant (S2) into 10 mL ultra centrifuge tubes. Add 20 mL 30 mM Tris-HCl, pH 7.4. /g brain wt to pellet (P2) and stand on ice for at least 15 min. Centrifuge S2 and P2 at 250,000g for 60 min at 4°C.
6. Collect supernatant of S2 (S3) into cold measuring cylinder. Combine supernatant of P2 (P2S) with S3. This is considered the cytosolic fraction (*see* **Note 6**).
7. Add 2 mL of 30 mM Tris-HCl/g wet brain wt to pellet (P2M) and rehomogenize. Make P2M (membrane fraction) to 10 mL/g wet brain wt with 30 mM Tris-HCl (*see* **Note 6**).

Table 1
Percoll Gradient Solutions

Percent percoll	4x Gradient buffer, mL	Percoll, mL	25 mM DTT, mL
3	6.25	0.75	0.25
10	6.25	2.50	0.25
15	6.25	3.75	0.25
23	6.25	5.75	0.25

3.1.2. Isolation of Synaptosomes

The procedure described here is based on the method developed by Dunkley et al. (*see* **Note 7**). These quantities are for one rat forebrain (approx 1.3 g wet weight).

1. Filter percoll with AP15 pre-filters (*see* **Table 1**). Make to about 20 mL with d.d.H$_2$O and then buffer to 7.4 at 4°C. Add protease inhibitors and make to 25 mL.
2. All steps to be performed at 4°C. Add 2 mL of each Percoll solution in seven 10-mL centrifuge tubes. Gradient layer using a fine long needle attached to technicon tubing on a peristaltic pump. Deliver the percoll very slowly as to not disturb the interfaces.
3. Using a 30-mL glass and Teflon homogenizer, homogenize brain in Gradient Buffer in 3 mL/g wet wt, with ten up and down strokes at 700 rpm. Centrifuge at 2000g for 5 min (*see* **Note 5**).
4. Make supernatant (S1) to 12 mL/g wet wt with Gradient Buffer. Ideally this should give a protein concentration of about 5 mg/mL. Layer 2 mL of S1 onto each gradient using peristaltic pump.
5. Centrifuge gradients at 32,500g for 5 min. Time does not include acceleration or deceleration. Decelerated with no brake on centrifuge (*see* **Note 8**).
6. Remove desired fractions with Pasteur pipet and pool in 50 mL centrifuge tubes. Avoid collecting percoll between factions. Fraction F4 is the most enriched for synaptosomes and F3 is the next most enriched for synaptosomes. Wash with at least 3x the volume with Krebs buffer or 0.87% NaCl. Centrifuge at 21,500g for 10 min.
7. Discard supernatant and gently resuspend pellets to approximately 5 mg protein/mL using a Pasteur pipet. Repeat wash and resuspend in either isotonic solution (for intact synaptosomes), or in double volume lysis buffer (for lysed synaptosomes). For complete lysis, final ionic concentration should be < 90 mM (*see* **Notes 9** and **10**).

3.1.3 Developmental Windows

As discussed in **Subheading 1.2.**, it is possible to use chicken brain from birds of defined post-hatch age to obtain samples representative of the synaptic formation and maturation phases *(9)*. Synapse formation is essentially complete by 14 d posthatch, whereas synapse maturation occurs from wk 3 to wk 8. Thus for samples representative of the changes occurring during synapse formation, use chicken brains of 2 and 14 d posthatch. For samples representative of the changes occurring during synapse maturation, use chicken brains of 21 d and >56 d posthatch.

3.2. Standard ^{32}P-phosphorylase Assay for PP1 and PP2A

3.2.1. ^{32}P Labeling of Glycogen Phosphorylase

1. To two scintillation vials add 1 mL scintillant. Label both with the date and one "ATP" and the other "TOTAL." Add 495 μL d.d.H$_2$O to a microfuge tube and label "ATP 1/100" and store on ice. Add 300 μL reaction buffer to a separate microfuge tube and label "R". Add appropriate amount of d.d.H$_2$O to "R" tube to make the final volume (after adding all the other reagents) to 500 μL and store on ice.
2. Add 0.5 mCi ^{32}P-ATP (<100 μL) (*see* **Note 11**),100 μL glycogen phosphorylase (100 mg/mL) and 10 μL phosphorylase kinase to "R" tube (*see* **Note 12**). Vortex well and incubate for at least 30 min at 30°C.
3. After the 30 min incubation of "R" tube, mix and remove 5 μL and add to "ATP 1/100" microfuge tube. Take 5 μL of "ATP 1/100" and add to vial labeled "ATP." Vortex and keep for counting later (*see* **Note 13**).
4. Stop reaction by adding 100 μL 75% saturated ammonium sulfate (stored at 4°C) to "R" tube. Mix and repeat four more times, so that a total of 500 μL of 75% saturated ammonium sulfate has been added to "R" tube. Place in fridge for 1 h.
5. To remove excess ^{32}P-ATP, spin in microfuge for 10 min at 4°C at 15,000*g*. Discard supernatant, resuspend pellet in 1 mL 45% saturated ammonium sulfate and spin in microfuge for 10 min at 4°C. Repeat twice more (*see* **Note 14**).
6. To remove ammonium sulfate, resuspend pellet in 1.5 mL solubilization buffer (room temperature) and stand at room temperature for 10 min.
7. Dialyze against 2 L solubilization buffer for 1 h at room temperature.
8. Collect labeled phosphorylase into a vial, and increase volume to a total of 3 mL with fresh solubilization buffer.
9. To avoid possible radioactive contamination using dialysis to remove excess ATP, it can also be removed using filtration centrifugation. Resuspend pellet in a total of 1.5 mL solubilization buffer and stand at room temperature for 10 min. Transfer to an Amicon 30 concentrator, increase volume to 2.5 mL and spin at 5000*g* for 30 min at 22°C. Resuspend pellet in a total of 2.5 mL, solubilization buffer and centrifuge at 5000*g* for 30 min at 22°C. Repeat. Collect labeled phosphorylase from upper reservoir into a scintillation vial, and wash reservoir with 5 × 200-mL aliquots of solubilization buffer, adding washes to vial marked "^{32}P-phosphorylase."
10. To measure degree of ^{32}P-Phosphorylation of glycogen phosphorylase, add 10 mL ^{32}P-phosphorylase to vial marked "TOTAL." Vortex and count with the vial marked "ATP." Store all three vials in a lead container at 4°C.

11. Calculate stoichiometry:
    a. ATP Concentration = Reaction Buffer Volume (255 μL) × [ATP](1.96 m*M*)/ Total Assay Volume (500 μL)
    b. nmoles ATP concentration in aliquot = 5 μL of 1/100 = (a)/100 × 5
    b. = 0.05 nmoles ATP
    c. μg Phosphorylase in aliquot = Protein conc. as determined by Bradford assay mg/μL × 10 mL × 1000 = Pro. Conc. × 10/1000 mL
    d. nmoles Phosphorylase in aliquot = (c) × MW (96000) / 1000 = (c)/96
    e. nmoles ^{32}Phosphate per aliquot = CPM "TOTAL" × (b)/CPM "ATP"
    f. Moles ^{32}Phosphate per mole phosphorylase = (e)/(d)—(*see* **Note 15**)

3.2.2. ^{32}P-PHOSPHORYLASE PP1 AND PP2A ASSAY

1. Perform Bradford assay to determine protein concentration of samples.
2. For every sample label 3 × 1.5 mL microfuge tubes and 3 scintillant vials, adding 1 mL of scintillant to vials. Ensure water bath is at 30°C.
3. Prepare 6 n*M* okadaic acid and store on ice.
4. Dilute tissue extracts with dilution buffer to 0.06 mg/mL. Store on ice (*see* **Note 16**).
5. Add 10 mL Assay Buffer with or without okadaic acid to microfuge tubes. Add 10 mL of appropriate tissue extract or 10 mL dilution buffer as blank.
6. Start reaction by adding 10 mL ^{32}P-labeled substrate. Stagger the start of each reaction (usually by 10 s) as they are placed in 30°C water bath. Incubate for a total of 20 min (*see* **Note 17**).
7. Stop reaction by adding 90 mL of 20% TCA. Vortex and stand on ice for 10 min.
8. Centrifuge in microfuge for 4 min at 15,000*g*. Carefully remove 100 mL supernatant and add to 1 mL of scintillant. Vortex until clear (*see* **Note 18**).
9. Count the radioactivity in the samples; the original ATP aliquot to account for radioactive decay; and TOTAL in scintillation counter, to determine degree of dephosphorylation. Calculate phosphatase activity (**Subheading 3.2.2.3.**) in units per milligram protein per minute where 1 U is equivalent to 1 nmole phosphate removed from the protein.
10. Calculate activity:

    [CPM sample – CPM blank/CPM "ATP"] × (nmoles ATP) × [(Total Vol./ Vol. Counted) × (1000/mL used) × (1/time) × (1/protein concentration)]

    Standard Assay : CPM Blank = mean of triplicate; [ATP] = 0.05 nmoles as determined from labeling procedure; total volume = 120 μL; volume counted = 100 μL; mL of sample used = 10 μL; time of incubation = 20 min.

In the absence of divalent cations, PP1 is the activity in the presence of 2 n*M* okadaic acid, whereas PP2A activity is the total activity minus the activity in the presence of 2 n*M* okadaic acid (*see* **Note 19**).

3.3. ^{32}P-Peptide PP2A Assay

3.3.1. Preparation of Peptide Substrate

1. Into two 1.5-mL microcentrifuge tubes, labeled "R" add: 100 μL peptide (5 mg/mL), 90 μL assay buffer, 50 μL 40 μ*M* ATP, enough d.d. H_2O to make the final volume in the tube, (after adding ^{32}P-ATP and 10 mL of PKA), 500 μL. Store on ice (*see* **Note 20**).
2. Put 245 μL d.d.H2O into two 1.5-mL microcentrifuge tubes labeled "1/50 ATP."
3. Add ^{32}P-ATP to "R" tubes. Start reaction with 10 μL of PKA. Mix and incubate for 6 h at 30°C (*see* **Notes 20** and **21**).
4. Prior to stopping the reaction, remove 5 μL from the reaction tubes and add to the "1/50 ATP" tubes.
5. Stop the reaction with 500 μL of 30% acetic acid.
6. To remove excess radiolabeled ATP, mix and load both reaction mixtures onto a 2 mL AG1X8 column. Wash with 1 mL of 30% acetic acid and collect wash into empty scintillation vial labeled "1st." Wash again with 500 μL of 30% acetic acid and collect second wash into a second scintillation vial labeled "2nd." (*see* **Note 23**).
7. To measure the degree of labelling, take 5 μL of each of the two "1/50 ATP" tubes and add to two empty scintillation vials and label "TOTAL."
8. Count the two washes and the "TOTAL".
9. Dry down the combined eluates into one tube to a minimal volume.
10. Recount the final concentrated eluate and make to appropriate volume (~5 mg/mL) with neutralization solution.
11. Dry down and resuspend in solubilization buffer to 1 mg/mL.
12. Calculate degree of phosphorylation:
    a. ATP Concentration in assay = 400 μ*M*
    b. nmoles ATP concentration in aliquot = 5 μL of 1/50 = (a) / 50 × 5/1000 (b) = 0.04 nmoles ATP
    c. μg Peptide in aliquot = [(Protein conc. of Peptide 5 mg/mL × 100 μL) × 1000 = 500 μg/1000 μL]
    d. nmoles Peptide in aliquot = (c) × 1000/MW(=672) = X nmoles
    e. nmoles ^{32}Phosphate per aliquot = CPM "TOTAL" × 0.04 nmoles ATP/CPM "ATP"
    f. Moles ^{32}Phosphate per mole peptide = (e)/(d) (*see* **Note 15**)

3.3.2. Assay Procedure

1. Perform Bradford assay to determine protein concentration of samples.
2. For every sample label three microtubes and three scintillant vials. Ensure water bath is at 30°C. Make up 6 n*M* okadaic acid and store on ice. Dilute tissue extracts with dilution buffer to 0.06 mg/mL. Store on ice.
3. Add 50 μL assay buffer with or without okadaic acid to microtubes. Add 50 μL of appropriate tissue extract or 50 μL dilution buffer as blank.
4. Start reaction by adding 50 μL ^{32}P-labeled substrate (1 mg/mL). Stagger the start of each reaction (usually by 10 s) as they are placed in 30°C water bath. Incubate for a total of 20 min.

5. Stop reaction by adding 450 μL of 30% acidic acid. Vortex and stand on ice for 10 min.
6. Load onto AG1X8 columns as in peptide labelling procedure. Wash with 1 mL of 30% acetic acid and collect wash into empty scintillation vial labeled "1st." Wash again with 500 μL of 30% acetic acid and collect second wash into a second scintillation vial labeled "2nd."
7. To calculate results, count eluates, samples, the original ATP aliquot to account for radioactive decay and the TOTAL to determine degree of dephosphorylation. Calculate phosphatase activity in units per mg protein per minute where 1 U is equivalent to 1 nmole phosphate removed from the protein.

(CPM sample – CPM blank/CPM "ATP") × (nmoles ATP) × (Total vol./vol. counted) × (1000/mL use) × (1/time) × (1/protein concentration)

CPM Blank = mean of triplicate; (ATP) = 0.04 nmoles as determined from labeling procedure; Total volume = 600 μL; Volume counted =xμL; μL of sample used = 50 μL, Time of incubation = min.

3.4. Nonradioactive Peptide Assay for PP2A

3.4.1. Design of Peptide Substrate

This technique can be applied to virtually any peptide substrate and is, therefore, not restricted to PP2A assays. It is possible during chemical synthesis to incorporate the phosphorylated amino acid, thereby providing 100% phosphopeptide and avoiding the need for in vitro phosphorylation prior to use. It is also possible to incorporate a range of different dyes at the N-terminus. As it has been shown that attaching methyl red at the N-terminus of the PP2A substrate RRATVA, has little effect on its dephosphorylation by PP2A, this cannot be guaranteed for every substrate.

3.4.2. Assay Procedure

1. Prepare assay tubes in duplicate for each condition containing 25 μL assay buffer and 25 μL sample including blank.
2. Prepare 1 mL chelating sepharose columns by applying sequentially 10 mL water, 1 mL $FeCl_2$, 2 mL 100 mM sodium acetate. This is known as the iron chelate column (*see* **Note 24**).
3. Start reaction by adding 25 μL peptide (2 mg/mL) and incubate for 10 min at 30°C.
4. Stop the reaction by applying sample to Iron chelate columns.
5. Wash column with 1.5 mL 100 mM sodium acetate and collect flow through.
6. Elute bound phosphopeptide with 1.5 mL 20 mM sodium phosphate and collect.
7. Measure the absorbance at 450 nm of both wash and eluted samples.
8. Measure the absorbance of a total tube containing 25 L peptide, 25 μL assay buffer, 25 μL dilution buffer made up to 1.5 mL with elution buffer.

3.4.3. Calculation of Results (see **Note 25**)

Since the peptide is synthetically phosphorylated to 1 mol/mol:
Total absorbance =50 μg peptide = 50/MW(672) × 1000 nmols = nmols phosphate (a) nmols phosphate released in sample (b) = sample absorbance/total absorbance × (a)

$$\text{Activity (c)} = \text{(b)}/[\text{Mg protein in sample X time (minutes)}]$$

3.5. Dephosphorylation of Intact Tissue

3.5.1. ^{32}P-incubation

1. Prepare synaptosomes, wash and resuspend in Krebs buffer to >5 mg/mL protein.
2. Incubate 150 μL synaptosomes with sufficient ^{32}Pi to a concentration of 0.75 mCi/mL for 45 min at 37°C (*see* **Note 25**).
3. Shake sample frequently to maintain in suspension.

3.5.2. Depolarization

1. Incubate samples with appropriate concentration of phosphatase inhibitors (*see* **Note 27**) for 10 min.
2. Incubate 70 μL aliquots of preincubated synaptosomes with an equal volume of prewarmed control or depolarization buffer.
3. Stop reaction at appropriate time points (*see* **Note 28**) by addition of 70 μL sodium dodecyl sulfate (SDS) sample buffer.
4. Run samples on SDS-polyacrylamide gel elctrophoresis (PAGE), dry, and autoradiograph.
5. Dephosphorylation is measured by monitoring the time-dependent decrease in the intensity of bands visualized by autoradiography and quantitated by densitometry.

3.6. Quantitation of Phosphatase Concentration

3.6.1. Western Blotting

1. Turn on cooling unit 2 h before transferring and ensure that transfer buffer is also at 4°C (*see* **Note 29**).
2. Cut two pieces of filter paper the size of the fibre pads. Cut 14 × 15 cm pieces of nitrocellulose membrane per medium sized gels and 10 × 11.5 cm per minigel (*see* **Note 30**).
3. Cut off stacking gel off gel and lower left hand corner for orientation later. Equilibrate gel by immersion in transfer buffer for 30 min prior to transfer. Wet filter paper, fibre pads, and nitrocellulose in transfer buffer, placing nitrocellulose in buffer at a 45° angle to avoid trapping air into the membrane.
4. Prepare sandwich of fiber pad, filter paper, gel (placing cut corner on the right hand side), membrane, filter paper, and fiber pad, beginning on the cathode (black) side of the gel-cassette. Roll out bubbles with Pasteur pipet.

5. Place cartridges in tanks. Fill tank with cooled transfer buffer. Ensure that the black lead is also attached to the black electrode and negative powerpack output.

6. For a 10% gel set on 0.6 Amps and run for > 5.5 h (with magnetic stirrer on). Higher acrylamide concentration gels will need longer transfer times. For mini 10% gels set on 0.25 Amps for 1 h and 11% gels for 1.5 h (*see* **Note 31**).

7. Immerse membrane/s in first blocking solution at room temperature for 45 min. Ensure that the side of the membrane that was next to the gel is facing up (*see* **Note 32**).

8. Wash membrane/s with TBST three times for 5 min.

9. Immerse membrane/s in second blocking solution at room temperature for 45 min (*see* **Note 33**).

10. Wash membrane/s with TBST three times for 5 min.

11. Place membrane/s on trays only a little larger than the membrane, and incubate with primary antibody (2 mL/ track for 14 × 15 cm gels and 1.5 mL/track for minigels) for appropriate time (as previously determined) on rocking platform.

12. Wash membrane/s with TBST three times for 5 min, or longer if the primary antibody weakly binds to other proteins not of interest (*see* **Note 34**).

13. Incubate membrane/s for 1 h with 1/3000 dilution secondary antibody at room temperature on rocking platform (*see* **Note 35**).

14. Wash membrane/s with TBST two times and once with TBS for 5 min.

15. For color development, make Alkaline Phosphatase Color Development Solution. Pour over membrane/s and place on orbital shaker until required color intensity is achieved or before background begins to appear. Stop reaction by discarding development solution and rinse with d.d.H_2O for about 10 min. Dry membranes between filter paper.

3.6.2. Sample Preparation

1. For direct comparison of the protein phosphatase concentration in different samples, the samples must be run on SDS-PAGE at identical protein concentrations.

2. To avoid the possible loss of antigenic recognition via proteolysis, samples should be prepared in the presence of protease inhibitors.

3.6.3. Specific Blotting Conditions

1. In **Subheading 1.5.2.**, the effects of posttranslational modifications of PP2A were discussed in relation to recognition by certain C-terminal antisera. In particular, differential levels of carboxylmethylation of Leucine309 will result in differential antibody recognition, which may be interpreted as differential protein levels. Since carboxylmethylation prevents antibody binding, this feature may be carboxylmethylation of PP2A. However, results must be compared with results obtained using fully demethylated protein or antisera unaffected by carboxylmethylation.

2. To demethylate PP2A for maximum antibody binding, incubate the transfer for >30 min in 0.1 *M* NaOH prior to primary antibody incubation and wash extensively with TBS.

3.6.4. Raising PP1 and PP2A Specific Antisera

3.6.4.1. IMMUNIZATION OF RABBITS

The most commonly used rabbit is the New Zealand White. The Dutch rabbit is also becoming popular for its small size and attractive temperament. A large number of different techniques are used. The present method represents the procedure used routinely in the authors' laboratory:

1. For initial immunization prepare a 4 mg/mL solution of conjugated peptide in TBS (*see* **Note 36**).
2. Emulsify 1:1 with Freund's complete adjuvant immediately prior to injection using two 3-mL Luer-Lock syringes with 18-gage needles connected by Tygon microbore tubing (0.3 × 0.09). The immunogen and adjuvant mixture is emulsified by passing the mixture back and forth between the syringes until the sample is thick.
3. As for all animals, approach the animal quietly and confidently. Before attempting to restrain the rabbit allow the animal to become acquainted with your smell by stroking the rabbit's head and ears (*see* **Note 37**). With one hand, scruff a large fold of skin behind the ears and with the other hand scoop up the rabbits hindquarters and lift. To carry the animal, tuck its head into the bend of the elbow of the supporting arm. Restrain by wrapping a towel around the rabbit. If it becomes restless, drape the end of the towel over the rabbits head and eyes and wait quietly for it to settle. Wrapping the rabbit in a towel will also help maintain body heat and, therefore, reducing the risk of the animal going into shock. Transportation of animals over any distance at all should be in a carrier box .
4. Shave rabbit around thigh muscle of left hind limb.
5. Inject 100 μL antigen (200 μg) intramuscularly using an 18-gage needle.
6. Repeat immunization at after 7 d using right hind limb (*see* **Table 2**).
7. Boost immunization by subcutaneous injection on the back at six sites with a total of 80 μg peptide emulsified with Freunds incomplete adjuvant.

3.6.4.2. BLEEDING AND PROCESSING SERUM

1. Gently pull out one ear from under the towel (*see* **Note 38**). Shave the ear to better maintain asepsis. The ear should be rinsed with ethanol and dried. Generously apply petroleum jelly on the marginal edge of the ear and on the region just under the ear. Do not apply where the cut is to be made. Once the ear is prepared, a razor blade is used to cut the vein using a firm, flat stroke perpendicular to the vein. Be careful not to cut through the ear or to tear the vein. For a few seconds the animal may not bleed. Collecting blood in glass tubes is preferable since clotting will be more efficient, and more serum can be obtained with minimal hemolysis. The animal should stop bleeding immediately or very shortly after the ear is released. Wash ear thoroughly to remove xylene. Naturally, if the rabbit begins to bleed uncontrollably, then the towel should be removed to reduce body temperature and apply pressure. As soon as the blood clots in the tube, rim the edges of the clot with a wooden applicator or glass pipet. Allow the blood to clot

Table 2
Table for Immunization

Day	Immunogen dose total	Carrier	Site of injection
1	200 μg	Freund's complete adjuvant: immunogen 0.1 mL total volume	Intramuscular, left hind limb
7	200 μg	Freund's complete adjuvant: immunogen 0.1 mL total	Intramuscular, right hind limb
35	80 μg	Freund's incomplete adjuvant: immunogen 4 × 0.1 mL total	Subcutaneous
45	Test bleed		

     for 2–4 h at room temperature, then decant serum and centrifuge at 1000*g* for 10 min to remove blood cells.

2. Perform antibody titre using antirabbit IgG and check for specificity by SDS-PAGE and Western blotting using a range of sera dilutions between 1/3000 to 1/20,000 and range of incubation times, between 0.5–1.5 h. Optimal dilution is the one that gives the maximum signal to background ratio.

3.7. Measurement of PP1 and PP2A Activity/Mole Enzyme

1. In order to measure the specific activity of PP1 and PP2A (i.e., activity per mole enzyme as opposed to activity per mg protein) from different subcellular fractions, it is essential to take into account the relative amounts of enzyme in each fraction (*see* **Subheadings 1.1.** and **1.5.1.**). This is done by measuring both the activity per mg protein and the concentration of PP1 or PP2A in that fraction by Western blotting and densitometry. Activity and concentration are both determined at the same protein concentration, and concentration measurements are standardized relative to all other fractions. The specific activity/mole protein by the concentration (arbitrary U/mg protein).

3.8. Purification of Neuronal PP1 and PP2A Binding Proteins

3.8.1. Preparation of Aminoethanethiol-Microcystin Sepharose

1. Rehydrate 0.5 g of activated CH Sepharose 4B (0.25 g in 2 × 10 mL tubes) in excess 1 m*M* HCl for 10 min. Spin and aspirate supernatant. Add 10 mL aliquots of 1 m*M* HCl, spin and aspirate supernatant four more times.
2. Combine beads while concurrently adding a total of 10 mL 50 m*M* Na bicarbonate, pH 9.2. Spin and aspirate supernatant. Repeat this step.
3. Equilibrate in a little Na bicarbonate and allow to settle.
4. For derivatization of microcystin, set water bath in fume hood to 50°C. Purge 5 mL of stock solutions of H_2O, DMSO, and 5 N NaOH with nitrogen. Dissolve a fresh 0.5 mg vial of microcystin in a total volume of 0.5 mL ethanol, and transfer to a 50-mL Falcon tube, labeled "Micro." Add 0.5 mL of ethanol in another

50-mL tube, labeled "Ethanol." Dissolve 1 g of aminoethanethiol in 1.833 mL of purged H_2O (in fume hood).

5. To both falcon tubes add 1 mL purged DMSO, 0.335 mL purged 5 N NaOH and 0.5 mL aminoethanethiol. Purge with Nitrogen, seal tubes with paraffin and incubate for 60 min at 50°C.

6. While tubes are incubating add 10 mL of 0.1% TFA to two C18 Sep-paks. Then 10 mL of 0.1% TFA in 100% acetonitrile and 20 mL 0.1% TFA. Label one Sep-pak "Micro" and the other "Ethanol."

7. After 60-min incubation, cool tubes on ice. Add 3.1 mL glacial acetic acid to each tube. Add 31 mL of 0.1% TFA, pH to 1.5 dropwise with 100% TFA. (60–70 drops). Apply microcystin to Sep-pak labeled "Micro" and the ethanol solution to the "Ethanol" Sep-pak. Wash both Sep-paks with 20 mL 0.1% TFA in 10% acetonitrile—collect wash.

8. Elute derivatized microcystin with 0.1% TFA in 100% acetonitrile in 10 1-mL fractions. Freeze dry. Redissolve in 0.02 mL methanol to approx 25 mM stock.

9. To monitor the extent of the reaction, remove 1 µL from stock and add 249 µL 0.1% TFA = 100 mM stock. Run 20 µL aliquots on HPLC using an appropriate acetonitrile gradient and compare with blank and microcystinstandards.

10. From results of above, calculate degree of labeling and stock concentrations (*see* **Note 38**).

11. To bind derivatized microcystin to sepharose, while transferring combined fractions of derivatized microcystin to a 10 mL yellow capped tube, add 1 mL of Na bicarbonate buffer.

12. Add 1 mL packed CH-Sepharose and mix end over end at room temperature for a minimum of 4 h.

13. Spin for 10 min, collect supernatant and wash three times with Na bicarbonate.

14. Block excess reactive groups by mixing with 10 mL, 50 mM Tris-HCl, pH 8.0 end over end for 1 h—spin and collect S/N and note volume. Wash three times with 10 mL, 50 mM Tris-HCl, pH 8.0.

15. Equilibrate all material to 4°C and pour microcystin-sepharose into 2-mL mini-column. Wash with 10 mL, 0.5 M NaCl = 50 mM NaAc, pH 4.0. Collect flow through and note volume. Wash with 10 mL 0.5 M NaCl. Collect flow through and note volume. Repeat this step four times.

16. Store column in column storage buffer.

17. Freeze dry down flow throughs. Resuspend in 0.5 mL of 0.1% TFA.

18. To monitor the extent of linkage, run 20–50 µL flow through on HPLC. Any unbound microcystin will be detected from which the extent of linkage can be determined.

19. Compare with standards and calculate amount in linkage (*see* **Note 39**).

3.8.2. Purification of PP1 and PP2A Binding Proteins from Crude Synaptic Plasma Membranes by Microcystin Affinity Chromatography

1. Equilibrate two batches of 5 g of SM-2 beads in SM-2 column buffer and microcystin column with microcystin column buffer.

2. Prepare crude synaptic plasma membranes (P2M) of 1 rat forebrain as per **Subheading 3.1.** Add fraction to NaCl soluble eluting buffer at a ratio of 1:2. Final NaCl concentration should be 1 M. (Keep 200 μL sample.) Mix end over end for 30 min at 4°C.

3. Centrifuge for 30,000g for 15 min. Collect supernatant (1st S/N) and rehomogenize pellet (1st Pell). Make 1st Pell to 35 mL with 30 mM Tris-HCl, pH 7.4. (Keep 200 μL of 1st Pell.)

4. Add 1st Pell to 70 mL of detergent solubilization eluting buffer. Sonicate 15". Mix end over end for 30 min at 4°C.

5. Centrifuge 30,000g for 15 min. Collect supernatant (2nd S/N) and rehomogenize pellet (2nd Pell). Make to 35 mL with 30 mM Tris-HCl, pH 7.4. (Keep 1 mL 2nd S/N sample and 2nd Pell.)

6. Dialyze 1st and 2nd S/N overnight against 1/10 SM-2 column buffer.

7. Concentrate 10-fold by evaporative centrifugation. ([NaCl] = 0.5 M and [protein] >1 mg/mL to reduce nonspecific binding to SM-2 beads.) Keep 300 μL of both 1st and 2nd S/N.

8. To remove Triton, batch 2nd S/N with 5g SM-2 beads 2 h, mixing end over end at 4°C.

9. Load batch into 50 mL column and collect flow through at 0.3 mL/min.

10. Wash beads with SM-2 column buffer until no protein is detected by UV at 280 nm. Adding washes to SM-2 Flow Through.

11. Repeat Triton removal with either fresh or regenerated beads (*see* **Note 40**).

12. Keep 300 μL of SM2 Flow through.

13. Combine 1st S/N (NaCl eluting proteins) and SM2FT (detergent solubilized proteins) and load onto microcystin column with positive pressure at 2 mL/min. Wash column with microcystin column buffer until no protein is detected by UV. Add washes to microcystin flow through.

14. Elute protein phosphatases by adding one column volume of phosphatase eluting buffer under positive pressure and incubate at 4°C for at least 30 min. Collect 300-μL fractions into microfuge tubes until UV returns to baseline.

15. Regenerate column by loading 20 mL of microcystin column buffer.

16. Repeat **step 14** while collecting eluates in the same microfuge tubes.

17. Dialyze microcystin column eluate against 1/5 microcystin column buffer.

18. Concentrate fivefold on Speed vac (*see* **Note 41**).

19. Equilibrate microcystin column with 20 mL of column storage buffer.

20. To analyze the effectiveness of the purification, either dot blot or Western blot using appropriate antisera. For activity levels perform phosphatase assays.

4. Notes

1. Protease inhibitors are recommended to be added to buffers. PP1C is particularly susceptible to tryptic digestion, which alters its activity. Soybean trypsin inhibitor is an effective inhibitor, however, preparations are rarely pure, such that Soybean-derived proteins may contaminate purifications. PMSF is an irreversible protease inhibitor but does release F– in solution, which even at a concentration of 1 μM, inhibits PP1 *(18)*. Therefore, it is not recommended. Recommended protease inhibitors include EDTA, EGTA, Leupeptin, Arpotinin, and Pepstatin A.

2. Conditions for labeling glycogen phosphorylase are based on material supplied by Life Technologies. Other preparations of phosphorylase kinase are available, but may have different activity, requiring adjustment of the labeling conditions.

3. A number of different peptides can be dephosphorylated by PP2A *(19)*. RRATVA is routinely used, which is readily phosphorylated by PKA.

4. It is possible to use commercially available GS_{1-12} peptide which is already dye-tagged as part of a PKC assay kit supplied by Pierce. However, this substrate is not optimum for PP1 or PP2A and also requires prior phosphorylation by PKC. The peptide RRATVA is routinely synthesized with the Threonine replaced by Phosphothreonine during synthesis and methyl-red attached to the N-terminus. This provides a 100% phosphorylated peptide ready for phosphatase assay. This custom peptide from Chiron Mimotopes, but a number of commercial peptide is obtained suppliers should be able to provide this service.

5. To minimize heat generation during homogenization, it is essential to immerse the homogenizer in an ice bath during the homogenization.

6. The constituents of each fraction have been determined as follows: S_3, general cytosol, P_2S, occluded cytosol, P_3, small ER and plasma membrane fragments, and P_2M, crude synaptic plasma membranes.

7. Full details of the method developed by Dunkley et al. can be found in **ref. *20***.

8. This method has been developed and optimized using a Beckman J2-21 centrifuge with JA20 rotor. The yield and purity of synaptosomes obtained using different centrifuges and rotors may be different.

9. For a protein concentration of about 5 mg/mL, resuspend fractions in the following volumes: F1, 800 μL; F2, 2400 μL; F3, 990 μL; F4 600 μL; F5, 500 μL.

10. It is essential to wash off all the percoll from the synaptosomes.

11. Use ^{32}P-ATP of sufficiently high specific radioactivity to ensure that the volume ≤100 μL.

12. Whereas glycogen phosphorylase is available from a range of suppliers, highly purified glycogen phosphorylase is available from Life Technologies.

13. This tube is essential for calculating the degree of phosphorylation of the substrate.

14. This procedure must be rigorously applied to ensure that there is no carry over of ^{32}P-ATP, which will otherwise introduce high blank levels.

15. If the reaction has gone to completion then (F) should equal 1 mol/mol.

16. For rat/chicken brain, it is found that a sample concentration of up to 0.03 mg/mL keeps the reaction within linear coordinates. However, for each new experiment a concentration curve should be carried out to determine the optimal protein concentration to use (*see* **Subheading 1.4.4.**).

17. At a protein concentration of 0.02 mg/mL brain tissue, sufficient dephosphorylation is measured in 20 min. However, for each new experiment, a time course should be carried out to ensure linearity.

18. It is essential not to disturb the pellet, which will contain the majority of the radioactivity if dephosphorylation is restricted to <30%.

19. Using the conditions described, 2 n*M* okadaic acid is sufficient to distinguish between PP1 and PP2A. However, for new studies it is essential to calibrate the

assay such that 2 n*M* inhibits all the PP2A present. The results obtained are routinely compared using okadaic acid with results obtained using Inhibitor-1. However, it should be noted that when using inhibitor-1, preincubation of tissue for 10 min at 37°C with inhibitor-1 prior to the addition of the substrate is necessary. Inhibitor-1 is now available from a number of commercial suppliers.

20. The two separate incubations are carried out for convenient use of microfuge tubes.
21. Extended incubation times are necessary to ensure maximum labeling of the substrate.
22. Impure PKA may lead to proteolysis of the peptide substrate.
23. The majority of the pellet is washed from the column (which retains ATP) in the first wash. Where a substantial percentage of the peptide is released in the second wash, this should be combined with the first.
24. Chelating sepharose when charged with iron has an extremely high affinity for phosphopeptides. The principle of this method is the specific binding of the phosphopeptide and the nonbinding of the dephosphopeptide. The amount of dephosphopeptide is used as the measure of the dephosphorylation reaction.
25. It is possible to use the absorbance of both the washed through dephosphorylated peptide, and the eluted phosphopeptide to calculate the activity. However, since under linear conditions, the majority of material remains phosphorylation, use of the eluted phosphopeptide has greater potential to introduce more errors than using the wash through dephosphopeptide.
26. These conditions have been optimized for rat brain synaptosomes. It is not possible to achieve full equilibrium because of the limited viability of isolated synaptosomes. Therefore, the conditions represent a compromise between maximum incorporation of radioactivity and viability. It is, therefore, important to recognize that changes in phosphorylation/dephosphorylation observed in response to a particular treatment may represent a change in the specific labeling of ATP. Under these circumstances it may be necessary to measure the specific radioactivity of the endogenous ATP.
27. Detailed comparison of effects obtained with high and low concentrations of different cell permeable phosphatase inhibitors may be used to determine which phosphatases are responsible for a given dephosphorylation event (*see* **Subheading 1.4.2.**).
28. Under these conditions, phosphorylation is maximal by 30 s after which there is a time-dependent dephosphorylation, which returns to baseline phosphorylation levels by 5 min.
29. Tank apparatus is preferred to semidry blotting since quantitative results are more accurately obtained with the former.
30. At no stage should the nitrocellulose be touched with bare hands or exposed to other proteinaceous material. Use tweezers and gloves and ensure containers are clean.
31. To check effectiveness of transfer or for identifying individual tracks prior to further analysis, immerse nitrocellulose in Ponceau S stain for 10 min and then rinse in either d.d.H$_2$O or very weak acetic acid. Then remove the stain, wash with TBS. Transfer efficiency can also be checked by staining the post-transfer

gel. To check for "blow through," add an additional piece of nitrocellulose to the back of the first membrane during transfer and stain this with Ponceau S.

32. Nitrocellulose membranes can be rinsed in TBS and left wet for up to 1 wk or dry overnight at 4°C.

33. Ensure that gelatin has cooled to room temperature, but not so cold that it sets.

34. If the primary antibody binds weakly to other proteins nonspecifically, this can often be reduced by longer washing times.

35. In this case, secondary antibody linked to alkaline phosphatase is used.

36. A number of different antigenic conjugates are available. Diphtheria toxoid has been found to be the most effective.

37. Since the immune reaction can be dramatically affected by the animals state of stress, it is highly recommended that some attention is given to becoming acquainted with the animal several days before immunization.

38. Although monitoring the extent of derivatization is recommended, the reaction is sufficiently robust that >90% derivatization can be reasonably expected if the procedure described is adhered to.

39. Whereas determining the degree of linkage is recommended, the procedure described reliably produces >90% derivatized microcystin bound to sepharose.

40. A second detergent removal step has been shown to improve resolution.

41. Concentration of the sample is usually necessary to obtain sufficient material for SDS-PAGE analysis.

References

1. Asztalos, Z., von Wergerer, J., Wustman, G., Dombradi, V., Gausz, J., Spatz, H-C., and Friedrich, P. (1993) Protein phosphatase 1-deficient mutant drosophila is affected in habituation and associative learning. *J. Neurosci.* **3,** 924–930.

2. Mulkey, R. M., Herron, C. E., and Malenka, R. C. (1993) An essential role for protein phosphatases in hippocampal long-term depression. *Science* **261,** 1051–1055.

3. Trojanowski, J. Q. and Lee, V. M. (1995) Phosphorylation of paired helical filament tau in Alzheimer's disease neurofibrillary lesions: focusing on phosphatases. *FASEB J.* **9(15),** 1570–1576.

4. Sim, A. T. R., Lloyd, H. G. E., Jarvie, P., Morrison, M., Rostas, J. A. P., and Dunkley, P. R. (1993) Synaptosomal amino acid release: effect of inhibiting protein phosphatases with okadaic acid. *Neurosci. Lett.* **160,** 181–184.

5. Wang, L-Y., Salter, M. W., and Macdonald, J. F. (1991) Regulation of kainate receptors by cAMP-dependent protein kinase and phosphatases. *Science* **253,** 1132–1135.

6. Halpain, S., Girault, J-A., and Greengard, P. (1990) NMDA receptor activation induces dephosphorylation of DARPP-32 in rat striatal slices. *Nature* **343,** 369–372.

7. Sim, A. T. R., Dunkley, P. R., Jarvie, P. E., and Rostas, J. A. P. (1991) Modulation of synaptosomal protein phosphorylation/dephosphorylation by calcium is antagonised by inhibition of protein phosphatses with okadaic acid. *Neurosci. Lett.* **126,** 203–206.

8. Sim, A. T. R., Ratcliffe, E., Mumby, M. C., Villa-Moruzzi, E., and Rostas, J. A. P. (1994) Differential activities of protein phosphatse types 1 and 2A in cytosolic and particulate fractions from rat forebrain. *J. Neurochem.* **62,** 1552–1559.

9. Rostas, J. A. P (1991) Molecular mechanisms of neuronal maturation: a model for synaptic plasticity, in *Neural and Behavioural Plasticity.* (Andrew, R. J., ed.) Oxford University Press, pp. 177–211.

10. Sim, A. T. R., Collins, E., Mudge, L-M., and Rostas, J. A. P. (Manuscript in preparation.)

11. Agostinos, P., Goris, J., Pinna, L. A., Marchiori, F., Perich, J. W., Meyer, H. E., and Merlevede, W. (1990) Synthetic peptides as model substrates for the study of the specificity of the polycation-stimulated protein phosphatases. *Eur. J. Biochem.* **189,** 235–241.

12. Dunkley, P. R., Jarvie, P. E., and Sim, A. T. R. (1996) Protein phosphorylation and dephosphorylation in the nervous system, in *Neurochemistry, A Practical Approach* (Turner, A. J. and Bacehlard, H., eds) IRL.

13. Ishihara, H., Martin, B. L., Brautigan, D. L., Karaki, H., Ozaki, H., Kato, Y., Fusetani, N., Watabe, S., Hashimoto, K., Uemura, D., and Hartshorne, D. J. (1989) Calyculin A and Okadaic acid: inhibitors of protein phosphatase activity. *Biochem. Biophys. Res. Comm.* **159,** 871–877.

14. Dunkley, P. R. and Robinson, P. J (1986) Depolarisation-dependent protein phosphorylation in synaptosomes: mechanisms and signficance. *Prog. Brain. Res.* **69,** 273–293.

15. Bollen, M. and Stalmans, W. (1992) The structure, role and regulation of type 1 protein phosphatases. *Crit. Rev. Biochem. Molec. Biol.* **27,** 227–281.

16. Favre, B., Zolnierowicz, S., Turowski, P., and Hemmings, B. A. (1994) The catalytic subunit of protein phosphatase 2A is carboxylmethylated in vivo. *J. Biol. Chem.* **269,** 16311–16317.

17. Moorhead, G., MacKintosh, R. W., Morrice, N., Gallagher, T., and MacKintosh, C (1994) Purification of type 1 proein (serine/threonine) phosphatases by microcystin-sepharose affinity chromatography. *FEBS Lett.* **356,** 46–50.

18. Bollen, M. and Stalmans, W. (1988) Fluorine compounds inhibit the conversion of active type-1 protein phosphatases into the ATPMg-dependent form. *Biochem. J.* **255,** 327–333.

19. Pinna, L. A. and Donella-Deana, A. (1994) Phosphorylated synthetic peptides as tools for studying protein phosphatases. *Biochim. Biophys. Acta.* **1222,** 415–431.

20. Dunkley, P. R., Jarvie, P. E., Heath, J. W., Kidd, G. J., and Rostas, J. A. P (1986) A rapid method for the isolation of synaptosomes on percoll gradients. *Brain. Res.* **372,** 115–129.

8

PTPA Regulating PP2A as a Dual Specificity Phosphatase

Veerle Janssens, Christine Van Hoof, Wilfried Merlevede, and Jozef Goris

1. Introduction

There are multiple regulatory devices of protein phosphatase type 2A (PP2A), known on the biochemical level: association of regulatory subunits, interaction with other proteins, covalent modification by phosphorylation on tyrosyl and threonine residues, methylation of the carboxy terminus (*see* **ref. 1** for a recent review). In this chapter still another device will be described and discussed. Although PP2A is generally known as a phosphatase that specifically dephosphorylates seryl and threonyl residues, it can also operate as a phosphotyrosyl phosphatase, and this activity can be regulated independently. Historically, PP2A was the first phosphatase described that could remove phosphate from phosphotyrosyl residues *(2,3)* and is, therefore, a "dual specificity" phosphatase "avant la lettre." The in vitro characterization of the dual specificity of PP2A indicated that the phosphoseryl and phosphotyrosyl phosphatase activities exhibited distinct catalytic properties and thermostability *(2,4)*; they were either conversely affected by free ATP or pyrophosphate *(5)*, or concurrently stimulated by tubulin *(6)*. Further observations led to the isolation of a protein that specifically stimulates the phosphotyrosyl phosphatase activity of PP2A without affecting its Ser-P/Thr-P phosphatase activity *(3,7–9)*. This phosphotyrosyl phosphatase activator or PTPA was shown to be highly specific for the dimeric form of PP2A (PP2A$_D$) *(7)*. PTPA stimulates the PTPase activity of PP2A$_D$ in a time-dependent enzyme-like reaction requiring ATP, Mg^{2+} as essential cofactor that cannot be replaced by nonhydrolysable ATP analogs *(7)*. Although the requirement of ATP/Mg^{2+} strongly suggests the

From: *Methods in Molecular Biology, Vol. 93: Protein Phosphatase Protocols*
Edited by: J. W. Ludlow © Humana Press Inc., Totowa, NJ

involvement of a kinase reaction, no phosphorylation of the $PP2A_D$ subunits nor of PTPA could be observed, neither could a protein kinase activity be ascribed to PTPA using exogenous substrates *(7)*. The low ATPase activity detected appeared to be a consequence rather than a direct cause in the activation *(3,9)*. The mechanism of activation, therefore, remains to be resolved. It seems that PTPA induces a reversible conformational change in PP2A, so that the same catalytic site is now attainable by the larger tyrosyl phosphate.

The possible relevant physiological role played by PTPA is suggested by its ubiquity and abundance in differentiated and proliferating tissues in organisms ranging from yeast to man *(9)*. The local specificity determinants, as well as the recognition requirements for dephosphorylation of protein substrates is different for PP2A as compared to an authentic PTPase *(10)*. This was no surprise, since both phosphatases have a different mechanism of catalysis; all PTPases, including the distantly related dual specificity phosphatases have a "PTPase signature" motif in their catalytic site (I/V)HCXAGXGR(S/T)G, where the cysteinyl residue functions as a transient acceptor site for a phosphor transfer intermediate during catalysis *(11)*. No such sequence is found in $PP2A_D$ or PTPA, nor could a phospho-intermediate be detected. Moreover, opposite to PTPases in "sensu stricto," the $PP2A_D$ PTPase needs metal ions for activity *(3,7–9)*.

The significance of PP2A acting as a tyrosyl phosphatase in the cell is still largely unknown. Two criteria can help to distinguish this activity from other PTPases: its sensitivity to okadaic acid and its relative insensitivity to vanadate. Okadaic acid, a strong inhibitor of Ser/Thr phosphatase activity of PP2A, is also a potent inhibitor of the PP2A PTPase activity. I_{50} values are in the same nanomolar concentration range as for the phosphoseryl substrates and also dependent on the enzyme concentration. Since no other PTPase was shown to be sensitive to okadaic acid, this inhibitor is a useful tool in discriminating between the PP2A-PTPase and other PTPases *(12)*. Whereas most PTPases are inhibited by micromolar concentrations of vanadate, millimolar concentrations are necessary to inhibit PP2A-PTPase (I_{50} = 0.6 mM) *(7)*. Based on these criteria, it was shown that PP2A associated with mT/st displays an intrinsic PTPase activity *(13)*, and that it can autodephosphorylate its catalytic subunit if it was phosphorylated on tyrosyl residues *(14)*. The authors will describe how to assay for, and purify PTPA.

2. Materials

2.1. Buffers

1. Buffer 1: 50 mM Tris-HCl, pH 7.5, 2 mM MgCl$_2$, 1 mM ethylenediamine tetraacetic acid (EDTA), 0.5 mM dithiothreitol (DTT), 0.1 mM phenylmethylsulfonyl with fonyl fluoride (PMSF) (*see* **Note 1**).
2. Buffer 2: 100 mM Tris-HCl, pH 9.0, 10% glycerol, 1 mM EDTA, 1 mM DTT, 3% Triton X-100, and 50 µM PMSF.

3. Buffer 3: 25 mM HEPES, pH 7.5, 10% glycerol, 1 mM DTT, 50 µM PMSF, and 100 µM benzamidine (*see* **Note 1**).

4. Buffer 4: 25 mM Tris-HCl, pH 7.5, 20% $(NH_4)_2SO_4$, 10% glycerol, 0.1 mM EGTA, 0.1% β-mercaptoethanol, 0.1 mM PMSF, and 1 mM benzamidine (*see* **Note 1**).

5. Buffer 5: 10 mM HEPES, pH 7.4, 2% glycerol, 1 mM DTT, 1 mM EDTA, 0.5 mM PMSF, and 0.5 mM benzamidine (*see* **Note 1**).

6. Buffer 6: 20 mM Tris-HCl, pH 7.8, 1 mM EDTA, 0.5 mM DTT, and 2% glycerol.

7. Buffer 7: 10 mM Tris-HCl, pH 8.0, 1 mM EDTA, 1 mM EGTA, 0.5 mM DTT, and 0.5 mM benzamidine.

8. Buffer 8: 50 mM Tris-HCl, pH 8, 0.5 mM DTT, 1 mM EDTA, 1 mM EGTA, 0.1 mM TLCK, 0.1 mM PMSF, 0.5 mM benzamidine, and 250 mM sucrose (*see* **Note 1**).

9. Buffer 9: 20 mM Tris-HCl, pH 7.4, 0.5 mM DTT, 1 mM EDTA, 1 mM EGTA, and 0.5 mM benzamidine.

10. Buffer 10: 20 mM Tris-HCl, pH 7.4, 0.5 mM DTT, 1 mM EDTA, 1 mM EGTA, 0.5 mM benzamidine, and 2% glycerol.

11. Polybuffer 74 (Pharmacia, Uppsala, Sweden).

2.2. Chromatography Media

1. Sephadex G-25 and G-50 (Pharmacia).
2. DEAE-Sephacel (Pharmacia).
3. Tyrosine agarose (Sigma, St. Louis, MO).
4. Heparin-Sepharose (Pharmacia).
5. Mono Q (Pharmacia).
6. Dowex I-X8 (Bio-Rad, Hercules, CA).
7. Ultrogel ACA-34 (LKB, BioSepra).
8. Polylysine-Sepharose (Sigma).
9. Phenyl-Sepharose (Pharmacia).
10. Mono P (Pharmacia).
11. Phenyl-Superose (Pharmacia).

3. Methods
3.1. Preparation of Substrates

Because of the lack of abundant physiological phosphotyrosyl proteins, the use of artificial substrates is the only alternative. Several investigators have been using different types of substrates, including pNPP and analogs, phosphotyrosine, phosphorylated peptides, proteins, and chemically modified proteins. The substrate that are used as a "work horse" in the assays for tyrosyl phosphatase activity is reduced carboxamidomethylated and maleylated lysozyme (RCAM lysozyme), phosphorylated by a tyrosine kinase isolated from porcine spleen (*see* **Subheading 3.1.3.**) (*see* **Notes 2 and 3**).

3.1.1. Preparation of RCAM-Lysozyme (according to **ref.15**)

1. Dissolve 0.5 g (*see* **Note 3**) of lysozyme in 6.25 mL of 0.25 M Tris-HCl, pH 8.5, 6 M guanidinium chloride.
2. Add 3.8 mL of 3 M dithiothreitol to reduce the protein.
3. Incubate 90 min at 50°C in the dark under N_2 (bubble the gas through the solution for 30 s, close the test tube by a plastic or rubber closure, and wrap in aluminium foil).
4. Cool on ice.
5. Add 0.295 g of iodoacetamide and incubate 15 min at room temperature in the dark under N_2 to carboxamidomethylate the protein.
6. Quench the reaction by adding 38 µL of β-mercaptoethanol and dialyze against water.
7. A white precipitate will appear that is collected by centrifugation. The pellet is dissolved in 5 mL of 50 mM Tris-HCl, pH 8.5, 6 M guanidinium chloride.
8. Maleylation of the lysyl side chains is performed as follows: Dissolve 0.41 mg maleic acid anhydride in 5 mL dioxane and add dropwise to the lysozyme solution, keeping the pH at 8.7 with 1 N NaOH.
9. Leave on ice overnight.
10. The protein is finally chromatographed on Sephadex G-25 (18 cm × 2.5), equilibrated in 20 mM Tris-HCl, pH 7.5, 0.5 mM DTT. The protein peak, as detected by A_{280}, was pooled (~20 mg/mL) and stored in 5 mL aliquots at –20°C.

3.1.2. Selection of Peptide Substrates

Angiotensin II (DRVYIHPF) is widely used for assaying PTPases, but it is certainly not the best one. Remember that the substrate preference of PP2A-PTPase and a bona fide PTPase (PTP-1B) is different. After comparison of 15 different peptides, modeled according to the phosphoacceptor sites of some potential cellular targets of PTPases *(10)*, the best substrate for dephosphorylation by PP2A-PTPase that is found so far is EKIGEGTYGVVFK, a peptide representing the amino acids 8–20 of cdc2, the central regulator of the cell cycle. The native tyrosyl in position 19 was replaced by a phenylalanine to allow unambiguous kinetic conclusions. It must be stressed, however,that, although phosphorylation of model peptide substrates is handsome for detection of enzymatic activity and sometimes for distinction between different enzymes, extrapolation toward dephosphorylation of proteins should be made very cautiously because the micro-environment of the primary structure around the phosphorylated residue is not the only determining factor for dephosphorylation.

3.1.3. Purification of an src-Like Tyrosine Kinase from Pig Spleen

A tyrosine kinase was prepared from the particulate fraction of porcine spleen (*see* **Note 5**) based on published procedures *(16,17)*. These methods were adapted in order to separate the contaminating PTPases early in the purification.

1. Homogenize 1 kg of porcine spleen in 1 L of buffer 4 (this and all subsequent steps are carried out on ice, in the cold room or at 4°C).

2. Centrifuge for 10 min 1000g, filter the supernatant twice over glass wool, and recentrifuge for 90 min at 50,000g.

3. Resuspend the pellet in 300 mL Extraction Buffer 2 using a Dounce homogenizer and mix gently for 30 min on a magnetic stirrer.

4. Remove particulate material by centrifugation (90 min, 100,000g), adjust to pH 7.5, absorb batchwise to (800 mL DEAE-Sephacel, equilibrated in Buffer 3 and wash extensively (seven times 400 mL) with the same buffer, pour the gel-slurry into a column (5-cm diameter) and continue to wash after settling of the gel.

5. Elute the column with a 2 × 1000 mL 0-0.5 M NaCl gradient in the equilibration buffer, collect fractions of 10 mL and assay tyrosine kinase activity (*see* **Subheading 3.1.5.**) in every fifth fraction.

6. Combine the fractions containing the major tyrosine kinase activity (eluting at (230 mM NaCl) and make the solution 20% in $(NH_4)_2SO_4$ by adding solid salt. Alternatively, the volume of the kinase can be first reduced by precipitation with 75% $(NH_4)_2SO_4$ taking up the pellet in one tenth of the original volume of the tyrosyl agarose equilibration buffer (*see* **step 7**).

7. Load on a tyrosine agarose column (2.5 × 12 cm) equilibrated in Buffer 4, wash with 100 mL of the same buffer, and elute the column with a descending gradient of 2 × 275 mL 20–0% $(NH_4)_2SO_4$ in the same buffer.

8. Pool the active fractions and dialyze overnight in Buffer 5.

9. Load on a heparin-Sepharose column (1.5 × 10 cm) equilibrated in Buffer 5 and elute with a 2 × 100 mL 0–0.6 M NaCl gradient in the same buffer.

10. Pool the active fractions and load on a Mono-Q (HR 5/5) column, equilibrated in buffer 6. Wash with 5 mL of the same buffer and elute with a 25 mL total gradient of 0–0.5 M NaCl in Buffer 6.

3.1.4. Phosphorylation and Reisolation of Substrates

3.1.4.1. RCAM LYSOZYME

RCAM lysozyme is phosphorylated on a single tyr (Y53) by the tyrosine kinase (isolated as described in **Subheading 3.1.3.**) as follows:

1. 4 mg of RCAM lysozyme is incubated with 15 U tyrosine kinase, 200 μM [γ-^{32}P] ATP (200–1000 cpm/pmol), 20 mM $MgCl_2$, 50 μM vanadate, and 0.07% Triton X-100 in 50 mM Tris-HCl, pH 7.5 in a total volume of 500 μL for 4 h at 30°C.

2. The kinase reaction is stopped by adding 50 μL of 250 mM EDTA and the solution is gel filtered over a Sephadex G-50 column (0.9 × 15 cm) in 10 mM Tris-HCl, pH 7.5 to separate the protein from the unreacted ATP.

3.1.4.2. TYROSYL SYNTHETIC PEPTIDES

Tyrpsyl synthetic peptides (150 μM) are phosphorylated in the same conditions (*see* **Subheading 3.1.4.1.**):

1. To separate the unreacted ATP from the peptide, the reaction is stopped by adding an equal volume of 60% acetic acid and loaded on a 1 mL Dowex 1-X8 anion-exchange resin equilibrated in 30% acetic acid.

2. The flow through fractions contain the phosphopeptide, whereas ATP remains attached to the column.

3. The eluate containing the peptide is lyophilized, washed several times with water and dissolved in an appropriate amount of 20 mM Tris-HCl, pH 7.5, 0.5 mM DTT, to obtain 30 μM ^{32}P-labeled peptide.

4. The specific activity of the ^{32}P incorporated into the peptides is assumed to be equal to that of [γ-^{32}P] ATP (200–1000 cpm/pmol), and the concentration of the ^{32}P-labeled peptide is based on this assumption.

3.1.5. Enzyme Assays

1. Phosphorylase phosphatase is assayed in 30 μL 20 mM Tris-HCl, pH 7.4, 0.5 mM DTT, 1 mg/mL BSA, 15 mM caffeine, and 1 mg/mL ^{32}P labeled phosphorylase a (*see* **Note 6**) in the absence or presence of 3.3 μg/mL protamine and 15 mM (NH$_4$)$_2$SO$_4$ (*see* **Note 7**). After 10 or 20 min incubation at 30°C, the reaction is stopped by TCA and the TCA soluble liberated inorganic ^{32}P, measured as described in **Subheading 3.2.2.2.**.

2. Protein tyrosine kinase is measured in a volume of 50 μL containing 50 mM Tris-HCl, pH 7.5, 20 mM MgCl2, 100 μM [γ-^{32}P] ATP (500 cpm/pmol), 50 μM VO$_4$$^{3-}$, 0.07% Triton X-100, and 5 mg/mL poly-[glu:tyr] (4:1). After various incubation times at 30°C, a 10 μL sample is spotted onto a 2 × 1 cm section of Whatman 3 MM filter paper, dropped in 10% TCA at 0°C, washed two times with 5% TCA at room temperature (each time 20 min), and once in Acetone. Finally, the filter papers are dried and counted by liquid scintillation spectroscopy. One kinase unit (U) transfers 1 nmol of phosphate from ATP to the substrate per minute.

3.2. Assay of PTPA

So far, no direct assay for PTPA activity is available, and the only activity assay is indirect by measuring the increase in tyrosyl phosphatase activity of the dimeric form of PP2A:PP2A$_D$ (=PP2A$_2$ or PCS$_L$). This resulting tyrosyl phosphatase activity can be measured by the phosphorylation of the substrates mentioned in **Subheading 3.1.**, or by a p-nitrophenol phosphate (pNPP) (*see* **Note 8**). A procedure for purification of PP2A$_D$ from rabbit skeletal muscle is described here (*see* **Notes 9** and **10**).

3.2.1. Purification of PP2A$_D$
from Rabbit Skeletal Muscle (According to *ref.* 18)

1. Three normally fed adult rabbits are sacrificed by severing the neck and exsanguinated.

2. The hind leg and back muscle (1200 g) are removed, chilled in ice and quickly homogenized, using a Robot Minute SEB blender, in 2 vol of Buffer 8.

3. The homogenate is centrifuged for 60 min at 6000*g* and the supernatant filtered through glasswool. (This and all consecutive steps are carried out at 4°C.)

4. The supernatant is mixed batch-wise with 500 mL of DEAE-Sephacel equilibrated in Buffer 9 and washed with 1 L of Buffer 9 and 0.2 M NaCl to remove calcineurin and most of the different forms of PP-1 and PP2A$_0$.

5. The slurry is packed into a 5 × 36 cm column and washed further with the same solution (total volume ±0.5 L).
6. The proteins eluting from the column with 1500 mL of a 0.2–0.5 M NaCl linear gradient in Buffer 9 are collected in 12 mL fractions and assayed for phosphorylase phosphatase activity in the presence or absence of polycations (*see* **Subheading 3.1.5.**).
7. Three peaks of polycation stimulated activity are observed (*see* **Note 11**). The third activity peak, eluting at about 0.4 M NaCl, where the A_{280} is at its lowest or in the ascending part of the next A_{280} peak, represents $PP2A_D$.
8. This can easily be checked by a plus and minus PTPA assay of the PTPase activity (*see* **Subheading 3.2.2.** and **Note 13**).
9. The fractions containing $PP2A_D$ are pooled and solid $(NH_4)_2SO_4$ is added to obtain 60% saturation.
10. After leaving on ice for 15 min, with gentle stirring, proteins are precipitated by centrifugation (10 min at 8000g), taken up in a minimal volume of Buffer 9 and dialyzed 1–2 h against the same buffer.
11. The dialyzed material, containing $PP2A_D$, is recentrifuged (10 min at 8,000g), to remove any precipitate and applied to an Ultrogel ACA-34 column (2.5 × 90 cm) equilibrated in Buffer 9 plus 0.1 M NaCl.
12. $PP2A_D$ elutes as a single symmetrical peak of polycation stimulated phosphatase activity, is pooled, made 2% glycerol and applied to a polylysine-Sepharose 4B column (0.9 × 35 cm), equilibrated in Buffer 10.
13. The column is washed with 0.2 M NaCl in Buffer 10 and eluted with 130 mL of a linear 0.2–0.7 M NaCl gradient in Buffer 10. $PP2A_D$ elutes as a single symmetrical peak at about 0.38 M NaCl.
14. The active fractions are pooled and concentrated by dialyzing against 20% polyethylene glycol in Buffer 9 to a volume of 5–6 mL followed by dialysis overnight against 60% glycerol in the same buffer to a final volume of 2–3 mL.
15. For a last purification step, the enzyme is applied to a Mono Q HR 5/5 column equilibrated in Buffer 10 and after washing with 5 mL 0.25 M NaCl in Buffer 10, the phosphatase is eluted with 30 mL of a 0.25–0.45 M NaCl linear gradient in the same buffer at a flow rate of 1 mL per min.
16. The single symmetrical peak of the $PP2A_D$ phosphatase is pooled and concentrated by dialysis against 20% polyethylene glycol in Buffer 9 followed by dialysis against 60% glycerol in Buffer 9.
17. The activity of the purified $PP2A_D$ is stable for at least 1 yr when stored at −20°C in the presence of 60% glycerol. In this condition, ice crystals are not formed and enzyme samples can be taken without thawing.

3.2.2. PTPA Assay

PTPA activates the aryl phosphatase (PTPase or pNPPase) activity of $PP2A_D$ in the presence of ATP and Mg, without affecting the phosphoseryl phosphatase activity. The mechanism of this activation is not fully understood. Therefore, no direct evaluation of this conversion is possible, and only the change in aryl phosphatase activity can be followed to assay PTPA. The routine assay of PTPA is based on the increase in p-nitrophenyl phosphatase or phosphotyrosyl phosphatase activity under conditions where the increase in activity is directly proportional to the quantity of PTPA added. Since the activation is transient in the presence of aryl substrate, a short incubation time for the assay of the PTPase activity is mandatory.

3.2.2.1. PTPASE ASSAY WITH P-NITROPHENYL PHOSPHATE AS SUBSTRATE

1. Activation is carried out in a preincubation of 100 μL containing 2.5–10 U potential p-nitrophenyl phosphatase, different dilutions of PTPA, 10 mM MgCl$_2$, 2 mM ATP, 20 mM Tris-HCl, pH 7.5, 0.5 mM DTT, 0.002% Triton X-100, and 1 mg/mL bovine serum albumin, during 5–10 min at 30°C.
2. After addition of 100 μL containing 20 mM pNPP, 20 mM MgCl$_2$, 50 mM Tris-HCl, pH 8.5, 0.5 mM DTT, and incubation for 2 min at 30°C, the reaction is stopped by adding 500 μL 1 M Na$_2$CO$_3$ and the absorbance measured at 410 nm.
3. Controls with omission of PTPA are measured in duplicate and substracted; they never exceed 10% of the maximal activity (*see* **Note 12**).
4. A unit (U) of PTPA is defined as the amount which activates 50% of the potential phosphatase activity in these experimental conditions.

3.2.2.2. PTPASE ASSAY WITH RCAM LYSOZYME AS SUBSTRATE

1. When ^{32}P labeled RCAM lysozyme is the substrate, 40–100 mU of PP2A (*see* **Note 13**) are preincubated in 20 μL with 10 mM MgCl$_2$, 2 mM ATP, 20 mM Tris-HCl, pH 7.5, 0.5 mM DTT, 0.003% Triton X-100, 1 mg/mL bovine serum albumin, and different dilutions of PTPA during 5–10 min at 30°C.
2. Then 10 μL of 3 μM ^{32}P-RCAM lysozyme, in 20 mM MgCl$_2$, 50 mM Tris-HCl pH 7.5, 0.5 mM DTT is added, incubated for 2–5 min at 30°C, and the reaction is stopped by the addition of 200 μL ice-cold 20% trichloroacetic acid and 200 μL containing 6 mg/mL bovine serum albumin as a coprecipitant.
3. The solution is left on ice during 10 min, centrifuged 5 min 3000g and 380 μL of the supernatant are counted in 5 mL of a liquid scintillation fluid.
4. A unit of PTPase catalyzes the release of 1 nmol phosphate/min.

3.2.2.3. PTPASE ASSAY WITH PHOSPHOPEPTIDES

1. When phosphotyrosyl peptides are used as substrate, a similar assay is used but the reaction is stopped according to **ref. 19** by 0.8 mL 5 mM silicotungstate in 1 mM H$_2$SO$_4$, 0.16 mL of 50 mg/mL ammonium-heptamolybdate in 2 M H$_2$SO$_4$ and 1.2 mL isobutyl alcohol/toluene (1:1, by vol). (*see* **Notes 14** and **15**).
2. The formed phosphomolybdate complex, opposite to the phosphopeptide, is extracted in the organic phase: after vigorous mixing, the organic phase is separated from the water phase by a short centrifugation (5 min 2000g) and 0.5 mL of the organic (upper) phase is counted in a liquid scintillation fluid.

3.3. Purification of PTPA from Several Tissues and Species (9)

1. Rabbit skeletal muscle (300 g) (*see* **Note 16**) is homogenized in 2 vol homogenization buffer (Buffer 8 without sucrose).

2. After a low-speed centrifugation (60 min at 5000g), the homogenate is filtered twice over glass wool.

3. The filtrate is acidified to pH 5.2 with 1 M acetic acid and centrifuged for 20 min at 10,000g.

4. The supernatant is adjusted to pH 7.8 by adding 1 M Tris-HCl, pH 7.8 and after filtration over glass wool, brought to 10% ammonium sulfate.

5. The solution obtained is absorbed onto 125 mL phenyl-Sepharose equilibrated in Buffer 9 plus 10% $(NH_4)_2SO_4$ by filtration on a Büchner filter.

6. The filtrate is reabsorbed and filtered on the same hydrophobic resin.

7. The gel is subsequently washed six times with 500 mL of equilibration buffer 9 plus 10% $(NH_4)_2SO_4$ and six times with Buffer 9.

8. Elution is achieved using 50% etheleneglycol in a batch-wise manner.

9. The eluate obtained with 50% ethyleneglycol in the previous batch-wise procedure, is loaded on a DEAE-Sephacel column (2.5 × 16 cm) equilibrated in Buffer 9.

10. The column is rinsed with 200 mL Buffer 9 and PTPA is eluted with a linear 0–0.3 M NaCl gradient (2 × 400 mL) in the same buffer.

11. Fractions of 6 mL are collected and those containing PTPA activity are pooled and concentrated by consecutive dialysis against 50% polyethyleneglycol and 60% glycerol in Buffer 9.

12. The concentrate obtained is diluted 10–20-fold in 25 mM Bistris, pH 7.1 (pH was adjusted with iminodiacetic acid) and further purified by chromatography on a Mono P HR 5/20 column equilibrated in the same buffer.

13. The column is eluted with 10 mL polybuffer 74, pH 4, creating a pH gradient.

14. Fractions of 0.5 mL are collected in tubes that contained 100 µL 0.5 M Tris-HCl, pH 7.8 to neutralize the acidic eluate immediately (*see* **Note 17**).

15. Fractions containing PTPA activity are pooled and made 20% $(NH_4)_2SO_4$ before loading onto a phenyl-Superose column (HR 5/5), equilibrated in Buffer 9 plus 20% ammonium sulphate.

16. The column is developed with a descending salt gradient (20–6% ammonium sulphate) of 6 mL in Buffer 9, and PTPA is eluted with 3 mL 6% ammonium sulphate in Buffer 9.

17. The PTPA activity is pooled and concentrated against 60% glycerol in Buffer 9.

18. The final purification step is a Mono Q HR 5/5 chromatography, and pure PTPA is obtained by elution with a 20-mL linear gradient of 0-0.25 M in Buffer 7.

19. The active fractions are pooled and concentrated two- to three-fold by dialysis against 60% glycerol in Buffer 9. As such, they can be stored for several months without any loss of activity.

4. Notes

1. Phenylmethylsulfonyl fluoride (PMSF) and 1-chloro-3-tosylamido-7-amino-2-heptanone-hydrochloride (TLCK) are solubilized in 1 mL ethanol just before the buffer is used and added immediately after solubilization.

2. Chemical denaturation is a necessity to expose tyrosyl side chains in proteins such as BSA or lysozyme that are normally not phosphorylated by tyrosine kinases, but become substrates after such treatment. Whereas in principle, any protein that can be obtained in large quantities is a suitable candidate for chemical modification, lysozyme is the protein of choice because it does not suffer the many difficulties encountered in the phosphorylation of these substrates: it is only phosphorylated at one site, and can be phosphorylated up to reasonable stoichiometries. Moreover, the maleylation of lysyl side chains significantly improves the solubility of the denatured lysozyme. Poor solubility of the modified derivative (remember the consistency of boiled egg albumin!) was indeed a serious problem, and harsh treatments with acid or alkali are not required to prepare soluble RCAM lysozyme.

3. One might think of using the commercially available poly[glu-tyr](4:1), normally used for tyrosine kinase routine assays. This cannot be used to assay the PP2A-PTPase activity, since it is shown that this substrate is resistant for dephosphorylation by PP2A.

4. This quantity can easily been scaled up or down by keeping the ratios of the reagents constant. However 0.5 g is a quantity that can be appreciated as "huge" since it allows to do 25,000 assays as described in **Subheading 3.2.2.2.**, or 100 times a phosphorylation batch of 5 mg that can be used over a period of 2–3 wk each. Once prepared, the stock can be stored in batches at −20°C almost indefinitely (The authors are still using the batch that was prepared 5 yr ago, without any problem).

5. Described here is the purification of "a" tyrosine kinase that can be used for routine phosphorylations of the substrates mentioned in **Subheading 3.1.** Even the cdc2 peptide was phosphorylated exclusively on tyrosine 15 near stoichiometry (0.8–0.9 mol/mol) by this kinase, although it is known that in vivo this site is specifically phosphorylated by wee1, a dual specificity kinase that also phosphorylates threonine 14. So far any other tyrosine kinase could be used especially the commercially available pp60[c-src] or Abl.

6. For the preparation of ^{32}P radiolabeled phosphorylase a, incubate 10 mg/mL phosphorylase b with 0.2 mg/mL phosphorylase kinase for 1–5 h at 30°C in 100 mM Tris-HCl, PH 8.2, 100 mM glycerol phosphate, 0.1 mM CaCl$_2$, 10 mM Mg acetate, and 0.2 mM [γ-^{32}P] ATP.

7. In the absence of (NH$_4$)$_2$SO$_4$, protamine stimulation has an optimal concentration, typical for each holoenzyme. Above this optimum the activity decreases. This diminishing of activity by higher protamine concentrations is not observed when 15 mM (NH$_4$)$_2$SO$_4$ is included in the assay. As a result, greater stimulations of phosphorylase phosphatase activities are observed with higher protamine concentrations (above the optimal concentrations, in the absence of [NH$_4$]$_2$SO$_4$) (*see* **Fig. 2** in **ref.** *20* to illustrate this phenomenon). At 33 μg/mL, the maximal activity is observed for all PP2A holoenzyme forms described so far and is, therefore, used in the routine assay.

8. Although pNPP can be used as an alternative for measuring the PTPase activity of PP2A, the disadvantage is the larger quantity of PP2A$_D$ to cope with the larger assay volumes necessary for spectrophotometrical assays.

9. Since it is known that the "third" variable subunits are inhibiting (PR55) or preventing (PR72, PR61) PTPA's action, it is essential that the dimeric form is isolated. An alternative is the use of the catalytic subunit, but this assay is less sensitive.

10. In principle, other tissues such as porcine brain or bovine heart can be used as starting material. However, the mechanism(s) governing the dissociation of the "third" subunits are not known, and, therefore, the relative amounts of the dimeric form may vary. By following the method described here, 60–200 μg of dimeric PP2A can be isolated from 1 kg of rabbit skeletal muscle. Once one has purified PTPA, the location of PP2A$_D$ during the different purification columns, can easily be followed with a plus and minus PTPA, PTPase assay (*see* **Subheading 3.2.2.**).

11. Depending on the amount of PP-1, still present in the first peak, the stimulation by polycations of this first peak is variable. Indeed PP-1 activity is inhibited by polycations, and whether stimulation of even inhibition of this first peak by polycations will be observed, will depend on the ratio of the coeluting PP2A$_D$ and PP-1.

12. The activatability of PP2A$_D$ by PTPA, and ATP, Mg is in fact dependent on a previous inactivation. In the purification procedure of PP2A$_D$ as described in **Subheading 3.2.1.**, the PTPase activity of PP2A$_D$ is almost completely inactive from the DEAE-Sephacel onwards. Some conditions, such as polyoma small t can prevent this inactivation *(13)*, and it is not excluded that still undetected cellular homologs might have similar effects.

13. These units are measured as phosphorylase phosphatase activity units. As a rule of thumb, PTPA stimulated PP2A-PTPase is measured in a 10-times lower dilution than its phosphorylase phosphatase activity: whereas the specific activity of basal phosphorylase phosphatase activity of PP2A$_D$ is 2000–3000 U/mg, its activated PTPase, measured with RCAM is 300–800 U/mg, but the latter is measured in a shorter assay.

14. Since short phosphopeptides are soluble in 20% trichloroacetic acid, a TCA precipitation cannot be used to separate the liberated inorganic phosphate from the unreacted phosphoprotein as is performed in the RCAM lysozyme or phosphorylase phosphatase assay.

15. For these reactions, glass tubes are mandatory.

16. The purification method as described here for rabbit skeletal muscle has also been succesfully applied to porcine brain, dog liver, *Xenopus* oocytes, and the yeast *Saccharomyces cerevisiae*, it is a simplified procedure *(9)*, originally described in **ref.** *7*.

17. The procedure to collect the fractions in tubes that contain Tris buffer is important to obtain good PTPA yields. PTPA is indeed much less stable at acidic pH's. The elution position of the PTPA's from different species is somewhat different, dependent on the pI of the protein. For instance, the oocyte PTPA was a little less acidic (pI = 5.8) than rabbit skeletal muscle or dog liver PTPA (pI = 5). It sometimes happens, for unknown reasons, that most of the PTPA appears in the flow-through of the mono P columns. In these cases, the active fractions are pooled, concentrated with polyethyleneglycol and glycerol as on the end of the previous step, and the whole procedure of **step 4** is (this time successfully) repeated.

References

1. Wera, S. and Hemmings, B. (1995) Serine threonine protein phosphatases. *Biochem. J.* **311**, 17–29.
2. Chernoff, J., Li, H. -C., Cheng, Y. -S. E., and Chen, L. B. (1983) Characterization of a phosphotyrosyl protein phosphatase activity associated with a phosphoseryl protein phosphatase of Mr=95,000 from bovine heart. *J. Biol. Chem.* **258**, 7852–7857.
3. Van Hoof, C., Cayla, X., Bosch, M., Merlevede, W., and Goris, J. (1994) PTPA adjusts the phosphotyrosyl phosphatase activity of PP2A. *Adv. Prot. Phosphatases* **8**, 283–309.
4. Silberman, S.,R., Speth, M., Nemani, R., Ganapathi, M.,K., Dombradi, V., Paris, H., and Lee, E. Y. C. (1984) Isolation and characterization of rabbit skeletal muscle protein phosphatases C-I and C-II. *J. Biol. Chem.* **259**, 2913–2922.
5. Hermann, J., Cayla, X., Dumortier, K., Goris, J., Ozon, R., and Merlevede, W. (1988) Polycation-stimulated (PCSL) protein phosphatase from *Xenopus laevis* oocytes. ATP-mediated regulation of alkaline phosphatase activity. *Eur. J. Biochem.* **173**, 501–507.
6. Jessus, C., Goris, J., Cayla, C., Hermann, J., Hendrix, P., Ozon, R., and Merlevede, W. (1989) Tubulin and MAP2 regulate the PCS phosphatase activity: a possible new role for microtubular proteins. *Eur. J. Biochem.* **180**, 15–22.
7. Cayla, X., Goris, J., Hermann, J., Hendrix, P., Ozon, R., and Merlevede, W. (1990) Isolation and characterization of a tyrosyl phosphatase activating factor from rabbit skeletal muscle and *Xenopus laevis* oocytes. *Biochemistry* **29**, 658–667.
8. Cayla, X., Van Hoof, C., Bosch, M., Waelkens, E., Vandekerckhove, J., Peeters, B., Merlevede, W., and Goris, J. (1994) Molecular cloning, expression and characterization of PTPA, a protein that activates the tyrosyl phosphatase activity of protein phosphatase 2A. *J. Biol. Chem.* **269**, 15,668–15,675.
9. Van Hoof, C., Cayla, X., Bosch, M., Merlevede, W., and Goris, J. (1994) The phosphotyrosyl phosphatase activator of protein phosphatase 2A. A novel purification method, immunological and enzymic characterization *Eur. J. Biochem.* **226**, 899–907.
10. Agostinis, P., Donella-Deana, A., Van Hoof, C., Cesaro, L., Brunati, A. M., Ruzzene, M., Merlevede, W., Pinna, L. A., and Goris, J. (1996) A comparative study of the phosphotyrosyl phosphatase specificity of protein phosphatase type 2A and phosphotyrosyl phosphatase type 1B using phosphopeptides and the phosphoproteins p50/HS1, c-Fgr and Lyn. *Eur. J. Biochem.* **236**, 548–557.
11. Stone, R. L. and Dixon, J. E. (1994) Protein-tyrosine phosphatases. *J. Biol. Chem.* **269**, 31,323–31,326.
12. Goris, J., Hermann, J., Hendrix, P., Ozon, R., and Merlevede, W. (1989) Okadaic acid, a specific protein phosphatase inhibitor, induces maturation and MPF formation in *Xenopus laevis* oocytes. *FEBS Lett.* **245**, 91–94.
13. Cayla, X., Ballmer-Hofer, K., Merlevede, W., and Goris, J. (1993) Phosphatase 2A associated with polyomavirus small t or middle T antigen is blocked by okadaic acid sensitive tyrosyl phosphatase. *Eur. J. Biochem.* **214**, 281–286.

14. Chen, J., Martin, L. M., and Brautigan, D. L. (1992) Regulation of protein serine-threonine phosphatase type-2A by tyrosine phosphorylation. *Science* **257**, 1261–1264.
15. Tonks, N. K., Diltz, D. C., and Fischer, E. H. (1988) Purification of the major protein-tyrosine phosphatases of human placenta. *J. Biol. Chem.* **263**, 6722–6730.
16. Tung, H. Y. L. and Reed, L. J. (1987) Identification and purification of a cytosolic phosphotyrosyl protein phosphatase from bovine spleen. *Anal. Biochem.* **161**, 412–419.
17. Swarup, G., Dasgupta, J. D., and Garbers, D. L. (1983) Tyrosine protein kinase activity of rat spleen and other tissues. *J. Biol. Chem.* **258**, 10,341–10,347.
18. Waelkens, E., Goris, J., and Merlevede, W. (1987) Purification and properties of the polycation-stimulated phosphorylase phosphatases from rabbit skeletal muscle. *J. Biol. Chem.* **262**, 1049–1059.
19. Shacter, E. (1984) Organic extraction of Pi with isobutanol/toluene. *Anal. Biochem.* **138**, 416–420.
20. Zolnierowicz, S., Van Hoof, C., Andjelkovic, N., Cron, P., Stevens, I., Merlevede, W., Goris, J., and Hemmings, B. A. (1996) The variable subunit associated with protein phosphatase $2A_0$ defines a novel multimember family of regulatory subunits. *Biochem. J.* **317**, 187–194.

9

Microinjection and Immunological Methods in the Analysis of Type 1 and 2A Protein Phosphatases from Mammalian Cells

Patric Turowski and Ned J. C. Lamb

1. Introduction

Type 1 and 2A protein phosphatases (PPases) account for the majority of enzyme activity dephosphorylating phospho-serine and -threonines in living cells. Their action is opposed to that of a large number of serine/threonine protein kinases. In recent years, the enigma of how the action of many protein kinases is reversed by a limited number of protein phosphatases in a regulated manner has been solved by a combination of biochemical and molecular cloning procedures: Both PP1 and PP2A were found to exist as multimeric holoenzymes in vivo, where a limited number of catalytic subunits are complexed to a vast number of regulatory subunits. The latter are now considered to be the determinants of substrate specificity and intracellular localization of the catalytic subunits *(1,2)*.

Protein phosphatase 1 (PP1) holoenzymes usually exist as 1:1 complexes between a catalytic subunit of 37 kDa and one of a number of regulatory subunits, which target the catalytic subunit (PP1c) toward their substrates *(3,4)*: a myosin specific holoenzyme, PP1M, consists of PP1c associated to proteins of 20 and 130 kDa, where the latter seems to be responsible for the targeting. PP1G consists of PP1c complexed with a regulatory protein, called G-subunit, which directs this holoenzyme towards glycogen particles and the sarcoplasmic reticulum. A soluble form of PP1, termed PP1S, is a dimer of PP1c in complex with inhibitor-2 (I-2). More recently I-2 has been considered a modulator of PP1, since under certain conditions PP1S is a fully active holoenzyme *(5)*. PP1c is also found complexed to bona fide inhibitors, such as I-1, a brain

From: *Methods in Molecular Biology, Vol. 93: Protein Phosphatase Protocols*
Edited by: J. W. Ludlow © Humana Press Inc., Totowa, NJ

specific isoform DARPP-32, or a group of nucleus-specific proteins, termed NIPPs *(4,6)*. In cultured mammalian cells, PP1 has been reported to be cytoplasmic, nuclear, and associated with microfilaments *(7,8)*.

Protein phosphatase 2A (PP2A) exists as heterotrimeric complexes in vivo. The 36 kDa catalytic subunit (PP2Ac) is constitutively associated with a constant regulatory subunit of 65 kDa (PR65 or A subunit). Final regulation and targeting of the holoenzyme is conferred by differential association of a third variable (B) subunit *(2)*. B subunits so far identified can be grouped into 3 gene families and different isoforms or splicing variants within each family give rise to at least 17 gene-products *(9,10)*. PP2A has been reported to be cytoplasmic, nuclear, associated with microtubules and cytoplasmic membranes *(11–13)*. The B55 family of regulatory subunits appears to render PP2A specific for sites phosphorylated by cyclin-dependent kinases *(14)*, whereas the proteins of the B72 family would rather confer specificity for casein kinase II sites *(15)*. Significantly, the latter, together with B61 class proteins appear as good candidates for nuclear targeting of PP2A, since nuclear localization signals have been found in their primary structure *(9,16)*.

From this, it is clear that, in order to elucidate the action of PP1 or PP2A on a given process, it has to be analyzed in the context of their holoenzyme composition (i.e., the regulatory subunit(s) associated) and their intracellular localization. The authors have chosen to analyze these aspects by microneedle injection. Microneedle injection is a powerful technique for introducing macromolecules into mammalian cells. The technique has been used for a wide variety of different biological samples including DNA (plasmids and oligonucleotides), RNA, peptides and proteins including antibodies, enzymes, and structural molecules. In the present article the use of microinjection combined with immunochemical and biochemical analyses is described in deciphering the role of PP1 and PP2A during mammalian cell growth and proliferation.

2. Materials

2.1. Reagents

1. Biologically inert antibodies: Rabbit IgG (I 5006, Sigma, St. Louis, MO), Mouse IgG (I 5381, Sigma). Store dry at 4°C. For microinjection resuspend in buffer G or PBS at 1 mg/mL. Store in aliquots of 10 µL at −20°C.
2. Dimethylsulfoxide (DMSO), Acetone, Methanol (p.a. grade, Merck, Mannheim, Germany).
3. Hydroxyurea (H 8627, Sigma). Resuspend in highly purified water at 200 m*M*. Filter sterilize and store frozen in aliquots of 1 mL. Thawed aliquots can be refrozen.
4. Formalin (p.a. grade, Sigma). Store in a dark place.
5. Glutaraldehyde (EM-grade, TAAB). Store at 4°C in a dark place.

6. Secondary antibodies for immunofluorescence: rhodamine and fluorescein coupled antirabbit/-mouse/-sheep (Cappel), biotinylated antibodies (antirabbit, RPN 1004; antimouse, RPN 1001; antisheep, RPN 1025; Amersham, Arlington Heights, IL). Store in aliquots of 1–5 μL according to the manufacturer's recommendations. Streptavidin-fluorescein (RPN 1232, Amersham), streptavidin-Texas Red (RPN 1233, Amersham). Store at 4°C.
7. Bisbenzimide (Hoechst, B 2883, Sigma). Make up at 100 μg/mL and store frozen in aliquots of 1–5 μL.
8. AirVol 205 (Airproducts Inc.).
9. [^{32}P]-orthophosphate (PBS 13, Amersham).
10. [^{35}S]-methionine (SJ 235, Amersham).
11. DNase, RNase (protease-free, Worthington) at 1 U/μL in 50 mM Tris-HCl, pH 7.5, 50 mM KCl, 10 mM MgCl$_2$, 250 mM sucrose.

2.2. Solutions

1. Sucrose/PBS: To approx 1 kg of sucrose add 500 mL of cold water. Agitate overnight in the coldroom. Decant the supernatant and add 11% (v/v) of PBS. This solution can be reused. Store at 4°C.
2. Cell culture solutions: Phosphate-buffered saline (PBS), Dulbecco's modified Eagle Medium (DME), fetal calf serum (FCS), 2.5% trypsin in Hanks' balanced salt solution (all from Gibco or Imperial). Cell freezing solution: 90% (v/v) FCS, 10% (v/v) DMSO. Filter sterilize and store frozen in aliquots of 10 mL.
3. Formalin fixation solution: 3.7% (w/v) formaldehyde in PBS. Store at room temperature for no longer than 1 mo.
4. Glutaraldehyde prefix solution: DME containing 10% (v/v) FCS, 0.1% (w/v) Triton X-100, and 0.025% (w/v) glutaraldehyde. Fixation solution: 0.25–1% (w/v) glutaraldehyde in PBS. Always prepare both solutions just before usage.
5. PBS-BSA: 0.5% (w/v) BSA, 0.05% (w/v) NaN$_3$ in PBS. Keep at room temperature. For longer periods store at 4°C.
6. Mounting medium: 15.2% (w/v) AirVol 205, 15% (w/v) glycerol in PBS. Store in a dark bottle at 4°C.
7. Buffer G *(17)*: Resuspend K-glutamate (final concentration 100 mM) and K-citrate (39 mM) in highly purified water. Adjust the pH to 7.2 by dropwise addition of 100 mM citric acid. Filter sterilize and store at 4°C.
8. Lysis buffer: 50 mM Tris-HCl, pH 7.4, 0.5% (w/v) SDS, 2% (w/v) Nonidet P-40, 5 mM dithiotreiol (DTT) containing approx 10^6 cells/sample as carrier.

2.3. Cell Lines

1. HS68 (CRL-1365, ATTC).
2. REF-52 *(18)*.

2.4. Other Equipment

1. Dialysis tubing (Spectra/Por 1; 6.4 mm).
2. 15 and 50 mL Falcon tubes; Eppendorf tubes; 35, 60, 100 mm Petri dishes; 25 cm^2 T-flasks.

3. Planar glass cover slips, 12 mm diameter (Schütt Labortechnik, Göttingen, Germany).
4. Diamond-cut square glass chips (approx 1×1 mm^2).

3. Methods

3.1. Generation of Antisera for Immunocytology, Biochemistry, or Microinjection

3.1.1. Choice of Epitopes (see Note 1)

For catalytic subunits, the choice of epitope is limited by the high levels of subunit conservation. Both peptides and whole subunits have been successfully used as immunogens. Regulatory subunits are less conserved and, therefore, choice of epitopes may be somewhat easier. *See* also **refs. 8,11,12,19–21.** In the following some of the most important steps that should be taken to start off the generation of specific antibodies as successfully as possible are summarized.

3.1.1.1. ANTIPEPTIDE ANTIBODIES

1. Choose your peptide according to the species of the cell model to be studied.
2. Run your peptide of choice against the various databases. Use TBLASTN *(22)* for the most accurate assessment.
3. Keep in mind that most PPase subunits exist as multiple isoforms. If your cell model contains several different isoforms of the subunit to be studied (to be checked by Northern blot analysis) the choice of epitopes depends on whether all isoforms should be studied together or separately.
4. Some of the posttranslational modification of PPase subunits are already known. When choosing epitopes from those regions, problems of epitope masking may arise (*see* **Note 2**).

3.1.1.2. ANTIPROTEIN ANTIBODIES

1. In light of the high conservation amongst serine/threonine PPases catalytic subunits, cross-reactivity with members from other classes may be encountered. A reliable control should exist (*see* **Note 3**).
2. In the case of subunits with multiple isoforms, the generation of antibodies against whole proteins leads very often to antisera specific for several members of the isoform family.
3. The purification of PPases (and especially of specific subunits) is very laborious. However, very good antibodies have been generated against recombinant (even insoluble) PPase subunits (*see* **Note 4**).

3.1.2. Immunization

Antibodies should be produced by standard immunization protocols *(23)*. Because of the extensive conservation of PPases, high titer antisera against

PPase subunits are difficult to obtain. In the authors' hands a number of simple rules might facilitate the generation of satisfactory antisera.

1. Peptides should be coupled to keyhole limpet hemocyanin.
2. Insoluble protein aggregates usually result in a good immune response.
3. The antigen should be taken up in Freund's adjuvant and not inverse. The ratio between antigen and adjuvant should be approx 4:6 (*see* **Note 5**).
4. Immunizations intervals should be at least 4 wk (*see* **Note 6**).
5. Sample the blood in either glass tubes or into plastic tubes containing a glass rod (*see* **Note 7**).

3.1.3. Storage of Antibodies

3.1.3.1. SERUM

Store serum at −20°C to −80°C.

3.1.3.2. PURIFIED ANTIBODIES

1. Supplement with a carrier protein, either BSA or biologically inert immunoglobulins, to a final concentration of 0.1–0.2 mg/mL (*see* **Note 8**).
2. Transfer the antibodies to a 6.4 mm dialysis tubing.
3. Concentrate by overnight dialysis against sucrose/PBS at 4°C.
4. Remove most of the sucrose from the outside of the dialysis tubing with a tissue.
5. Using two fingers, concentrate your solution at the bottom of the tubing.
6. Clamp the tubing so that the solution is under maximal pressure.
7. Remove sucrose by dialysis against PBS at 4°C.
8. Cut open the dialysis tubing, invert it into a 15-mL Falcon tube, and fix the tubing by screwing the tube's top.
9. Collect the antibodies by centrifugation at 5000*g* for 5 min.
10. Antibodies treated in this way can usually be stored in small (10–100 µL) aliquots at −20°C to −80°C (*see* **Note 9**).

3.2. Culture and Synchronization of Nontransformed Mammalian Fibroblasts

3.2.1. Cells lines

Biochemical, immunochemical, and microinjection experiments can be performed with essentially any cell line. However, care should be taken in respect to the questions asked and the methods used (*see* **Notes 10** and **11**). Two nontransformed mammalian fibroblast cell lines are used: human HS68 and rat embryo fibroblasts (REF-52).

3.2.2. Cell Culture

3.2.2.1. MAINTENANCE OF HS68 AND REF-52 FIBROBLASTS

1. Aspirate the growth medium from the cells of a 25 cm² T-flask.
2. Wash with PBS.

3. Add 2 mL of warmed trypsin solution.
4. Aspirate the trypsin as soon as you see the cells detaching on their edges.
5. Gently dislodge the cells into 5 mL of warmed serum containing DME.
6. Collect cells by centrifugation at 900*g* for 1 min.
7. Resuspend the cells in 5 mL of fresh serum containing DME.
8. Seed 0.5 to 1 mL of the cell suspension per 25 cm^2.
9. Pass cells in this way once every week (*see* **Note 12**).
10. Cells for biochemical analysis are passed into 35, 60, or 100 mm Petri dishes.

3.2.2.2. FREEZING AND THAWING HS68 AND REF-52

1. Proceed as in **Subheading 3.2.2.1.**
2. At **step 7** take up the cells from a single 25 cm^2 T-flask in 1 mL of 90% FCS, 10% DMSO solution.
3. Transfer to freezing vial.
4. Freeze over night at –20°C, then 24 h at –80°C.
5. Store cells in liquid nitrogen.
6. Quickly thaw cells by warming them in your hand.
7. Take them up into 10 mL of warmed DME containing 10% FCS.
8. Seed the cells into two 25 cm^2 T-flasks.
9. Change the growth medium 1–12 h later.

3.2.2.3. SUBCULTURING HS68 AND REF-52
FOR IMMUNOFLUORESCENCE ANALYSIS AND MICROINJECTION

1. Flame sterilize planar glass coverslips (c/s) and put 6–8 c/s into a 60 mm Petri dish.
2. Trypsinize your cells as described in **Subheading 3.2.2.1.**
3. Put 100 µL (or two drops from a 5-mL pipet) of the cell suspension onto each c/s and incubate for 3–5 min. Depending on when the cells should be further treated use higher dilutions.
4. Check that most cells have adhered by microscopy (*see* **Note 13**).
5. Add 5 mL of medium per dish and allow cells for at least 24 h to recover and become fully flattened before further manipulation.

3.2.3. Cell Synchronization

Cells can be synchronized by many different methods *(24)* (*see* **Note 14**). However, two simple methods are sufficient to efficiently synchronize nontransformed fibroblasts: serum deprivation to obtain quiescent and G$_1$ cells and resynchronization using hydroxurea for cells synchronous in S, G$_2$, and mitosis (*see* **Note 15**).

3.2.3.1. SYNCHRONIZATION BY SERUM DEPRIVATION

1. Heat denature 10 mL of FCS in a 15-mL Falcon tube in a boiling water bath under gentle agitation for 15 min.
2. Pass cells as described in **Subheading 3.2.2.** and leave for at least 24 h to recover.

3. Wash the cells once with serum free DME to remove serum remnants.
4. Replace the culture media with DME supplemented with 0.5% heat-denatured FCS (*see* **Note 16**).
5. Incubate for at least 30 h, but maximally 48 h (*see* **Notes 17** and **18**). These cells are now in G_0.
6. Refeed cells by exchanging the media with DME supplemented with 10% FCS. These cells immediately enter G_1 and will remain in this phase for the next 15 h.

3.2.3.2. SYNCHRONIZATION FOR S, G_2, OR MITOSIS BY A HYDROXUREA (HU) BLOCK

1. Produce quiescent cells by serum deprivation as described in **Subheading 3.2.3.1.**
2. Activate cells into proliferation by serum addition.
3. After 3–12 h add HU to a final concentration of 2 mM for REF-52 and 0.5–1 mM for HS68 (*see* **Note 19**).
4. Incubate cells until at least 20 h have elapsed since serum addition (*see* **Note 20**). Cells are now blocked just before the G_1/S transition.
5. To release cells into S phase, aspirate off the HU containing growth media and wash cells in prewarmed solutions as follows: PBS for 30 s, then 5 min, followed by serum-supplemented DME for 10 min, then 15 min. Cells enter S phase during the next 10 min.
6. Cells in S phase should be further manipulated during the 2 h that follow.
7. Cells enter G_2 approx 4 h and mitosis 7–8 h after HU removal.

3.3. Immunolocalization of Serine Threonine Protein Phosphatases in Mammalian Cells

3.3.1. Fixation and Extraction Methods

Three fixation techniques are used for the immunological analysis of mammalian cells: fixation in 3.7% formalin followed by permeabilization with acetone, direct fixation/permeabilization with –20°C methanol, or detergent permeabilization followed by fixation in glutaraldehyde. Care has to be taken since any fixation/extraction procedure can lead to artifactual relocalization of a given protein (*see* **Note 21**).

3.3.1.1. FORMALIN FIXATION

1. Transfer the cells on the coverslip into a 35-mm Petri dish.
2. Fix cells by adding 3.7% formalin in PBS. Leave for 10 min at room temperature (*see* **Notes 22** and **23**).
3. Discard the formalin solution and add –20°C acetone. Leave for 30–60 s.
4. Discard the acetone and rehydrate the cells by incubation in PBS-BSA for 10 min (*see* **Notes 24** and **28**).

3.3.1.2. METHANOL FIXATION (*SEE* **NOTE 25**)

1. Transfer the cells on the c/s into a 35-mm Petri dish.
2. Fix cells by adding –20°C methanol. Leave for 5 min at room temperature (*see* **Note 26**).

3. Hold the dish with the c/s over a sink and under gentle agitation add PBS dropwise. Add PBS until the methanol solution is completely replaced (*see* **Note 27**).
4. Replace the PBS by PBS-BSA and incubate for 10 min (**Notes 24** and **28**).

3.3.1.3. GLUTARALDEHYDE FIXATION

1. Transfer the cells on the c/s into a 35 mm-Petri dish containing freshly prepared prefix solution. Incubate for 30 s.
2. Replace this prefix with 0.25–1% glutaraldehyde in PBS and incubate for 15–30 min.
3. Extract cells in –20°C acetone or methanol for 30 s.
4. Rehydrate in PBS.
5. Incubate the cells with 0.5 mg/mL $NaBH_4$ in PBS three times for 5 min (*see* **Note 29**).
6. Incubate the cells in PBS-BSA for 10 min (*see* **Notes 24** and **28**).

3.3.2. Alkali Pretreatment
to Study PP2Ac Carboxymethylation In Situ *(11)*

1. Fix cells with fresh formalin as described in **Subheading 3.3.1.1.** (*see* **Note 23**). Extract with acetone.
2. Wash the cells with PBS.
3. Replace with 100 m*M* NaOH and incubate for 10 min at room temperature.
4. Wash cells with several changes of PBS until the solution is no longer alkaline (check with pH paper).
5. Incubate the cells in PBS-BSA for 10 min (*see* **Notes 24** and **28**).

3.3.3. Immunostaining

1. Manipulate the c/s by carefully holding it by its edges with Dumont #5 forceps.
2. Remove excess PBS-BSA by touching a soft tissue with the edge of the c/s for a few seconds (*see* **Note 30**).
3. Place the c/s, cell side up, on the parafilm sheet of a humidified chamber (*see* **Note 31**).
4. Add 30–50 µL of the primary antibody(ies) diluted in PBS-BSA (*see* **Note 32**). If more than one primary antibodies are used on a single cell sample, make sure they are derived from different species.
5. Incubate for 60 min at 37°C in the humidified chamber.
6. Wash the c/s for 5 min by placing them into a 60-mm Petri dish filled with PBS-BSA (*see* **Note 33**).
7. During the wash appropriately dilute the secondary antibody(ies) in PBS-BSA. For more than one primary antibody use the appropriate secondary reagent each coupled to a different fluorochrome. If signal amplification is desired a biotinylated secondary antibody should be used. (*see* **Notes 34** and **35**).
8. Proceed as in **steps 2** and **3**.

9. Add 50 μL of the secondary antibody solution and incubate for 30–40 min at 37°C in the humidified chamber.
10. Wash cells for 5 minutes in PBS-BSA as in **step 6**.
11. (Optional) When a biotinylated secondary antibody was used in **step 9, steps 7–9** should now be repeated with a fluorochrome coupled to streptavidin (*see* **Note 36**).
12. (Optional) If DNA staining is desired the solution of the last immunodecoration step should be supplemented with 1 μg/mL Hoechst.
13. Wash cells in PBS for 5 min.
14. Wash cells in water for 2 min (*see* **Note 37**).
15. Mount cells by inverting onto a drop of mounting medium.
16. Air-dry and observe.
17. Make sure to perform all necessary controls (*see* **Notes 38–41**).
18. The immunofluorescence signal is more stable if specimen are stored in a dark and cold place.

3.4. Microinjection as a Means to Assess Protein Phosphatase Function in Mammalian Cells

3.4.1. Characteristics of the Sample to be Microinjected

Microinjection of either proteins or nucleic acids may significantly alter the intracellular environment. Unspecific effects that arise from the environment of your sample are unwanted but common. In the following the authors summarize what they think should help to successfully microinject your sample.

3.4.1.1. THE MICROINJECTION BUFFER (*see* ALSO **NOTE 42**)

One of the most important aspects of microinjection must be the choice of the correct microinjection buffer. This solution should be as close as possible to the intracellular environment. Samples are routinely injected in buffer G (100 mM K-glutamate, 39 mM K-citrate pH 7.2) (*see* **Note 43**). If your sample is incompatible with this buffer try to observe the following set of rules.

1. Avoid the use of strong artificial buffers like Tris, HEPES, MES, PIPES, MOPS, or if necessary, at very low concentrations (5 mM).
2. Do not use chelating agents.
3. Microinjected samples should be free of any traces of glycerol.
4. Avoid any detergent even at very low concentrations (*see* **Note 44**).
5. Always perform a control experiment where only the microinjection buffer is injected alone.

3.4.1.2. INJECTION OF EXPRESSION CONSTRUCTS

1. Subclone the target cDNA in either sense or antisense orientation into an appropriate eukaryotic expression vector. Avoid using CMV driven promoters (*see* **Note 45**).

2. Isolate the expression construct by two consecutive purifications on CsCl gradients (*see* **Note 46**).
3. Resuspend the DNA in highly purified water at a final concentration of 1–3 mg/mL.
4. Store at 4°C until microinjection.
5. Spin at 20,000*g*, 4°C for at least 20 min, mix in the marker antibodies and microinject.
6. Always perform control injections with the empty vector in the context of the phenomena to be studied.

3.4.1.3. INJECTION OF ANTIBODIES

1. Injection of crude antisera is not recommended.
2. Isolate IgGs by standard methods or affinity purify the antibodies.
3. Concentrate the antibodies as described in **Subheading 3.1.3.2**. Add inert marker antibodies as carriers in the case of affinity purified antibodies. The specific antibodies should reach a final concentration of at least 0.1 mg/mL (*see* **Note 47**).
4. Do not add azide as preservative.
5. Spin at 20,000*g*, 4°C for at least 20 min before microinjection.
6. Control injections comprise marker antibodies or even better the flow-through of the affinity purification.

3.4.1.4. INJECTION OF PROTEIN PHOSPHATASES, REGULATORY SUBUNITS, OR PROTEIN INHIBITORS

1. Purify the protein from either tissue, cells, or recombinant source.
2. Check purity by gel electrophoresis and/or analytical HPLC. Check purity by two-dimensional gel electrophoresis.
3. Concentrate the protein by ion exchange chromatography in a buffer that convenes to the rules given in **Subheading 3.4.1.1.**
4. Concentrate PP1c and PP2Ac by ethanol precipitation, inhibitor 1 and 2 by TCA precipitation (*see* **Note 48**).
5. Resuspend/dilute the protein sample in buffer G to a final concentration of at least 0.1–0.2 mg/mL (*see* **Note 47**).
6. Spin at 20,000*g*, 4°C for at least 20 min, mix in the marker antibody and microinject.
7. Parallel to the microinjection check the activity of the protein (after **step 4**) by appropriate in vitro assays.

3.4.2. Injection of Agents that Reduce the Intracellular Activity of PPases

Reducing the intracellular activity of either PP1 or PP2A can be achieved by: the microinjection of neutralizing antibodies, catalytic subunit antisense expression constructs, overexpression from inhibitor-encoding constructs, or protein or polycarbon inhibitors. Each of these approaches has significant advantages and disadvantages in respect to the biological question asked. Unfortunately none of them combines high specificity with precise timing.

3.4.2.1. NEUTRALIZING ANTIBODIES

1. Control neutralizing activity and specificity in phosphatase assays and immuno-precipitations.
2. Neutralizing antibodies have the advantage of instantaneous action at a chosen cell stage.
3. Neutralizing antibodies are usually not specific for a particular holoenzyme configuration.
4. Their in vivo action should be monitored by metabolic labeling experiments (*see* **Subheading 3.4.5.**).

3.4.2.2. INHIBITION OF PPASE ACTIVITY BY MICROINJECTION OF ANTISENSE-CATALYTIC-SUBUNIT EXPRESSION VECTORS

1. Highly specific for a given class of PPases.
2. Not specific for particular holoenzyme configurations.
3. Action depends on the turn-over rate of the catalytic subunit (*see* **Note 51**).
4. Monitor the knock-out by immunofluorescence.

3.4.2.3. OVEREXPRESSION OF INHIBITOR PROTEINS FROM PLASMID INJECTION VECTORS

1. Highly specific for a given class of PPases.
2. Not specific for any particular holoenzyme configuration.
3. Timing may be imprecise due to slow or weak expression. Monitor expression by immunofluorescence.
4. For the interpretation of the result, bear in mind that a protein recognized as a bona fide inhibitor of PPases in vitro might not act the same in vivo (*see* **Note 49**).

3.4.2.4. INJECTION OF PPASE INHIBITORS

1. Highly specific for a given class of PPases.
2. Precise timing possible since they act instantaneously.
3. Not specific for a particular holoenzyme configuration.
4. Generally it is recommended not to use polycarbon inhibitors like okadaic acid since nothing is known about their in vivo selectivity.
5. For the interpretation of the result, bear in mind that a protein recognized as a bona fide inhibitor of PPases in vitro might not act the same in vivo (*see* **Note 50**).

3.4.3. Increasing the Intracellular Activity of PPases Through Microinjection

The strategies are similar to those for the down-regulation of PPase activity (*see* **Subheading 3.4.2.**): Injection of purified enzyme preparations or the overexpression of catalytic subunits. The general features described in **Subheading 3.4.2.** (precise timing when injecting proteins rather than DNA expression construct) apply as well. On top of that the following should be observed:

1. Never quantitate the PPase injected as activity per volume but as real protein concentration. This concentration should be at least in the order of 0.1–0.2 mg/mL (*see* **Note 47**).
2. The injection of isolated catalytic subunits of PP1 or PP2A may produce similar results due to the absence of regulators.
3. Remember that PP2A subunits (in particular PP2Ac) cannot be overexpressed in eukaryotic cells (**Note 50**).

3.4.4. Deciphering the Role of a Specific Subunit

To address these questions the authors favor either specific knock-out or selective overexpression of a subunit through injection of antisense or sense constructs, as appropriate.

3.4.4.1. GENERAL STRATEGY

1. Knock-out or overexpress the subunit of interest (*see* **Subheading 3.4.1.2.**). Monitor this process by immunofluorescence. The injection area should be well marked.
2. Incubate injected cells for 18–36 h (*see* **Note 51**).
3. Analyze injected cells in intervals of 4–8 h by phase contrast microscopy for changes in morphology or cell cycle arrest.

3.4.4.2. ANALYSIS IN RESPECT TO A GIVEN TARGET PROTEIN

1. Proceed as in **Subheading 3.4.4.1.**
2. Analyze target protein by either immunofluorescence (*see* **Subheading 3.3.**) or metabolic labeling (**Subheading 3.4.5.**) (*see* **Note 52**).

3.4.5. Measuring Changes in Phosphorylation State and/or Protein Level in Injected Cells

Most applications of microinjection are followed by phenomenological observations (i.e., events monitored by immunocytochemistry). The authors have developed the means to analyze in vivo changes in cellular protein synthesis and phosphorylation state of specific proteins. With respect to protein phosphatases, changes in both steady-state protein phosphorylation and phosphate turnover can be measured by this technique.

3.4.5.1. METABOLIC LABELING OF INJECTED CELLS

1. Plate and culture cells on 1×1 mm^2 glass chips. Synchronize as necessary.
2. Lyophilize 1 mCi [^{32}P] orthophosphate (*see* **Note 53**). Resuspend in 5 µL phosphate-free DME. containing 10% FCS dialyzed against phosphate-free DME.
3. Microinject each cell on the chip while counting the number of cells injected (approx 300–400 cells), but stop after 15 min.
4. Scrape uninjected cells from the chip using fine forceps.
5. Transfer the chip immediately to 5 µL of warmed phosphate-free DME and wash twice.

6. Spot the 5 μL of DME containing the [^{32}P]-orthophosphate. Incubate for 15–45 min in a humidified incubator.
7. Rinse cells three times in 10 μL of PBS.
8. Lyse the cells by transferring the chip into an Eppendorf tube containing 20–50 μL of lysis buffer.
9. Boil for 2 min. Freeze or process for electrophoretic analysis.
10. This should be done for each sample to be microinjected including the control (buffer only). Begin with the control sample which should set the number of cells to be injected.
11. Essentially the same procedure can be performed to analyze changes in protein levels. Label the cells with 500 μCi of [^{35}S]-methionine in 5 μL of methionine-free DME.
12. Analysis of phosphate turnover involves essentially the same technique except-ing that cellular phosphate pools are preloaded with [^{32}P] by incubation in 5 μL of DME containing orthophosphate prior to injection. Subsequently, cells are injected during 5 min in phosphate-free medium and then returned to the radiolabel.

3.4.5.2. SAMPLE PREPARATION FOR TWO-DIMENSIONAL GEL ELECTROPHORESIS

1. Lyophilize the labeled protein samples generated by the method in **Subheading 3.4.5.1.**
2. Resuspend in 10 μL of protease-free DNAse, RNase solution (10 U of each).
3. Incubate for 15 min at 30°C.
4. Precipitate the proteins by the addition of 90 μL of –20°C acetone. Leave on ice for 10 min.
5. Centrifuge at 15,000g, 4°C for 10 min.
6. Air-dry or lyophilize the pellets.
7. Resuspend in IEF sample buffer and perform standard two-dimensional gel elec-trophoresis for 20,000 Vh as described *(25)*.
8. Stain and dry the gel and expose on film or phosphor-imager.
9. Alternatively transfer the gel to nitrocellulose or polyvinylidene difluoride (PVDF) membrane. Subsequently, the protein of interest can be studied by both immunoblot and autoradiography.

4. Notes

1. An intriguing property of PPases is their existence as multimeric holoenzymes in vivo. Usually PPases are very difficult to immunoprecipitate in their native state (e.g., so far no antisera has been generated against PP2A, which quantitatively immunoprecipitates the native enzyme). This is probably a result of surfaces occupancy in subunit association.
2. PP2Ac is reversibly methylated at its carboxyl terminus. Antibodies against the 9–12 C-terminal amino acids are sensitive to this modification *(11)*. This obser-vation is based on several anti-C-terminal antisera from different laboratories and commercial suppliers.

3. Testing the antisera in immunoblots of related PPase subunits (may be produced as recombinant proteins), as described in *(19)*, is an appropriate control.

4. Recombinant, insoluble PPase subunits can easily be partially purified as inclusion bodies and directly used for immunization. The authors have successfully used antibodies prepared in such way against PP2Ac and the PP2A B subunit of 55 kDa *(11,19,21)*.

5. This leads to an emulsion of water in oil, which remains as a deposit under the skin of the animal much longer then an emulsion of oil in water.

6. Desensitization of the animal to the antigen should be avoided.

7. Sampling of serum should be performed with great care to avoid blood cell lysis, which usually titrates out and thus reduces the amount of specific antibodies, unfortunately, for obvious reasons, those which will recognize native holoenzymes. In general, the use of glass tubes (or the addition of a glass rod into Falcon tubes) minimizes this process since silicate increases the rate of blood clotting. Serum should always be light brown, possibly orange, but never reddish.

8. Add immunoglobulins of the same species (*see* **Subheading 2.**) for those antibodies that are used in microinjection experiments. Add BSA if the antibodies are to be used in immunoblots. Some of the carrier immunoglobulins, although biologically inert result in an increase of background in immunoblots.

9. The resistance to freezing of the antibody should be tested with a small aliquot. During this time keep the rest on ice in the coldroom.

10. For most purposes, adherent cells are easier to manipulate in microinjection and immunolocalization experiments. Here, the authors concentrate on fibroblast cells lines. However, epithelial cells have been used for the immunolocalization of different phosphatase subunits *(12)*.

11. Observations made in a specific cellular background may not automatically be generalized. A typical example found in the literature is the question of which PPase dephosphorylates the transcription factor CREB. In mammalian fibroblasts, PP1 appears to be the PPase responsible *(26,27)*, whereas in systems derived from liver, PP2A has convincingly been shown to dephosphorylate CREB *(28,29)*.

12. Cells cultured in this way usually become confluent at d 5 or 6, but support very well being passed on d 7.

13. Problems may be encountered with cell adhesion to glass. This is because of impurities in the glass. Under optimal conditions most cells should be immediately attach with the c/s. If this is not the case, the coverslips should be acid-washed before use *(30)*.

14. Whereas some transformed cells can also be synchronized by treatment with microtubule depolymerising drugs (e.g., nocodazol), most nontransformed fibroblasts respond poorly to these agents and instead block at any point in the cell cycle.

15. Hydroxyurea blocks ribonucleotide reductase *(31)*.

16. Starvation in DME containing no FCS frequently fragilizes the cells. This is an important aspect if these quiescent cells are to be microinjected.

17. Cells enter G_0 only after the completion of a complete cycle (24–26 h in HS68 or REF-52). Once cells pass a specific checkpoint in G_1, they will complete the

cycle even in serum-free medium. Therefore, cells should be starved for sufficient time to complete a single cell cycle.

18. REF-52 and HS68 are stable in G_0 for at least 48 hours.

19. HU should be added when all cells are in G_1 to prevent effects on DNA repair as cells leave G_0.

20. Sufficient time must elapse for cells to reach the G_1/S boundary and all be blocked before entering S.

21. No fixation/permeabilization procedure reflects the in vivo distribution of a protein in an absolute manner. Some fixation procedures are known to create significant artifacts. In particular, the use of detergent permeabilization before fixation may yield artifactual staining of cytoskeletal structures (unpublished observations and ref. *32*). In order to determine the localization of a given protein reliably, the authors recommend to perform immunofluorescence with all three procedures described. The distribution should be comparable between any two techniques.

22. Formalin is very toxic. Avoid breathing it and handle the stocks in a fume hood.

23. Fresh formalin should be used for all experiments. In general, the working dilutions should not be stored longer than a month. The cross-linking action of formalin should be fast, highly contrasted surface blebs should be seen at the cell borders by phase contrast microscopy within 15–30 s after formalin/PBS is added. In particular, when studying PP2Ac carboxyl methylation with anticarboxyl-terminal antibodies *(11)*, the formalin to be used should be fresh. Bottles not older than 6 mo after production are recommended. Once opened, the bottle should be stored in a dry and dark place and may be used for maximally 3–4 wk. Dilutions into PBS should be made fresh before every experiment.

24. This step serves to block free or unreacted aldehydes or other crosslinking agents and prevent non-specific antibody adhesion to the cells or coverslip.

25. Nucleosolar proteins of high solubility (e.g., PP2Ac) are often extracted by this method and, therefore, under-represented, the amount of soluble, cytosolic proteins is reduced similarly. However, in the case of proteins associated with subcellular structures methanol removes the nonassociated "background." Best examples are microtubule associated proteins or the association of PP1c with the condensed chromatin during mitosis *(8)*.

26. Methanol acts instantaneously. Therefore, it is not necessary to keep the c/s at −20°C for the 5 min.

27. This is the critical step of this fixation procedure and should be performed with greatest care. If the methanol is replaced too abruptly, cells undergo an osmotic shock.

28. Sometimes it is necessary to store fixed c/s (e.g., during a cell-cycle experiment where all immunofluorescence should be performed in parallel). Store formalin fixed c/s before the acetone extraction in PBS at 4°C in a dark place. Acetone extract the stored c/s just before starting the immunodecoration. Methanol and glutaraldehyde-fixed c/s are stored in PBS-BSA (supplemented with azide!) at 4°C. Alternatively methanol-fixed c/s can be air-dried directly after the methanol treatment and stored in a dry and dark place. Before usage they should be rehydrated in PBS-BSA.

29. NaBH$_4$ blocks unreacted aldehydes, which exhibit strong autofluorescence.
30. Avoid the c/s drying out at any time during the immunostaining since this strongly enhances unspecific background staining.
31. The authors use 10 × 10 cm^2 square Petri dishes containing a 9 × 9 cm^2 parafilm sheet. To keep up a high humidity, a moist tissue is placed at the edge and the dish is closed with a top.
32. Crude sera can usually be used at 1:100. If you have already determined an optimal dilution of your antibody for Western blots, dilute the antibodies 2–3 times less for immunofluorescence.
33. This washing method is very efficient because of the comparatively large volume of PBS-BSA used. Sometimes a large number of 60-mm dishes are needed. However, they can be rinsed in distilled water and reused.
34. The authors use the secondary reagents listed under **Subheading 2.** at the following dilutions: fluorescein- and rhodamine-labeled antibodies at 1:30 to 1:50; biotinylated antibodies at 1:100 to 1:200. If secondary reagents from other suppliers are used, controls should be performed using these antibodies alone. Keep in mind that higher dilutions eliminate unspecific background, but also weaken the detection of the primary signal.
35. In double or triple immunofluorescence experiments the weakest signal should always be combined with Texas Red or rhodamine, and not fluorescein detection since the latter has the highest rate of photobleaching. For example, when staining cells for a PPase subunit using a rabbit serum and for microtubules using a commercial monoclonal antibody, use fluorescein-labeled antimouse and rhodamine/Texas Red-labeled antirabbit antibodies.
36. The authors use streptavidin-Texas Red and streptavidin-fluorescein at 1:200.
37. This step eliminates salt crystals which can destroy your microscope lens.
38. Significantly, all immunocytochemistry should be accompanied by biochemical studies. Simple extraction and subcellular fractionation followed by immunoblotting usually help to exclude artifactual localization. A simple protocol using mechanical fractionation to separate cytoplasm from nuclei, avoiding the use of detergent and possible leaking from the nuclear compartment, is described in **ref. *11***.
39. As a simple rule of thumb, immunocytochemistry of serine/threonine protein phosphatases should be performed with at least two antisera against different epitopes of the molecule to be studied. In addition different fixation, extraction procedures should be used (*see* **Note 21**).
40. Epitope competition should be performed by preincubating the primary antibody with the antigen (1–50 μ*M*) for 30–60 min. Antibodies raised against insoluble, recombinant proteins can be efficiently competed with membrane strips of blotted protein.
41. Antisense expression is undoubtedly the most accurate method to control for unspecific staining patterns. For example, one of the anti-PP2Ac antibodies (highly specific in immunoblots) suggested an association of PP2Ac with intermediate filaments. The immunofluorescence signal was competed by the anti-

genic peptide. Other anti-PP2Ac antibodies suggested a ubiquitous distribution of PP2A in the cytoplasm and the nucleus. Only the latter signal was abolished by antisense expression and therefore considered to reflect the in vivo distribution *(11)*.

42. The microinjection buffer is the only component (apart from the biological molecule to be injected) to enter the cell and, therefore, the contribution of this buffer should be understood to minimize creation of artifacts. By far, the most suitable injection buffer is one in which the component of the injection solution have little or no buffering capacity, and are rapidly degraded by common cellular metabolism. Although many types of buffer have been reported, injection solutions that contain either strong buffers or chelating agents should be avoided. A quick survey of the injection solutions most commonly used shows that the general tendency is toward using the buffer in which the sample is most stable at the expense of modifying the intracellular environment.

43. The components of buffer G are rapidly metabolized by the citric acid cycle and, thus, quickly disappear from the intracellular environment.

44. The use of detergents during the purification should be avoided (especially during dialysis steps) since a lot of proteins tend to "soak" the detergent. The bound detergent is, nevertheless, released within the cell and this might often lead to severe damage.

45. Empty eukaryotic expression vectors can have strong effects on cell viability. It has been found that the nuclear injection of CMV-driven plasmids can lead to S phase arrest.

46. Using plasmids purified in this way the highest and most reliable ratio of expression per injected cell is observed. This is probably because of the fact that plasmids of this purity are stable at 4°C, since the authors remarked that plasmids stored frozen usually express at an inferior rate. Take care to remove well ethidum bromide after CsCl centrifugations.

47. Assuming a total protein concentration of 200 mg/mL in living cells, intracellular concentrations of PP1 and PP2A might be as high as 0.1 mg/mL *(11,33)*. Solutions are diluted 10–30-fold during microinjection. Injection of an antibody at 0.1 mg/mL may, therefore, only inhibit 10% of the PPase present. Likewise, injection of a PPase at 0.1 mg/mL might change the intracellular PPase content by only 10%.

48. Such precipitations are the easiest way to get rid of components like glycerol or Brij that are often used during purification and storage of these proteins.

49. Inhibitor-2 was initially thought to be a protein whose only action was to inhibit PP1. Now this protein is rather considered a modulator of PP1 activity *(5)*.

50. Whereas PP2Ac mRNA can be significantly overexpressed, protein and activity levels remain unchanged. It is possible that PP2Ac regulates its own expression (Turowski, P., Mayer-Jaekel, R., and Hemmings, B. A., unpublished observations). *See* also **ref. *34***.

51. Long incubation times are often necessary for effects to become visible. Two reasons appear likely. First, overexpression or especially knock-out of proteins

may take a considerable amount of time depending on the expression rate and/or the half life of the protein studied *(11)*. Second, the cell has to reach a certain stage of its cycle before the loss or the overabundance of the protein causes any effects *(8,34)*.

52. Start initial analyses at time points where overexpression or down-regulation are significant, but not maximal (as corroborated by immunofluorescence). At time points of maximal down-/up-regulation, cell metabolism might be altered in a general manner and changes in the target protein might be of secondary nature.
53. When working with radioactivity take the necessary safety precautions.

Acknowledgments

The authors would like to thank Anne Fernandez, with whom all their work has been carried out, for critical comments on the manuscript. We gratefully acknowledge the contribution of the laboratories of David Brautigan and Brian Hemmings, which have and continue to provide us with innumerable reagents for our studies of PP1 and PP2A over the years. This work was in part supported by the Association Francaise contre les Myopathies and the Association pour la Recherche contre le Cancer (contrat 1306). P.T. is the recipient of a post-doctoral fellowship from the Roche Research Foundation.

References

1. Hubbard, M. J. and Cohen, P. (1993) On target with a new mechanism for the regulation of protein phosphorylation. *Trends Biochem. Sci.* **18,** 172–177.
2. Mayer-Jaekel, R. E. and Hemmings, B. A. (1994) Protein phosphatase 2A—a 'menage a trois'. *Trends Cell Biol.* **4,** 287–291.
3. Bollen, M. and Stalmans, W. (1992) The structure, role, and regulation of type 1 protein phosphases. *Crit. Rev. Biochem. Mol. Biol.* **27,** 227–281.
4. Shenolikar, S. (1994) Protein serine/threonine phosphatases- new avenues for cell regulation. *Annu. Rev. Cell Biol.* **10,** 55–86.
5. Bollen, M., DePaoli-Roach, A. A., and Stalmans, W. (1996) Native cytosolic protein phosphatase-1 (PP1S) containing modulator (inhibitor-2) is an active protein. *FEBS Lett.* **344,** 196–200.
6. Beullens, M., Van Eynde, A., Stalmans, W., and Bollen, M. (1992) The isolation of novel inhibitory polypeptides of protein phosphatase 1 from bovine thymus nuclei. *J. Biol. Chem.* **267,** 16,538–16,544.
7. Fernandez, A., Brautigan, D. L., Mumby, M., and Lamb, N. J. C. (1990) Protein phosphatase type-1, not type-2A, modulates actin microfilament integrity and myosin light chain phosphorylation in living nonmuscle cells. *J. Cell Biol.* **111,** 103–112.
8. Fernandez, A., Brautigan, D. L., and Lamb, N. J. C. (1992) Protein phosphatase type 1 in mammalian cell mitosis: chromosomal localization and involvement in mitotic exit. *J. Cell Biol.* **116,** 1421–1430.
9. Zolnierowicz, S., Van Hoof, C., Andjelkovic, N., Cron, P., Stevens, I., Merlevede, W., Goris, J., and Hemmings, B. A. (1996) The variable subunit associated with

protein phosphatase 2A0 defines a novel multimember family of regulatory subunits. *Biochem. J.* **317**, 187–194.

10. Csortos, C., Zolnierowicz, S., Bako, E., Durbin, S. D., and DePaoli-Roach, A. A. (1996) High complexity in the expression of the B' subunit of protein phosphatase 2A0. Evidence for the existence of at least seven novel isoforms. *J. Biol. Chem.* **271**, 2578–2588.

11. Turowski, P., Fernandez, A., Favre, B., Lamb, N. J. C., and Hemmings, B. A. (1995) Differential methylation and altered conformation of cytoplasmic and nuclear forms of protein phosphatase 2A during cell cycle progression. *J. Cell Biol.* **129**, 397–410.

12. Sontag, E., Nunbhakdi-Craig, V., Bloom, G. S., and Mumby, M. C. (1995) A novel pool of protein phosphatase 2A is associated with microtubules and is regulated during the cell cycle. *J. Cell Biol.* **128**, 1131–1144.

13. Pitcher, J. A., Payne, E. S., Csortos, C., DePaoli-Roach, A. A., and Lefkowitz, R. J. (1995) The G-protein-coupled receptor phosphatase: a protein phosphatase type 2A with a distinct subcellular distribution and substrate specificity. *Proc. Natl. Acad. Sci. USA* **92**, 8343–8347.

14. Mayer-Jaekel, R. E., Ohkura, H., Ferrigno, P., Andjelkovic, N., Shiomi, K., Uemura, T., Glover, D. M., and Hemmings, B. A. (1994) *Drosophila* mutants in the 55 kDa regulatory subunit of protein phosphatase 2A show strongly reduced ability to dephosphorylate substrates of p34^{cdc2}. *J. Cell Sci.* **107**, 2609–2616.

15. Cegielska, A., Shaffer, S., Derua, R., Goris, J., and Virshup, D. M. (1994) Different oligomeric froms of protein phosphatase 2A activate and inhibit simian virus 40 replication. *Mol. Cell. Biol.* **14**, 4616–4623.

16. Hendrix, P., Mayer-Jaekel, R. E., Cron, P., Goris, J., Hofsteenge, J., Merlevede, W., and Hemmings, B. A. (1993) Structure and expression of a 72-kDa regulatory subunit of protein phosphatase 2A. *J. Biol. Chem.* **268**, 15,267–15,276.

17. Saxton, W. M., Stemple, D. L., Leslie, R. J., Salmon, E. D., Zavortink, M., and McIntosh, J. R. (1984) Tubulin dynamics in cultured mammalian cells. *J. Cell Biol.* **99**, 2175–2186.

18. McClure, D. B., Hightower, M. J., and Topp, W. C. (1982) Effect of SV40 transformation on the growth factor requirements of the rat embryo cell line REF52 in serum free medium. *Cold Spring Harbor Conf. Cell Prolif.* **9**, 345–364.

19. Hendrix, P., Turowski, P., Mayer-Jaekel, R. E., Goris, J., Hofsteenge, J., Merlevede, W., and Hemmings, B. A. (1993) Analysis of subunit isoforms in protein phosphatase 2A holoenzymes from rabbit and *Xenopus*. *J. Biol. Chem.* **268**, 7330–7337.

20. Zolnierowicz, S., Mayer-Jaekel, R. E., and Hemmings, B. A. (1995) Protein phosphatase 2A- purification, subunit sructure, molecular cloning, expression, and immunological analysis. *Neuroprotocols* **6**, 11–19.

21. Michelson, S., Turowski, P., Picard, L., Goris, J., Landini, M. P., Topilko, A., Hemmings, B. A., Bessia, C., Garcia, A., and Virelizier, J. L. (1996) Human cytomegalovirus carries serine/threonine protein phosphatases PP1 and a host-cell derived PP2A. *J. Virol.* **70**, 1415–1423.

22. Altschul, S. F., Gish, W., Miller, W., Myers, E. W., and Lipman, D. J. (1990) Basic local alignment search tool. *J. Mol. Biol.* **215,** 403–410.
23. Harlow, E. and Lane, D. (1988) *Antibodies: A Laboratory Manual,* Cold Spring Harbor Laboratory Press, Cold Spring Harbor, New York.
24. Krek, W. and DeCaprio, J. A. (1995) Cell synchronization. *Methods Enzymol.* **254,** 114–115.
25. O'Farrell, P. H. (1975) High resolution two-dimensional electrophoresis of proteins. *J. Biol. Chem.* **250,** 4007–4021.
26. Hagiwara, H., Alberts, A., Brindle, P., Meinkoth, J., Feramisco, J. R., Deng, T., Karin, M., Shenolikar, S., and Montminy, M. (1992) Transcriptional attenuation following cAMP induction requires PP-1-mediated dephosphorylation of CREB. *Cell* **70,** 105–113.
27. Alberts, A. S., Montminy, M., Shenolikar, S., and Feramisco, J. R. (1994) Expression of a peptide inhibitor of protein phosphatase 1 increases phosphorylation and activity of CREB in NIH 3T3 fibroblasts. *Mol. Cell. Biol.* **14,** 4398–4407.
28. Wadzinski, B. E., Wheat, W. H., Jaspers, S., Peruski, L. F., Lickteig, R. L., Johnson, G. L., and Klemm, D. J. (1993) Nuclear protein phosphatase 2A dephosphorylates protein kinase A-phosphorylated CREB and regulates CREB transcriptional stimulation. *Mol. Cell. Biol.* **13,** 2822–2834.
29. Wheat, W. H., Roesler, W. J., and Klemm, D. J. (1994) Simian virus 40 small tumor antigen inhibits dephosphorylation of protein kinase A-phosphorylated CREB and regulates CREB transcriptional stimulation. *Mol. Cell. Biol.* **14,** 5881–5890.
30. Parker, R. C. (1961) *Methods of Tissue Culture,* 3rd ed. (Hoeber, P., ed.), New York, p. 31.
31. Adams, R. L. P. and Lindsay, J. G. (1967) Hydroxyurea. Reversal of inhibition and use as a cell-synchronizing agent. J. Biol. Chem. 242, 1314–1317.
32. Melan, M. A. and Sluder, G. (1992) Redistribution and differential extraction of soluble proteins in permeabilized cultured cells. Implications for immunofluorescence microscopy. *J. Cell Sci.* **101,** 731–743.
33. Ruediger, R., van Wart Hood, J. E., Mumby, M., and Walter, G. (1991) Constant expression and activity of protein phosphatase 2A in synchronized cells. *Mol. Cell. Biol.* **11,** 4282–4285.
34. Wera, S., Fernandez, A., Lamb, N. J. C., Turowski, P., Hemmings, M., Mayer-Jaekel, R. E., and Hemmings, B. A. (1995) Deregulation of translational control of the 65 kDa regulatory subunit (PR65a) of protein phosphatase 2A leads to multinucleated cells. *J. Biol. Chem.* **270,** 21,374–21,381.

10

Use of Immunocomplexed Substrate for Detecting PP1 Activity

John W. Ludlow, Deirdre A. Nelson, and Nancy A. Krucher

1. Introduction

Numerous studies have been reported that demonstrate differences in protein function dependent upon the phosphorylation state of the protein. For many of these studies, a logical next step involves studying the enzyme(s) responsible for placing the phosphates on, as well as taking them off. When performing enzyme studies, it is advantageous to have the substrate protein in as pure a state as possible, and in a form that readily lends itself to manipulation and subsequent analysis. In situations where eukaryotic cell lysate is both the source of the enzyme and the protein substrate, multiple steps may be required to isolate the substrate to facilitate biochemical analysis of the reaction.

Here, a procedure is described for using immunocomplexed cellular protein as a substrate for in vitro phosphatase assays. This system relies on the specificity of the antibody to ensure immunopurification of the substrate to as close to homogeneity as possible. As is shown in **Fig. 1**, the substrate is easily manipulated, also remaining immunocomplexed with protein-A Sepharose. Successful use of immunocomplexed substrate may hinge on the antibody recognition site being different from the phosphorylation site(s) on the substrate protein. As such, the antibody does not sterically hinder the phosphatase from catalyzing phosphate removal. This assay system was developed to identify and characterize the serine/threonine protein phosphatase responsible for dephosphorylating the negative growth regulatory protein pRB, the product of the retinoblastoma susceptibility gene (1). This method has also been successfully used to demonstrate in vitro dephosphorylation of other proteins, such as the SV40 viral T-antigen (2).

From: *Methods in Molecular Biology, Vol. 93: Protein Phosphatase Protocols*
Edited by: J. W. Ludlow © Humana Press Inc., Totowa, NJ

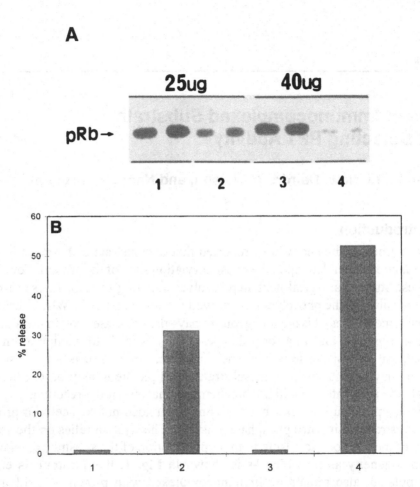

Fig. 1. *Detection of pRB-specific phosphatase activity*. **(A)** Aliquotes of ^{32}P-labeled pRB immunocomplexed to Protein A-Sepharose were incubated with mitotic cell extracts and then separated on an 8% SDS-polyacrylamide gel, fixed, dried, and then exposed to X-ray film overnight. Each reaction was performed in duplicate, with lanes 1 and 3 being the negative control reactions containing boiled cell lysate. Two different concentrations of mitotic cell lysate were used (25 µg and 40 µg), as indicated above the lanes. The location of pRB is indicated by the arrow to the left of the figure. **(B)** Quantitation of the band intensities shown in (A) was accomplished by exposing the dried gel to a phosphorimager screen and utilizing Imagequant software. Percent release was determined by comparing band intensity of the control reaction (designated 0% release) to that of the experimental for each protein concentration.

For this method, the authors' procedure for in vivo radiolabeling of pRB with ^{32}P is included, and subsequent immunopurification of the pRB substrate. Also provided are the procedures for synchronizing cells for optimum pRB-directed phosphatase activity, extraction of this activity, and detection of pRB dephosphorylation following gel electrophoresis.

2. Materials

Unless otherwise stated, all reagents were purchased from Sigma Chemical Company (St. Louis, MO) and are of the highest grade possible.

1. TBS: 25 mM Tris-HCl, pH 8.0, 150 mM NaCl. Store at 4°C.
2. EBC: 50 mM Tris-HCl, pH 8.0, 120 mM NaCl, 0.5% (v/v) Nonidet P-40. Store at 4°C.
3. NET-N: 20 mM Tris-HCl, pH 8.0, 100 mM NaCl, 1 mM EDTA, 0.5% (v/v) Nonidet P-40. Store at 4°C.
4. NET-N/4% BSA : NET-N containing 4% (w/v) bovine serum albumin (BSA), fraction IV. Store at 4°C.
5. Phosphatase wash buffer : 20 mM imidazole, pH 7.0, 150 mM NaCl. Store at 4°C.
6. Phosphatase assay buffer: Same as phosphatase wash buffer (20 mM imidazole, pH 7.0, 150 mM NaCl), containing 14 mM β-mercaptoethanol, 50 μM TLCK, 50 μM TPCK, 50 μM leupeptin, 2 μM benzamidine, and 1 μM pepstatin. Prepared fresh just before use. Store at 4°C until use.
7. 3X SDS sample buffer: 6% (w/v) sodium dodecyl sulfate (SDS), 3% (v/v) glycerol, 100 mM dithiothreitol (DTT), 0.03% (w/v) bromophenol blue. Store at −20°C in 1-mL aliquots. May be frozen-thawed at least four times before loss of potency.
8. Phenylmethylsulfonylfluoride (PMSF): 1 mg/mL (w/v) stock in 100% ethanol. Store at 4°C.
9. Leupeptin: 1 mg per mL (w/v) stock in distilled water. Long-term storage at −20°C in 100 μL aliquotes. Store at 4°C after thawed. Use within 1 wk after thawing.
10. Leupeptin hemisulfate (Calbiochem-Novabiochem International, La Jolla, CA.): 10 mM stock in distilled water. Long-term storage at −20°C in 100 μL aliquotes. Store at 4°C after thawed. Use within 1 wk after thawing.
11. Aprotinin: Purchased in aqueous solution having a protein concentration between 1 mg and 2 mg per mL. Store at 4°C.
12. Benzamidine: 1 mM stock in distilled water. Store at 4°C.
13. Pepstatin: 1 mM stock in 100% ethanol. Long-term storage at −20°C in 100 μL aliquotes. Store at 4°C after thawed. Use within 1 wk after thawing.
14. N-α-p-tosyl-L-lysine chloromethyl ketone (TLCK): 10 mM stock in distilled water. Prepare fresh just before use. Store at 4°C until use.
15. N-tosyl-L-phenylalanine chloromethyl ketone (TPCK): 10 mM stock in methanol. Store at 4°C.
16. Methyl-(5[2-thienyl-carbonyl}-1H-benzimidazol-2yl)carbamate (Nocodazole): 4 mg/mL stock solution in dimethylsulfoxide (DMSO). Store in 10 μL aliquotes at −20°C. Do not refreeze aliquotes after use.

17. Cell lysis buffer: EBC containing 10 µg/mL each of PMSF, leupeptin, and aprotinin. Prepare fresh just before use. Store at 4°C until use.
18. Dulbecco's Modified Eagle's Medium (DMEM): Purchased in liquid form, ready to use, from Gibco-BRL Life Technologies (Grand Island, NY).
19. Phosphate-free Dulbecco's Modified Eagle's Medium: Purchased in liquid form, ready to use, from Gibco-BRL Life Technologies.
20. Protein-A Sepharose: Purchased in dehydrated form. Swell the beads in NET-N/4% BSA. Store at 4°C. Before use, wash at least five times with NET-N/4% BSA. Resuspend the beads 1:1 (v/v) with NET-N/4% BSA.
21. ^{32}P-orthophosphate: Purchased from DuPont New England Nuclear Research Products (Boston, MA) in the form of phosphoric acid. Typical concentration is 10 mCi per mL.
22. Bovine calf serum (BCS): (Gibco-BRL Life Technologies). Keep frozen (−20°C) until use.
23. G3-245: antibody to PRB (Pharmingen, San Diego, CA). Store at 4°C.

3. Methods

3.1. Radiolabeling

1. Plastic dishes (100 mm diameter) containing actively growing CV-1P cells at approx 70–80% confluency are drained of medium and then rinsed and drained with 3 mL of warm phosphate-free DMEM. This is repeated a total of three times. After draining of the final rinse, 5 mL of phosphate-free medium is then added and the plate incubated for 30 min at 37°C in a humidified, 5% CO_2 containing atmosphere. The purpose here is to deplete inorganic phosphate stores within the cell (*see* **Note 1**).
2. After the 30 min incubation, the phosphate-free medium is drained and then replaced with 3 mL of phosphate-free medium freshly supplemented with 1 mCi ^{32}P-orthophosphate. The plate is then incubated for 3–4 h at 37°C in a humidified, 5% CO_2 containing atmosphere, with intermittent rocking of the plate every 15 min (*see* **Note 2**). The relatively high mCi amount of radiolabel used is because approx 90% will appear in nucleic acid, whereas only 10% appears in protein products.
3. Quickly drain and rinse the plate with 5 mL of ice cold TBS. Repeat a total of three times. Keep the dish on ice (*see* **Note 3**). This will remove the majority of unincorporated radiolabel from the cell surface.
4. After thoroughly draining the last rinse, add 1 mL of ice-cold cell lysis buffer per 100 mm dish.
5. Scrape the adhered cells using a rubber policeman or other cell scraping device into the lysis buffer. Transfer lysate to 1.5 mL Eppendorf tube and rock for 15 min at 4°C. This will facilitate further solubilization of the total cellular proteins.
6. Centrifuge at 10,000g for 10 min at 4°C. Transfer clarified supernate to fresh 1.5 mL Eppendorf tube and maintain on ice (*see* **Note 4**).

3.2. Immunoprecipitation

1. For each mL of clarified supernate, prepared from a single radiolabeled dish of cells, 5 μg of G3-245 antibody is added and continuously rocked overnight at 4°C (*see* **Note 5**).
2. The following day, 15 μL of anti-mouse IgG is added and continuously rocked for 2 additional h (*see* **Note 6**). The reason for this addition is that G3-245 is an IgG$_1$ species of mouse monoclonal, and requires an antimouse IgG for immunocomplex isolation on Protein-A Sepharose.
3. 100 μL of Protein-A Sepharose beads, prepared as a 1:1 slurry in NET-N/4% BSA, is then added to each tube and continuously rocked 1 h.
4. Centrifuge for 2 s at 10,000*g* to pellet the beads and discard the supernate (*see* **Note 7**).
5. Resuspend the beads in 1 mL of NET-N, centrifuge for 2 s at 10,000*g*, and discard supernate. Repeat for a total of five times.
6. Resuspend the beads in 1 mL of phosphatase assay buffer, centrifuge for 2 s at 10,000*g*, and discard supernate (*see* **Note 8**).

3.3. Cell Synchronization

1. CV-1P cells are passaged into DMEM containing 10% BCS and incubated for 3 h at 37°C in a humidified, 5% CO_2-containing atmosphere (*see* **Note 9**). During this incubation, the most healthy and vigorous cells will adhere to the surface of the plastic dish.
2. At the end of this 3-h incubation, nonadhered cells are removed by draining the culture medium and rinsing the adhered cells three times with 5 mL of warm, serum-free DMEM (*see* **Note 10**). This removes dead and less robust cells.
3. After draining the final wash of serum-free DMEM, the cells are maintained for 18 h in DMEM containing 0.4 mg/mL nocodazole supplemented with 10% BCS. This compound destabilized microtubule formation, thus arresting cell proliferation at the entrance to mitosis.
4. Mitotic cells will round up and are removed from the culture dish by vigorous pipeting of the culture medium used to maintain the cells for the 18-h time period onto the surface of the dish three to five times (*see* **Note 11**).
5. Mitotic cells are pelleted by centrifuging the culture medium at room temperature for 5 min at 2000g.
6. Culture medium is discarded, and the cell pellet resuspended in ice-cold serum-free DMEM, centrifuged at room temperature for 5 min at 2000g, and the supernate medium discarded. This washing procedure is repeated for a total of three times (*see* **Note 12**).

3.4. Preparation of Cell Lysates for use in Phosphatase Assay

1. Wash cells three times with ice cold phosphatase wash buffer.
2. Cells are lysed in phosphatase assay buffer by 15 times passage through a 28-gage syringe needle.
3. Centrifuge for 2 s at 10,000*g*.

4. Transfer supernate to a new Eppendorf tube and store on ice until use. Discard pellet (*see* **Note 13**).

3.5. Phosphatase Assay

1. Add 270 µL of phosphatase assay buffer to the Protein-A Sepharose beads. Resuspend and aliquot 30 µL into nine separate Eppendorf tubes (*see* **Note 14**).
2. Add 30 µL minimum, 80 µL maximum, of soluble cell supernate in phosphatase assay buffer to Protein-A sepharose beads (*see* **Note 15**).
3. Incubate at 30°C for 40 min with intermittent, gentle mixing of the beads with the lysate every 10 min (*see* **Note 16**).
4. Add half of the total bead/lysate volume of 3X sample buffer and boil. Samples may now be stored at −20°C or immediately analyzed on an 8% SDS-PAGE gel (*see* **Note 17**).
5. Quantify the intensity of the RB band by phosphorimaging or scanning densitometry of an autoradiogram.
6. Negative control reactions consist of ^{32}P-labeled pRB immunocomplexed to protein A-Sepharose mixed with boiled cell lysate or lysate preincubated for 15 min with 100 nM oakadic acid.

4. Notes

1. To ensure that the cells are actively growing, passage approximately 18 h before labeling is recommended. "Starving" the cells of phosphate during the 30-min incubation with phosphate-free DMEM seems to result in a greater ^{32}P incorporation of the cellular proteins.
2. Periodic rocking the plate to mix the medium during the incubation also appears to increase ^{32}P uptake by the cell, probably by eliminating microenvironmental areas of radionucleotide depletion.
3. Keeping the cells on ice will impede the off-rate of the radiolabeled phosphate groups as well as protein degradation.
4. At this stage, a protein assay as well as liquid scintillation counting may be performed if specific incorporation/activity values are needed to monitor labeling efficiency.
5. This step may be reduced to a 1 h incubation time. In so doing, the amount of radiolabeled pRB immunoprecipitated may be reduced, however.
6. An alternative approach is to use Protein-G Sepharose, which will bind to the G3-245 IgG$_1$ species. It has been the authors' experience, however, that this reaction is less efficient using this particular pRB antibody than first complexing with antimouse IgG and then complexing with Protein-A Sepharose.
7. Avoid excessive centrifugation to pellet beads. This may cause beads to fragment, thus producing "fines" that bind labeled protein nonspecifically, thus increasing the background of the autoradiogram. An alternative approach is to let the beads settle by gravity, which takes 5–10 min.
8. Removal of virtually all of the supernate can be achieved by aspiration with a 28-gage syringe needle; the beads will not pass through the needle opening.

9. When starting from a 90–100% confluent 100 mm dish, the authors typically pass CV-1P cells 1:3.
10. It is important to remove all of the "floaters" at this stage, because many cells blocked in mitosis following nocodazole treatment will round up and float in the culture medium. This will ensure that the population of mitotic cells is not contaminated by dead cells or cells in another phase of the cell cycle.
11. Whereas many mitotic cells are floating in the medium following the 18-h incubation in nocodazole, an equivalent or greater number are rounded up and loosely adhered to the dish. Vigorous pipeting of the medium onto the cell surface coaxes them off so that they may be recovered.
12. Washing removes BCS with may give artificially high protein readings following cell lysis.
13. At this stage, a protein assay may be performed if it is desirable to standardize samples. The pellet may also be saved for phosphatase analysis if needed.
14. It is crucial that the beads remain suspended for aliquoting. Gentle mixing of the tube by hand before each and every aliquot removal ensures an equal amount of beads being delivered per tube. All assays are also performed in triplicate to further alleviate discrepancies caused by pipeting errors.
15. This range of lysate works the best. Less than the minimum and exceeding the maximum generally results in poor efficiency of the reaction. Although not rigorously tested, it is suspected that this may have something to do with the ratio of bead volume to lysate volume. Also, whatever the control reaction is, make sure that it has a comparable amount of lysate added to it as carrier before boiling for gel analysis.
16. Periodic mixing of the tube incubation also appears to increase the efficiency of the reaction, perhaps by eliminating microenvironmental areas of enzyme depletion.
17. The entire reaction mixture is boiled in sample buffer, all of which (beads and liquid) is loaded into one lane of an SDS gel for analysis. This ensures that any of the substrate that may have come off during incubation, or that may not have eluted from the beads during boiling is not lost.

Acknowledgments

Work in this laboratory is supported by NIH grant CA56940 (J. W. L.) and Cancer Center Core Grant CA11198. D. A. N. is the recipient of a predoctoral fellowship from the Oral and Cellular Molecular Biology Training Grant (DE07202). N. A. K. is supported by NIH Cancer Research Training Grant T32 CA09363.

References

1. Ludlow, J. W., Glendening, C. L., Livingston, D. M., and DeCaprio, J. A. (1993) Specific enzymatic dephosphorylation of the retinoblastoma protein. *Mol. Cell Biol.* **13,** 367–372.
2. Ludlow, J. W. (1992) Selective ability of S-phase cell extracts to dephosphorylate SV40 large T-antigen in vitro. *Oncogene* **7,** 1011–1014.

11

The Biochemical Identification and Characterization of New Species of Protein Phosphatase 1

Monique Beullens, Willy Stalmans, and Mathieu Bollen

1. Introduction

This chapter deals with a family of Ser/Thr protein phosphatases that are known as type-1 protein phosphatases or PP1 *(1,2)*. They all share one of five known isoforms of the same catalytic subunit (PP1$_C$) of 36–38 kDa, designated with the subscripts α1, α2, δ, γ1, and γ2. In addition, each PP1 holoenzyme contains one or two regulatory subunits. The latter are multifunctional and not only determine the substrate specificity and the activity of the holoenzyme, but also anchor the phosphatase to the cellular compartment that harbors its physiological substrate(s). The targeting role of the noncatalytic subunits has formed the basis for the nomenclature of the various PP1 holoenzymes (**Table 1**).

Together with the protein phosphatases of type-2A, -2B, and -2C, the type-1 protein phosphatases constitute nearly all the protein phosphatase activity that is measured in crude tissue fractions with common substrates *(1,2)*. Other, more recently discovered Ser/Thr protein phosphatases are quantitatively less important. The purpose of this chapter is to describe specific assays for PP1 in crude or partially purified phosphatase preparations and to illustrate some biochemical approaches for a rapid initial characterization of PP1 holoenzymes.

2. Materials

1. 25 mM Tris-HCl, pH 7.4 with 0.5 mM dithiothreitol (DTT). Store at 4°C.
2. 50 mM glycylglycine, pH 7.4 with 0.5 mM DTT. Store at 4°C.
3. 50 mM imidazole, pH 7.4 with 0.5 mM DTT. Store at 4°C.
4. 50 mM caffeine. Store at −20°C.
5. 50 mg/mL bovine serum albumin (BSA). Store at −20°C.
6. 50 mg/mL ^{32}P-labeled phosphorylase *a* (Gibco-BRL). Store at 4°C.

From: *Methods in Molecular Biology, Vol. 93: Protein Phosphatase Protocols*
Edited by: J. W. Ludlow © Humana Press Inc., Totowa, NJ

Table 1
Diversity of Mammalian Type-1 Holoenzymes

Name	Subcellular localization	Regulatory polyeptide(s)	Mass on SDS-PAGE
PP1S[a]	Cytosol (Soluble)	Inhibitor-2	31 kDa
PP1G[b]	Glycogen particles/	G-subunit	161 kDa (skeletal muscle)
	Endoplasmic reticulum	33 kDa (liver)`	130 kDa (smooth muscle)
PP1M[c]	Myofibrils	M130	20 kDa (smooth muscle)
		M20	
PP1N[d]	Nucleus	NIPP1	41–47 kDa
		R111	111 kDa
		sds22	44 kDa
		p53BP2	unknown
		Rb	110 kDa
		PSF	95 kDa
PP1R[e]	Ribosomes	RIPP1	22 kDa
		L5	35 kDa

[a]Inhibitor-2 can be isolated as a complex with $PP1_C$ (1,2). In addition, cytosolic proteins have been isolated, termed inhibitor-1 and its isoform DARPP-32, that inhibit $PP1_C$ after phosphorylation by protein kinase A.

[b]The G-subunit from skeletal muscle has separate glycogen- and membrane-binding domains (1,2). The hepatic G-subunit is not membrane-anchored (1,3).

[c]The smooth muscle myosin phosphatase is a heterotrimer of $PP1_C$, M130, and M20 (4).

[d]The nucleus contains two major PP1 holoenzymes, which are heterodimers of the catalytic subunit and the inhibitory polypeptides NIPP1 or R111 (5). sds22 is a mammalian homolog of a $PP1_C$-interacting protein in yeast (6). p53BP2, PSF and Rb have originally been identified as $PP1_C$-interacting polypeptides in the yeast two-hybrid assay, but they have also been shown to interact with the type-1 catalytic subunit using biochemical approaches (7–9).

[e]The ribsomal protein L5 is a cobalt-dependent activator of PP1 (10). RIPP1 has been purified as a ribosomal inhibitor of PP1 (11).

7. 10, 20, and 100% (w/v) trichloroacetic acid (TCA). Store at 4°C.
8. 6 mg/mL BSA. Store at 4°C.
9. 10 nM okadaic acid.
10. 1 μM inhibitor-2.
11. 1 mg/mL trypsin. Store at −20°C.
12. 5 mg/mL soybean trypsin inhibitor. Store at −20°C.
13. Ice-cold acetone.
14. $PP1_C$.
15. Sodium dodecyl sulfate polyacrylamide gel electrophoresis (SDS-PAGE) and blotting equipment.
16. DIG protein labeling and detection kit from Boehringer (Mannheim, Germany, catalog number 1336371).
17. 50 mM sodium borate, pH 8.5 with 1 mM DTT.

18. Dialysis buffer: 50 mM Tris-HCl, pH 7.4, 1 mM DTT without or with 60% (v/v) glycerol.
19. Blocking solution 1: 20 mM Tris-HCl, pH 7.4, 0.5 M NaCl and 3% (w/v) milk powder.
20. DIG-PP1$_C$-incubation solution: 20 mM Tris-HCl, pH 7.4, 150 mM NaCl, 1 mg/mL BSA with 1/1000 DIG-PP1C.
21. Blocking solution 2: 50 mM phosphate buffer, pH 8.5, with 0.1 g blocking reagent (kit Boehringer).
22. TBS: 20 mM Tris-HCl, pH 7.4, and 150 mM NaCl. Store at 4°C.
23. Developing solution: 10 mL 100 mM Tris-HCl, pH 9.0, 50 mM MgCl$_2$, 100 mM NaCl with 37.5 μL 5-bromo-4-chloro-3-indolyl-phosphate and 50 μL 4-nitroblue tetrazolium chloride, as present in the DIG protein labeling and detection kit.
24. 5 mg/mL myelin basic protein, 10 mg/mL histone IIA and 10 mg/mL casein. Store at –20°C.
25. Catalytic subunit of protein kinase A (10 U/μL, Sigma, St. Louis, MO). Dissolve in 50 mM Tris-HCl, pH 7.4, with 60% (v/v) glycerol and store at –20°C.
26. Phosphorylation mixture: 1 mM [γ^{32}P]ATP (1000–10,000 cpm/pmol) and 20 mM MgCl$_2$.
27. Diethylether.
28. 10 mM EDTA, pH 7.4.
29. Isobutanol:toluol (1:1).
30. 5 mM silocotungstic acid in 1 mM H$_2$SO$_4$.
31. 5% (w/v) ammonium molybdate in 2 M H$_2$SO$_4$.

3. Methods

3.1. The Specific Assay of PP1

Protein phosphatase activities are classically determined from the rate of dephosphorylation of a ^{32}P-labeled substrate, which can easily be measured by counting the released (acid-soluble) radioactivity. With phosphorylase α as a substrate for the assay of protein phosphatases in crude tissue fractions, only PP1 and PP2A activities are measured. PP1 and PP2A can then be further differentiated by the use of specific inhibitors, e.g., inhibitor-2 or okadaic acid. The data can be extended by Western analysis with (commercially available) antibodies against the catalytic subunits of PP1 and PP2A.

1. Pipet 24 μL of a suitably diluted phosphatase sample in a test tube (*see* **Note 1**).
2. Start the dephosphorylation reaction by adding 6 μL substrate mixture, containing 3 μL of 50 mM caffeine, 2 μL of 50 mg/mL BSA, and 1 μL of 50 mg/mL ^{32}P-labeled phosphorylase *a* (*see* **Note 2**). Incubate the assay mixture for 10 min at 30°C.
3. Stop the phosphatase reaction by adding 250 μL of ice-cold 20% (w/v) TCA and 250 μL of an ice-cold solution of BSA (6 mg/mL). Keep the mixture during 10 min at 0°C (*see* **Note 3**).
4. Sediment the precipitated proteins by centrifugation during 5 min at 5000*g*.
5. Count the radioactivity of 350 μL supernatant in a liquid scintillation counter (*see* **Note 4**).

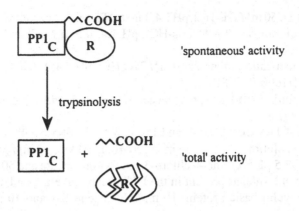

Fig. 1. Schematic illustration of the action of trypsin on a PP1 holoenzyme. 'R' represents a noncatalytic subunit. The 'spontaneous' and 'total' activities refer to the phosphorylase phosphatase activities before and after trypsinolysis, respectively.

6. The obtained radioactivities are diminished with the 'blank' value, i.e., a measure of free $[^{32}P]P_i$ in the substrate mixture, which is obtained by performing the assay with buffer instead of a phosphatase sample. Knowing the specific and total radioactivities of the substrate, the phosphatase activities can be calculated in U/mL. One unit of protein phosphatase liberates 1 nmol of phosphate per min under the specified assay conditions. The dephosphorylation of 1 nmol phosphorylase *a* subunit (97 kDa) results in the release of 1 nmol phosphate.

7. The assay can be rendered specific for PP1 or PP2A by inclusion of 10 n*M* okadaic acid or 1 μM inhibitor-2, respectively (*see* **Note 5**). The contribution of PP1 can also be calculated from the difference in activities before and after addition of inhibitor-2.

3.2. Identification of Regulatory Polypeptides of PP1

3.2.1. The 'Spontaneous' vs the 'Total' Phosphorylase Phosphatase Activity

Trypsinolysis of $PP1_C$ results in the proteolytic removal of its 30–40 carboxyterminal residues (**Fig. 1**). However, such a proteolysis barely affects the phosphorylase phosphatase activity of $PP1_C$, when assayed as described above (*see* **Section 3.1.**). In contrast, all known regulatory polypeptides of PP1 are destroyed by trypsin. The assay of PP1 holoenzymes before and after a trypsin treatment thus allows one to identify regulatory polypeptides/subunits that affect the phosphorylase phosphatase activity of $PP1_C$. Since nearly all known regulatory polypeptides of PP1 (inhibitor-1/DARPP-32, inhibitor-2, G-subunit, M130, NIPP1, R111, RIPP1) are inhibitory to the phosphorylase phosphatase activity of the catalytic subunit, the trypsin-revealed activity is referred to as the 'total' activity, as compared to the 'spontaneous' activity, i.e., the activity without prior proteolysis.

1. Incubate 20 µL PP1 holoenzyme during 5 min at 30°C with 2 µL dilution buffer or with 2 µL trypsin (1 mg/mL). Subsequently add 2 µL soybean trypsin inhibitor (5 mg/mL) to ensure complete inhibition of trypsin (*see* **Note 6**).
2. Assay the phosphorylase phosphatase activities as detailed in **Subheading 3.1.**

3.2.2. Assay for Heat- and Acid-Stable Inhibitors of PP1$_C$

Some inhibitory polypeptides of PP1 (inhibitor-1/DARPP-32, inhibitor-2) or fragments of inhibitory polypeptides (NIPP1) resist an incubation at 90°C and a treatment with trichloroacetic acid *(1,5)*. The finding of such an extreme resistance to denaturing procedures provides some attractive means for a rapid partial purification and allows for a complete control of (further) proteolytic degradation of the inhibitor.

3.2.2.1. HEAT-STABLE INHIBITORS

1. Incubate 30 µL of the inhibitory polypeptide of PP1 during 5 min at 90°C.
2. Precipitate the denatured proteins by centrifugation during 5 min at 15,000*g*. Incubate 4 µL PP1$_C$ (5 n*M*) with 20 µL dilution buffer or with 20 µL heat-stable supernatant. Add 6 µL substrate mixture and assay the spontaneous phosphorylase phosphatase activity, as detailed in **Subheading 3.1.** (*see* **Note 7**).

3.2.2.2. ACID-STABLE

1. Incubate 30 µL of the inhibitory polypeptide of PP1 during 10 min at 0°C with 5 µL 100% (w/v) TCA.
2. Sediment the precipitated proteins by centrifugation during 5 min at 15,000*g*.
3. Remove the TCA by washing the pellet three times with 200 µL ice-cold acetone. Resuspend the pellet in 30 µL dilution buffer and adjust, if necessary, the pH to 7.4 with 0.1 *M* KOH.
4. Precipitate the denatured proteins by centrifugation during 5 min at 15,000*g*. Assay the acid-stable supernatant for PP1$_C$-inhibitory activity, as described for the heat-stable supernatant (*see* **Subheading 3.2.2.1.**).

3.2.3. Inhibition vs Inactivation of PP1$_C$

Some regulatory subunits of PP1 (inhibitor-2 and RIPP1) not only instantaneously block the activity of PP1$_C$ (*inhibition*), but also convert slowly ($t_{1/2}$ = 10–60 min at 25°C) the catalytic subunit into an inactive conformation (*inactivation*). Inhibition and inactivation can be biochemically differentiated because the latter is not reversed by the removal of the inhibitor, e.g., by proteolysis, and because trypsinolysis destroys the inactivated, but not the inhibited catalytic subunit (**Fig. 2**) (*see* **ref. 1**).

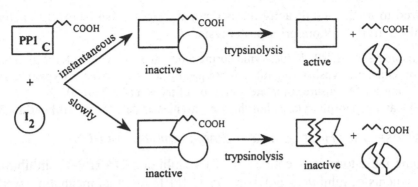

Fig. 2. The effect of trypsinolysis on the inhibitor-2-mediated inhibition and inactivation of $PP1_C$. I_2, inhibitor-2.

1. Add 5 µL $PP1_C$ (5 nM) to 45 µL dilution buffer or to 45 µL of a dilution of the inhibitor that causes a nearly complete inhibition of the catalytic subunit.
2. At various times, up to 60 min at 25°C, 3-µL aliquots are taken for the immediate assay of the spontaneous and trypsin-revealed phosphorylase phosphatase activities, as described in **Subheading 3.2.1.**
3. The spontaneous and trypsin-revealed phosphorylase phosphatase activities of the free catalytic subunit should remain constant during the entire incubation period (*see* **Note 8**). The spontaneous activity should be blocked in the presence of the inhibitor, but this inhibition should be completely reversed by trypsinolysis at the beginning of the incubation period. A time-dependent decrease of the trypsin-revealed phosphorylase phosphatase activity is an indication of an inactivation of $PP1_C$ (*see* **Note 9**).

3.2.4. The Detection of $PP1_C$-Interacting Polypeptides by Far-Western Blot Analysis

The basic principles of the method are illustrated in **Fig. 3.** The method allows for a rapid and reproducible screening of polypeptides that bind with high affinity to the catalytic subunit, and at the same time yields a precise estimate of the mass of the interacting polypeptide(s). Regulatory polypeptides that have been visualized by this approach are the M-subunit *(4)*, the G-subunit *(3)*, NIPP1 *(5)*, and R111 *(5)*. However, some known regulators of $PP1_C$ (e.g., inhibitor-2) cannot be detected by this method, probably because the interaction with $PP1_C$ is measured after denaturing electrophoresis and blotting, which may destroy or mask $PP1_C$ interacting site(s).

3.2.4.1. Preparation of Digoxigenin-Labeled $PP1_C$

1. Dialyze 100 µL $PP1_C$ (20 µM) against 50 mM sodium borate, pH 8.5, 1 mM dithiothreitol.

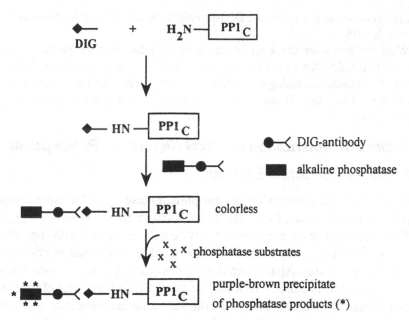

Fig. 3. Identification of PP1$_C$-interacting polypeptides by far-Western blot analysis.

2. Incubate the catalytic subunit during 1 h at room temperature with 10 μL digoxigenin-carboxymethyl-*N*-hydroxysuccinimide ester, as present in the DIG protein labeling and detection kit from Boehringer.
3. Remove the free digoxigenin-carboxymethyl-*N*-hydroxysuccinimide ester by extensive dialysis (2 × 2 h) against 50 m*M* Tris-HCl, pH 7.4, 1 m*M* dithiothreitol.
4. Dialyze the digoxygenin-labeled catalytic subunit (DIG-PP1$_C$) at least 5 h against 50 m*M* Tris-HCl, pH 7.4, 1 m*M* DTT, 60% (v/v) glycerol and store at –20°C.

3.2.4.2. DETECTION OF PP1$_C$-INTERACTING POLYPEPTIDES

1. Apply the samples of interest (a partially purified species of PP1 or even a crude tissue fraction) onto SDS-PAGE and blot the separated polypeptides onto a polyvinylidene difluoride membrane.
2. Incubate the membrane (8 × 5 cm) for 4 h at room temperature (or overnight at 4°C) in 20 mL blocking solution 1.
3. Incubate the blot during 4 h at room temperature (or overnight at 4°C) with 5 mL digoxigenin-labeled PP1$_C$ (*see* **Subheading 3.2.4.1.**) incubation solution (*see* **Note 10**).
4. Incubate the membrane during 30 min at room temperature in 20 mL of blocking solution 2.
5 Wash the blot three times for 10 min with TBS. Incubate the blot during 1 h at room temperature with 10 μL digoxigenin antibodies, conjugated with alkaline

phosphatase (as present in the DIG-protein detection kit), and diluted until up to 10 mL TBS.

6   Wash the blot three times for 10 min with TBS. Develop the blot by incubation with 10 mL developing solution (*see* **Note 11**). The staining is stopped after 1–60 min by exhaustive rinsing of the blot with water. The blot can be air-dried without loss of the stain. Upon prolonged storage (wk-mo), the stain will gradually fade away.

3.3. Initial Characterization of Novel Regulatory Polypeptides

3.3.1. Effect on Substrate Specificity

The noncatalytic subunits of protein phosphatase 1 are 'substrate-specifying' polypeptides. As has, for example, been shown for the G-subunit *(1,3)* and M130 *(4)*, their effect on the type-1 catalytic subunit varies with the substrate used. It is, therefore, essential to analyze the role of the noncatalytic subunits with various substrates. Apart from the physiologically relevant substrates, for which information is usually lacking, a first glance of the importance of the primary structure, size, and charge of the substrate can be easily obtained by comparing the substrate quality of phosphorylase *a* and the acid-stable proteins casein, myelin basic protein, and histone IIA. Assay for protein phosphatase activities with the latter substrates is essentially the same as described for phosphorylase phosphatase (*see* **Subheading 3.1.**). The final concentration of the acid-stable substrates in the assay is 0.1–0.5 mg/mL. Under these assay conditions, $PP1_C$ releases $[^{32}P]P_i$ about 10–20-fold faster from phosphorylase and myelin basic protein than from histone IIA and casein.

A kit for the preparation of ^{32}P-labeled phosphorylase *a* is commercially available (Gibco-BRL, Gaitherburg, MD). ^{32}P-labeled myelin basic protein, histone IIA and casein (Sigma) can be obtained by phosphorylation with the catalytic subunit of protein kinase A as follows:

1. Mix 80 μL substrate (5 mg/mL of either casein, histone IIA or myelin basic protein in dilution buffer, *see* **Note 1**) with 10 μL catalytic subunit of protein kinase A (10 U/μL, Sigma) and 10 μL phosphorylation mixture. Incubate for 1 h at 37°C (*see* **Note 12**).
2. Stop the phosphorylation reaction by addition of 10 μL of 100% (w/v) TCA and keep the mixture at 0°C for 10 min.
3. Sediment the precipitated proteins by centrifugation during 1 min at 15,000*g*.
4. Resuspend the pellet twice with 200 μL 10% (w/v) TCA to remove the remaining free $[\gamma^{32}P]ATP$.
5. Remove the TCA by washing the pellet multiple times with 200 μL diethylether, until the pH of the washing fluid is neutral (measured with pH papers). Dissolve the pellet in 100 μL buffer and adjust, if necessary, the pH to 7.4 with 0.1 *M* KOH.

3.3.2. Regulation by Phosphorylation?

Nearly all known regulatory polypeptides of PP1 are themselves regulated by reversible phosphorylation (1,5). In several cases, phosphorylation of a noncatalytic subunit has been shown to change the substrate specificity of the holoenzyme or the affinity of the noncatalytic subunit for the catalytic subunit. Described here is a test that allows one to check for phosphorylation by an exogenous protein kinase and to investigate the effects of such phosphorylation on the activity of the holoenzyme and on the affinity of the regulatory polypeptide(s) for the catalytic subunit. The description is for phoshorylation by protein kinase A, but can easily be expanded to include other kinases like the casein kinases-1 or -2, GSK-3, and protein kinases-C or -G.

1. Mix 10 µL (partially) purified PP1 holoenzyme with either 8 µL dilution buffer or with 8 µL catalytic subunit of protein kinase A (0.2 U/µL), and 2 µL phosphorylation mixture, containing 1 mM nonradioactive or [γ^{32}P]-labeled (1000–10,000 cpm/pmol) ATP and 20 mM MgCl$_2$ (*see* **Note 1**). A control for phosphorylation by endogenous protein kinases (contaminants of the protein phosphatase sample) is obtained by incubation without exogenously added protein kinase. A control for the presence of (auto)phosphorylating polypeptides and protein phosphatases in the kinase sample is obtained by incubation with dilution buffer instead of phosphatase sample.
2. Incubate for 1 h at 37°C.
3. Stop the phosphorylation reaction by adding either 10 µL 10 mM EDTA (reactions with nonradioactive ATP) or 10 µL SDS-PAGE sample buffer (^{32}P-labeled ATP).
4. Following SDS-PAGE and blotting of the ^{32}P-labeled samples, phosphorylation of the phosphatase regulatory subunit(s) is visualized by autoradiography. The same blot can also be used for far-Western blotting with DIG-PP1$_C$ (*see* **Subheading 3.2.4.**) to detect differences in the affinity of the regulatory polypeptide(s) for the catalytic subunit. Such differences should be abolished by preincubation of the blot with an excess of (commercially available) Ser/Thr protein phosphatases (5).
5. The samples incubated under phosphorylation conditions with unlabeled ATP are assayed (after dialysis, if necessary) for spontaneous and trypsin-revealed phosphorylase phosphatase activities (*see* **Subheading 3.2.1.**) to check for the effects of phosphorylation on the activity of the holoenzyme.

4. Notes

1. As dilution and assay buffers 25 mM Tris-HCl is used, or 50 mM glycylglycine, or 50 mM imidazole/HCl, at pH 7.4, and supplemented with 0.5 mM DTT. Phosphate or glycerophosphate buffers cannot be used because they are inhibitory to Ser/Thr protein phosphatases. For the same reason the salt concentration in the assay should not exceed 50 mM, unless the effect of 'physiological' ionic strength is under investigation.

2. Caffeine binds to the nucleoside binding site of phosphorylase and prevents the binding of AMP to the nucleotide binding site. AMP may be present in crude phosphatase samples and its binding to phosphorylase *a* renders this substrate inaccessible for dephosphorylation. Bovine serum albumin is added to decrease the proteolytic degradation of protein phosphatases by providing an excess of alternative protease substrate.

3. If in doubt, whether all the released radioactivity stems from P_i (the acid-soluble radioactivity may also represent proteolytic fragments), one can check whether the radioactivity can be extracted as a phosphomolybdate complex in isobutanol: arrest the phosphatase reaction by the addition of 0.8 mL isobutanol:toluol (1:1) and 1.2 mL of a mixture containing 5 mM silicotungstic acid and 1 mM H_2SO_4. Vortex vigorously and add 160 µL 5% (w/v) ammonium molybdate in 2 M H_2SO_4. Vortex and centrifuge during 5 min at 5000g. Count the radioactivity in 500 µL supernatant.

4. In order to have an assay that is linear with time and dilution of the phosphatase, care should be taken that the liberated radioactivity is less than 30% of the total radioactivity. When the value is higher, the assay should be repeated at higher phosphatase dilutions.

5. 10 nM okadaic completely blocks PP2A (unless the phosphatase concentration is higher than 10 nM), but is virtually without effect on PP1. The free catalytic subunit of PP1 is blocked completely by 10 nM inhibitor-2, but the PP1 holoenzymes are generally more resistant to inhibition by inhibitor-2 *(1)*, hence the proposal to use micromolar concentrations of the inhibitor. Moreover, the inhibition of some PP1 holoenzymes by inhibitor-2 is time-dependent. It is therefore, advisable to preincubate the protein phosphatase samples with inhibitor-2 (5 min at 30°C) before adding the substrate mixture.

6. Trypsin will cleave off an acid-soluble phosphopeptide from phosphorylase *a* if not completely arrested by soybean trypsin inhibitor, resulting in an overestimation of the protein phosphatase activity.

7. In case an inhibitory activity is detected, its proteinaceous nature can easily be checked by its likely destruction during trypsinolysis (*see* **Subheading 3.2.1.**).

8. In the authors' experience, the free catalytic subunit is most stable when the incubation is done at 25°C and the incubation buffer consists of 50 mM imidazole, pH 7.4, and freshly added dithiothreitol (0.5 mM) plus 2-mercaptoethanol (6 mM).

9. The inactivation induced by inhibitor-2 can be reversed by trypsinolysis in the presence of Mn^{2+} or by the transient phosphorylation of inhibitor-2 with protein kinase GSK-3 *(1)*.

10. The diluted DIG-PP1$_C$ can be stored at –20°C and reused up to five times.

11. As a positive control for the staining procedure a sample of digoxygenin-labeled PP1$_C$ can be used.

12. Such an incubation typically results in the incorporation of 0.3–0.6 mol phosphate/mol substrate.

References

1. Bollen, M. and Stalmans, W. (1992) The structure, role and regulation of type 1 phosphatases. *Crit. Rev. Biochem. Mol. Biol.* **27**, 227–281.
2. Wera, S. and Hemmings, B. (1995) Serine/threonine protein phosphatases. *Biochem. J.* **311**, 17–29.
3. Moorhead, G., MacKintosh, C., Morrice, N., and Cohen, P. (1995) Purification of the hepatic glycogen-associated form of protein phosphatase-1 by microcystin-Sepharose affinity chromatography. *FEBS Lett.* **362**, 101–105.
4. Alessi, D., MacDougall, L. K., Sola, M. M., Ikebe, M., and Cohen, P. (1992) The control of protein phosphatase–1 by targetting subunits. The major myosin phosphatase in avian smooth muscle is a novel form of protein phosphatase-1. *Eur. J. Biochem.* **210**, 1023–1035.
5. Jagiello, I., Beullens, M., Stalmans, W., and Bollen, M. (1995) Subunit structure and regulation of protein phosphatase-1 in rat liver nuclei. *J. Biol. Chem.* **270**, 17,257–17,263.
6. Renouf, S., Beullens, M., Wera, S., Van Eynde, A., Sikela, J., Stalmans, W., and Bollen, M. (1995) Molecular cloning of a human polypeptide related to yeast sds22, a regulator of protein phosphatase-1. *FEBS Lett.* **375**, 75–78.
7. Helps, N. R., Barker, H. M., Elledge, S. J., and Cohen, P. T. W. (1995) Protein phosphatase-1 interacts with p53BP2, a protein which binds to the tumour suppressor p53. *FEBS Lett.* **377**, 295–300.
8. Hirano, K., Erdödi, F., Patton, J. G., and Hartshorne, D. J. (1996) Interaction of protein phosphatase type 1 with a splicing factor. *FEBS Lett.* **389**, 191–194.
9. Durfee, T. Becherer, K., Chen, P. L., Yeh, S.-H., Yang, Y., Kilburn, A. E., Lee, W.-H., and Elledge, S. J. (1993) The retinoblastoma protein associates with the protein phosphatase type 1 catalytic subunit. *Genes Dev.* **7**, 555–569.
10. Hirano, K., Ito, M., and Hartshorne, D. J. (1995) Interaction of the ribosomal protein, L5, with protein phosphatase type-1. *J. Biol. Chem.* **270**, 19,786–19,790.
11. Beullens, M., Stalmans, W., and Bollen, M. (1996) Characterization of a ribosomal inhibitory polypeptide of protein phosphatase-1 from rat liver. *Eur. J. Biochem.* **239**, 183–189.

12

The Relationship Between Insulin Signaling and Protein Phosphatase 1 Activation

Louis Ragolia and Najma Begum

1. Introduction

Insulin signaling is mediated by a cascade of phosphorylation/dephosphorylation reactions via the stimulation of specific serine/threonine (ser/thr) kinases and phosphatases (1–7). Dephosphorylations catalyzed by ser/thr phosphatases result in the activation or inhibition of several intracellular enzymes and proteins that govern the final steps of insulin action (2). The best examples are glycogen synthase (GS), pyruvate dehydrogenase (PDH), acetyl-CoA carboxylase, hormone-sensitive lipoprotein lipase, and Glut-4 (the insulin regulatable muscle/fat glucose transporter) (1–10). Dephosphorylation increases the activity of GS, PDH, and Glut-4, and inhibits phosphorylase kinase (2,8–10). Protein phosphatase 1 (PP1) appears to be the key enzyme responsible for connecting the insulin-initiated phosphorylation cascade with the dephosphorylation of insulin-sensitive substrates (2). The exact in vivo molecular mechanism by which insulin regulates the activity of this enzyme remains unclear. A large proportion of the spontaneously active PP1 is found associated with the particulate fraction of the cell, notably with glycogen, membranes, and myofibrils (11). Therefore, it appears that the catalytic activity of PP1 is determined by its interaction with regulatory subunits that target the enzyme to particular locations in the cell and alter its substrate specificity (11).

The glycogen-associated form of PP1 is the best characterized phosphatase to date. It consist a 37 kDa catalytic subunit (C) complexed to 160 kDa regulatory subunit ($PP1_{RG}$). $PP1_{RG}$ is believed be involved in the hormonal regulation of glycogen metabolism (11). This subunit also directs PP1 to the sarcoplasmic reticulum ($PP1_{SR}$) (12). In vitro studies from Cohen's laboratory suggest that PP1 catalytic activity is regulated by site-specific phosphorylation of the gly-

From: *Methods in Molecular Biology, Vol. 93: Protein Phosphatase Protocols*
Edited by: J. W. Ludlow © Humana Press Inc., Totowa, NJ

cogen-associated regulatory subunit, PP1$_{RG}$ *(11)*. An insulin-stimulated protein kinase (a homologue of mammalian S6 kinase II) phosphorylates PP1$_{RG}$ at site-1 in vitro, causing activation of PP1 *(13)*. cAMP dependent protein kinase A causes phosphorylation at site-2, resulting in PP1 inhibition because of the dissociation of PP1 catalytic subunit from its regulatory subunit *(2,14–16)*. Treatment of rabbits with insulin increases the phosphorylation levels of site-1, but not site-2, however, no increase of phosphatase activity has been demonstrated *(11)*. Thus physiologic relevance of these site-specific phosphorylations to PP1 activation and insulin action are unknown. Therefore, the question remains whether the biochemical properties of PP1$_{RG}$ observed in vitro are relevant to its physiological role in vivo.

Using cultured rat skeletal muscle cells and freshly isolated rat adipocytes as model systems, the authors have recently demonstrated that insulin rapidly stimulates cellular PP1 activity and concomitantly inhibits PP2A activity *(2,17–21)*. Stimulation of PP1 is accompanied by an increase in the phosphorylation status of PP1$_{RG}$.

In this chapter, the authors describe in detail the experimental strategies that they have used to demonstrate: an in vivo insulin effect on PP1 activation in two insulin sensitive cells (rat skeletal muscle cell line, L6 and freshly isolated rat adipocytes), subcellular localization and immunoprecipitation of PP1 with antibodies directed towards the PP1 regulatory subunit, PP1$_{RG}$, phosphorylation of PP1$_{RG}$, and a potential mechanism of PP1 activation by insulin via its regulatory subunit.

2. Materials

1. Phosphate-buffered saline (PBS): 0.14 M NaCl, 1.5 mM KH$_2$PO$_4$, 8.1 mM Na$_2$HPO$_4$, pH 7.4. Store at 4°C.
2. Phosphatase Extraction/Assay Buffer: 20 mM imidazole HCl (pH 7.2), 2 mM ethylene diamine tetra-acetic acid (EDTA), β-mercaptoethanol (0.2%), 2 mg/mL glycogen (rabbit liver, type X, Sigma Chemicals, St. Louis, MO), 1 mM benzamidine, 1 mM phenylmethylsulfonyl fluoride (PMSF), 10 µg/mL each of aprotinin, leupeptin, antipain, Soy trypsin inhibitor (STI), and pepstatin A. Store in 5–10 mL aliquots at –20°C.
3. Phosphorylation reaction buffer: 250 mM Tris-HCl, pH 8.2, 16.7 mM MgCl$_2$, 1.67 mM ATP, 0.83 mM CaCl$_2$, and 133 mM β-glycerophosphate. Store in 300-µL aliquots at –20°C.
4. Substrate/kinase: glycogen phosphorylase b, Cat. no. 3189SA and phosphorylase kinase, Cat. no. 3190SA (Life Technologies, Gaithersburg, MD).
5. Ammonium sulfate solution: 90% saturated.
6. Solubilization buffer: 50 mM Tris-HCl, pH 7.0, 0.1 mM EDTA, 15 mM caffeine, 0.1% (v/v) β-mercaptoethanol. Store at –20°C.
7. Concentrators: Centricon-30 (Amicon, Danvers, MA).

8. Trichloroacetic acid: 20% (w/v). Store at 4°C.
9. Okadaic acid (Moana chemicals, Honolulu, HA). Prepare a 2 μ*M* stock solution in Dimethyl formamide (DMF). Store at –20°C in small aliquots.
10. Protein assay reagents: Bradford Reagent (Bio-Rad Labs, Hercules, CA), BCA Reagent (Pierce, Rockford, IL).
11. Lysis buffer for immunoprecipitation: 20 m*M* triethanolamine, pH 7.2, 0.5 m*M* ethylene glycol-bis [β-amino ethyl ether] - N,N,N′,N′ - tetra-acetic acid (EGTA), 1 m*M* EDTA, 1 m*M* benzaminidine, 0.1 m*M* PMSF, 10 μg/mL each of leupeptin, aprotinin, antipain, Soy trypsin inhibitor (STI), and pepstatin A, 100 m*M* NaCl, and 1% Triton X-100.
12. Lysis buffer (w/ phosphatase inhibitors): Add 2 m*M* sodium orthovanadate, 100 m*M* sodium pyrophosphate, 100 m*M* sodium fluoride, 40 m*M* β-glycero-phosphate and 1.0 μ*M* microcystin, adjust pH to 7.6.
13. Laemmli Sample Buffer (2X): 0.004% bromophenol blue, 20 m*M* sodium phosphate, pH 7.0, 20% glycerol, 4% SDS. Add 15 mg dithiotheitol (DTT) to each 1.0 mL of LSB (1X) prior to use.
13. TBS: 20 m*M* Tris-HCl, 137 m*M* NaCl; pH 7.6
14. Protein A Sepharose CL-6B (Sigma): Swell 1.5 g of Sepharose beads in 6 mL of buffer containing 25 m*M* HEPES; pH 7.4, 2 m*M* EDTA, 100 m*M* NaCl. Transfer beads to a sterile 15-mL tube, centrifuge at 2000*g* for 2 min. Resuspend the beads in 6 mL of buffer and centrifuge for 2 min. Wash the beads an additional four times. Resuspend the beads in the aforementioned buffer containing 0.1% sodium azide (v/v) and store upright at 4°C in 1-mL aliquots.
15. Krebs-Ringer Phosphate (KRP) buffer: 25 m*M* phosphate buffer, pH 7.4, 118 m*M* NaCl, 4.7 m*M* KCl, 1.3 m*M* CaCl$_2$, and 1.2 m*M* MgSO$_4$.
16. Staining solution: 0.25% Coomassie Brilliant Blue R 250 (Bio-Rad Labs) in 20% Trichloroacetic acid.
17. Destaining solution: 10% acetic acid, 45% MeOH.
18. Transfer Buffer: 25 m*M* Tris-base, 0.2 *M* glycine, and 20% MeOH.

3. Methods

3.1. Insulin Effect on PP1 Activation In Vivo

3.1.1. Preparation of Cells and Insulin Treatment

1. Serum starve rat skeletal muscle cells (*see* **Note 1**) overnight or use freshly isolated rat adipocytes (*see* **Note 2**).
2. The next day add 1 mL of fresh serum-free medium. Incubate the cells with 10 μl of the appropriate insulin dose (*see* **Note 3**). For controls, add an equal volume of 0.1% BSA. Stagger additions 1–2 min between dishes (*see* **Note 4**).
3. Incubate dishes with insulin for 5–10 min (*see* **Note 4**).
4. Terminate the reaction by quickly aspirating the medium.
5. Transfer the dish to an ice bucket and rinse three times with ice-cold PBS. Remove any traces of PBS from the last wash (*see* **Note 5** for adipocyte washing).
6. Add 0.3–0.5 mL of ice-cold phosphatase extraction buffer (*see* **Note 6**).

3.1.2. Extraction of Phosphatases (All steps performed at 4°C)

1. Scrape cells immediately with a sterile plastic spatula.
2. Transfer cells to clean, prechilled microcentrifuge tube.
3. Sonicate for 10 s (see Note 7).
4. Centrifuge cell extract at 2000g for 5 min to remove any cell debris.

3.1.3. Subcellular Fractionation

1. Centrifuge cell extract at 100,000g for 20 min at 4°C in a miniultracentrifuge. Save an aliquot of cell extract prior to centrifugation.
2. Transfer the supernatant to new a tube (cytosolic fraction; CF).
3. Wipe the sides of the centrifuge tube with a tissue and reconstitute the pellet to the original volume (particulate fraction; PF) (see Note 8).
4. Assay 10 μL aliquots of the extracts, particulate and cytosolic fractions for protein content.

3.1.4. Assay of Phosphatase Activity

3.1.4.1. CHOICE OF SUBSTRATES (SEE NOTE 9)

3.1.4.2. PREPARATION OF RADIOLABELED GLYCOGEN PHOSPHORYLASE A (SEE NOTE 10)

1. Add 0.5 mL of [τ-^{32}P]-ATP to a tube containing 300 μL of phosphorylation reaction buffer.
2. In a separate microcentrifuge tube, mix 100 μL of phosphorylase b (100 mg/mL) and 1 μL of phosphorylase kinase. (This is the kinase/substrate mix.)
3. Start the reaction by adding the phosphorylation reaction buffer from **step 1** (see **Note 11**). Vortex well.
4. Incubate at 30°C for 1 h.
5. Terminate the reaction with the addition of 0.5 mL of a 90% ammonium sulfate solution.
6. Vortex and incubate the tube on ice for 30 min.
7. Centrifuge at 12,000g for 10 min at 4°C.
8. Decant the supernatant (caution radioactive waste).
9. Resuspend the pellet in 1 mL of 45% ice-cold ammonium sulfate, centrifuge as in **step 7**. Repeat the washings four times.
10. Dissolve the pellet in 1 mL of solubilization buffer and transfer to the concentrator (see **Note 12**).
11. Rinse the tube with 1 mL of solubilization buffer and add it to the concentrator.
12. Spin the concentrator at 5000g for at least 25 min at 20°C (see **Note 13**). The remaining volume in the upper chamber of the concentrator should be approx 400 μL.
13. Bring the volume back up to 2 mL with solubilization buffer and centrifuge as in **step 12**.
14. Transfer the retentate from the concentrator to a 4-mL tube. Rinse the concentrator repeatedly with small volumes of solubilization buffer and bring the volume up to 3.1 mL (=3 mg/mL phosphorylase a).
15. Store in 0.5 mL aliquots at 4°C.

3.1.4.3. DIFFERENTIAL PP1 ENZYMATIC ASSAY USING OKADAIC ACID

1. On ice aliquot equal amounts of each protein extract (1–5 μg/assay, from control or insulin treated cells) into six prechilled microcentrifuge tubes.
2. Bring the volume to 40 μL with either ice-cold assay buffer alone (tubes 1–3) or assay buffer containing 2–3 n*M* okadaic acid (tubes 4–6) (*see* **Note 14**).
3. To the Blank tube add 40 μL assay buffer without enzyme.
4. Start the reaction by adding 20 μL of [^{32}P]-labeled phosphorylase a to each tube (stagger the additions by 15 s), vortex and incubate in a water bath at 30°C for 10 min (*see* Note 15).
5. Terminate the reaction by adding 180 μL of ice-cold 20% TCA to each tube (staggering additions as in **step 4**). Add 10 μL of 3% BSA, vortex and incubate on ice for 10 min.
6. Centrifuge the tubes at 12,000*g* for 3 min at 4°C and count 200 μL of the clear supernatant to determine the amount of radioactivity released as ^{32}P$_i$ (*see* **Note 16**).

3.1.5. Immunoprecipitation of PP1 from Cell Lysates

1. Extract control and insulin treated dishes as described in **Subheading 3.1.1.**, with 0.5 mL lysis buffer (*see* **Subheading 2.**). Rock the dishes at 4°C for 20 min. Transfer the lysates to microcentrifuge tubes.
2. Centrifuge at 16,000*g* for 10 min at 4°C. Assay 10 μL of clear lysates for protein content using the bicinchoninic acid (BCA) reagent.
3. Preclear lysates (100 μg of protein in 1 mL of lysis buffer) with rat IgG (5 μg/mL, coupled to protein A-Sepharose) for 1 h at 4°C with gentle agitation.
4. Centrifuge at 5000*g* for 2 min and transfer the supernatants to new tubes. Add 5 μg of affinity purified PP1$_{RG}$ site-1 peptide antibody precoupled to protein A sepharose (*see* **Note 17**). Incubate at 4°C with continous mixing for 1 h.
5. Centrifuge briefly as in **step 4**. Remove the supernatant and save.
6. Wash the sepharose beads four times with 1 mL of lysis buffer being sure to completely remove any traces of buffer from the last wash.

3.1.5.1. ASSAY OF PP1 ACTIVITY IN THE IMMUNOPRECIPITATES

1. Resuspend the sepharose beads with 1 mL of lysis buffer containing 0.1% β-mercaptoethanol and 15 μg/mL of site-1 antigenic peptide. Incubate at 4°C for 1 h with constant mixing to release the bound enzyme from the immuno-complex.
2. Assay an aliquot of the supernatant from **step 1**, the original lysate, and the immunodepleted supernatant (**Subheading 3.1.5.**, **step 5**) for PP1 enzymatic activity using [^{32}P]-labeled phosphorylase a as a substrate (*see* **Subheading 3.1.4.3.**).

3.1.5.2. IDENTIFICATION OF THE PP1 CATALYTIC AND REGULATORY SUBUNITS

1. Add 40 μL of 2X Laemmli sample buffer to the immunecomplex from **Subheading 3.1.5.**, **step 6**.
2. Incubate in a boiling water bath for 5 min.

3. Centrifuge briefly, collect the supernatant and load onto a 7.5% SDS polyacryla-mide gel at 120 V using Bio-Rad minigel apparatus.

4. Transfer the proteins to a PVDF membrane at 100 V for 1 h in transport buffer using the Bio-Rad mini transblot transfer apparatus.

5. Block the membrane in TBS containing 0.05% Tween-20 (TBST) with 5% non-fat dry milk.

6. Cut the membrane in half horizontally along the 60 kDa molecular weight marker.

7. Probe the upper half of the membrane with the PP1$_{RG}$ subunit site-1 antibody (1:500 in TBST containing 1% milk protein) for 1 h at room temperature with constant shaking.

8. Incubate the lower half of membrane with PP1 C- subunit antibody (δ isoform) for 1 h.

9. Wash blots three times each for 5 min with TBST.

10. Incubate blots for 1 h with [^{125}I]-labeled protein A (0.05 μCi/mL) in TBST con-taining 1% milk protein.

11. Wash blots three times with TBS containing 0.5% Tween-20 for 5 min each.

12. Expose the blots to Kodak-XAR film overnight with an intensifying screen at −70°C.

3.2. Experimental Strategy to Demonstrate a Potential Mechanism of PP1 Activation by Insulin

3.2.1. Identification and Quantification of PP1$_{RG}$ by Western Blot Analysis

1. Prepare and load equal amounts of control and insulin treated cell extracts (20 μg protein), PF and CF (see **Subheading 3.1.3.** on subcellular fractionation) on duplicate 7.5% SDS gels (see **Note 18**).

2. After electrophoresis, transfer proteins and probe one blot with the site-1 PP1$_{RG}$ anti-body as detailed in **Subheading 3.1.5.2.**, followed by detection with [^{125}I]-protein A.

3. Probe the duplicate blot with the site-1 PP1$_{RG}$ antibody in the presence of com-peting site-1 peptide (15 μg/mL). The remaining protocol is identical to **Sub-heading 3.1.5.2.**.

4. Quantitate the amount of protein in the 160 kDa band with a densitometer or cut out the band and quantitate the radioactivity with a τ- counter.

3.2.2. The Role of PP1$_{RG}$ Phosphorylation

3.2.2.1. METABOLIC LABELING OF CELLS AND INSULIN TREATMENT

1. Phosphate starve serum-depleted cells for 1 h in phosphate free DMEM.

2. Add [^{32}P]-orthophosphoric acid (0.5 mCi/mL) and incubate cells for 4 h (see **Notes 11** and **19**).

3. Treat cells with insulin (10–100 nM) or 0.1% BSA (controls) for 5–10 min at 37°C.

4. Quickly aspirate the medium (see **Note 11**) and rinse dishes four times with ice-cold PBS containing phosphatase inhibitors (see **Note 20**).

5. Add 0.5 mL of lysis buffer containing phosphatase inhibitors (see **Subheading 2.**).

6. Freeze dishes in liquid nitrogen.

7. Thaw dishes on ice, collect cell lysate and centrifuge at 16,000g for 10 min to remove cell debris. Assay the protein content using the BCA reagent.

3.2.2.2 IMMUNOPRECIPITATION OF PHOSPHORYLATED PP1$_{RG}$ SUBUNIT

1. Use 100 μg of cell lysate from **step 7** for immunoprecipitation (*see* **Subheading 3.1.5.** for details) (*see* **Note 21**).
2. Separate the immunoprecipitates on duplicate gels, using one gel to quantitate the amount of PP1RG by Western blot analysis (*see* **Subheading 3.1.5.**). and the other gel to analyze total protein by incubation with 20% TCA for 5 min, followed by rapid staining with Coomassie Blue for 5 min.
3. Destain the gel with destaining solution, transfer to 3 MM Whatman filter paper and dry under vaccum for 1 h at 80°C.
4. Expose the dried gel to Kodak-XAR film overnight at –70°C.

3.2.2.3. QUANTITATION OF PP1$_{RG}$ PHOSPHORYLATION AND CALCULATION OF SPECIFIC ACTIVITY

1. Mark the area surrounding the 160 kDa band on the dried gel by aligning with the autoradiogram. Cut out the 160 kDa band and count radioactivity, or quantitate the ^{32}P content of the 160 kDa band by optical density scanning of the autoradiogram.
2. Calculate the relative specific activity by dividing the values of the ^{32}P peaks by those of nonradioactive protein peaks.

3.2.3. USE OF SPECIFIC INHIBITORS TO BLOCK INSULIN-DEPENDENT PP1$_{RG}$ PHOSPHORYLATION

1. Incubate [^{32}P]-labeled cells (*see* **Subheading 3.2.2.1.**) with either wortmannin (50 nM), calfostin (100 nM), GDPβS (1 mM), or rapamycin (12 ng/mL) for 20 min followed by the incubation in the absence or presence of insulin for 5–10 min (*see* **Note 22**).
2. Immunoprecipitate PP1$_{RG}$ as detailed in **Subheading 3.2.2.2.** Assay PP1 enzymatic activity on nonradioactive preparations, perform a sodium dodecyl sulfate polyacrylamide gel electrophoresis, autoradiography and quantitate as detailed in **Subheading 3.2.2.3.**

4. Notes

1. Insulin's effect on PP1 activation can only be observed in fused rat skeletal muscle cells. Therefore, use of L6 myotubes (10–14 d cultures) is strongly recommended (*see* **ref. *17***).
2. Isolated rat adipocytes are prepared by collagenase (CLS-1, Worthington) digestion of rat epididymal fat pads according to the method of Rodbell *(22)*. Briefly, 1 g of adipose tissue is minced and digested in polypropylene vials, with collagenase (3 mg/g of fat pad in 3-mL Krebs Ringer-phosphate (KRP) buffer, pH 7.4, containing 3% BSA (Fraction V, RIA grade, Sigma), and 5 mM glucose, prepared fresh each day) for 45 min in a shaking waterbath at 37°C. Each lot of

collagenase must be tested and comparedwith a previous lot by measurement of basal and insulin stimulated glucose uptake on identical fat cells isolated from each of the epididymal fat pad of a single rat.

3. Prepare insulin stocks (Porcine crystalline insulin, Eli Lilly) of 5 mg/mL in 0.005 N HCl (pH 2.5). Filter sterilize and store in sterile polypropylene tubes at 4°C. Do not freeze insulin stocks (stable for 1 yr when kept sterile). Dilute stocks with sterile KRP buffer containing 0.1% BSA just before use.

4. Each experiment should be performed on triplicate dishes plated at the same cell density using the same batch of cells from the same passage to yield similar results. Exposure time to insulin should be determined in a preliminary experiment by performing a time course of insulin effect on PP1 activation.

5. For adipocytes, an equal volume of packed cells (0.5 mL) suspended in KRP-3% BSA are exposed to insulin, as described for L6 cells. At the end of incubation, medium is removed after centrifugation for 5 s at 2000g. Cells are rinsed four times with warm PBS and then transferred to prechilled 4 mL plastic tubes for homogenization and processed as detailed in **Subheading 3.**

6. The presence of EDTA, β-mercaptoethanol, glycogen and a mixture of protease inhibitors in the extraction buffer are needed for the preservation of active phosphatases. Removal of EDTA or glycogen results in a significant loss of insulin-stimulated response.

7. It is important to sonicate L6 cells with a microtip and a pulse cycle. The authors use 0.1 s on/off cycle. Do not touch the tip to the walls or bottom of the tube. Samples should be kept on ice during sonication. Avoid frothing to prevent enzyme inactivation. Monitor cell disruption under a light microscope. Adipocyte cell extracts are prepared using a polytron at setting 3 for 10 s at 2°C (Tissue Miser, Fisher Scientific, Pittsburgh, PA) in a cold room.

8. To demonstrate the distribution of PP1 between PF and CF, it is important that the PF is not contaminated with the CF. The PF should be reconstituted with assay buffer to the original extract volume. Use a sonicator for reconstitution.

9. Although glycogen synthase and phosphorylase kinase have been used as substrates by Dent et al. *(11)*, to demonstrate the activation of purified PP1$_{RG}$, these substrates are phosphorylated at multiple sites and are highly unstable. The authors have, therefore, used glycogen phosphorylase a for the measurement of PP1 activation by insulin for two reasons: glycogen phosphorylase is phosphorylated at a single serine site by phosphorylase kinase, thereby making it substantially less complicated to analyze the dephosphorylation kinetics of a single phosphoserine residue on phosphorylase a than the dephosphorylation of multiple serine sites of glycogen synthase, and both substrates are dephosphorylated by the same class of phosphatase in most tissues.

10. It is very important to use highly pure (> 95%) phosphorylase b and phosphorylase kinase. Check the purity by running a sample on SDS/PAGE. If impure, purify according to the method of Fischer and Krebs *(23)*. We have used commercial preparations from Gibco/BRL.

11. Prior to starting the reaction, make sure that the work area is properly shielded and protected with adsorbent paper. Wear rubber gloves, a lab coat, and protective plastic eye glasses. Carefully monitor gloves, centrifuge, and rotors for radioactive contamination with a Geiger counter.

12. Premoisten the upper reservoir of the concentrator with solubilization buffer (this prevents loss caused by absorption), centrifuge briefly to remove residual buffer, do not allow the column to dry.

13. Adhere to the manufacturer's centrifugation speed to prevent any rupturing of the membrane. Centrifuge at 20°C to prevent denaturation of phosphorylase a during concentration.

14. Reconstitute okadaic acid (OA) with DMSF. Freeze 1000X stock solutions in small aliqouts. Dilute the stock OA with PP1 assay buffer just prior to use. Discard any remaining buffer after use. Using between 1–5 µg of protein/assay, the authors have found that a final concentration of 3 nM okadaic acid inhibits only PP2A, and the remaining activity represents the contribution of PP1. This should be confirmed emperically by assaying enzyme activity in the presence of Inhibitor-2 (2 µg, preincubate the enzyme with the inhibitor for 5 min before the addition of the substrate). The amount of activity inhibited by Inhibitor-2 should be equal to the activity assayed in the presence of okadaic acid. When comparing phosphatase activities in different extracts (i.e., control vs insulin) it is important that the enzyme assays be performed with identical protein concentrations.

15. [^{32}P]-labeled phosphorylase a should be brought to room temperature prior to use. Check the incubation time in a preliminary experiment for linearity. Typically, < 30% dephosphorylation is recommended.

16. Each assay tube contains 60 µg of [^{32}P]-labeled phosphorylase a, which is equivalent to 0.62 nmoles of protein bound phosphate. Calculate the specific activity of the substrate as follows:

$$SPA = \frac{(\text{total cpm/reaction tube} - \text{cpm blank})}{0.62 \text{ nmoles phosphate}}$$

Protein phosphatase activity (expressed as nmoles P_i released/min/mg protein) is calculated as follows:

$$\frac{cpm_{sample} - cpm_{blank} \times 1000}{SPA \times 10 \text{ min} \times \mu g \text{ protein/Rxn}}$$

17. PP1$_{RG}$ site-1 antibodies were generated by immunizing rabbits against a synthetic peptide corresponding to the sequence surrounding site-1 of rabbit skeletal muscle PP1$_{RG}$.

18. It is important to solubilize the PF with lysis buffer (*see* **Subheading 2.**) for complete extraction of PP1$_{RG}$.

19. Steady state equilibrium of cells with [^{32}P]-Orthophosphate is critical.

20. Add 1 mM of sodium orthovanadate, 1 M of microcystin, and protease inhibitors to PBS just prior to use.

21. To prevent dephosphorylation, all steps including immunoprecipation and SDS/ PAGE should be performed on the same day. Antibody concentration should be titrated in a preliminary experiment with different amounts of protein to achieve complete immunoprecipitation of $PP1_{RG}$.
22. These inhibitors are commercially available and should be reconstituted according to manufacturers instructions.

References

1. Saltiel, A. R. (1996) Diverse signaling pathways in the cellular actions of insulin. *Am. J. Physiol.* **270,** E375–E385.
2. Begum, N. (1995) Role of protein serine/threonine phosphatase 1 and 2A in insulin action. *Adv. Prot. Phosphatases* **9,** 263–281.
3. Nimmo, H. G. and Cohen, P. (1977) Hormonal control of protein phosphorylation. *Adv. Cyclic. Nucleotide Res.* **8,** 145–266.
4. Krebs, E. G. and Beavo, J. A. (1979) Phosphorylation-dephosphorylation of enzymes. *Annu. Rev. Biochem.* **48,** 923–959.
5. Cheatham, B. and Kahn, R. C. 1995. Insulin action and the insulin signaling network. *Endocr. Rev.* **16,** 117–142.
6. Rosen, O. M. (1987) After insulin binds. *Science* **237,** 1452–1458.
7. Keller, S. R., and Lienhard, G. E. (1994) Insulin signalling. *Trends Cell Biol.* **4,** 115–119.
8. Zhang, J-N, Hiken, J., Davis, A. E., and Lawrence, J. C. Jr. (1989) Insulin stimulates dephosphorylation of phosphorylase in rat epitrochlearis muscle. *J. Biol. Chem.* **264,** 17,513–17,523.
9. Mandarino, L. J., Wright, K. S., Vesity, L. S., Nicholas, J., Bell, J. M., Kolterman, O. G., and Beck-Nielsen, H. (1987) Effects of insulin infusion on human skeletal muscle pyruvate dehydrogenase, phosphofructokinase and glycogen synthase. *J. Clin. Invest.* **80,** 655–663.
10. Begum, N. and Draznin, B. (1992) Effect of streptozotocin-induced diabetes on GLUT-4 phosphorylation in rat adipocytes. *J. Clin. Invest.* **90,** 1254–1262.
11. Dent, P., Lavoinne, A., Nakielny, S., Caudwell, F. B., Watt, P., and Cohen, P. (1990) The molecular mechanism by which insulin stimulates glycogen synthesis in mammalian skeletal muscle. *Nature* **348,** 302–308.
12. Hubbard, M. J., Dent, P., Smythe, C., and Cohen, P. (1990) Targetting of protein phosphatase-1 to the sarcoplasmic reticulum of rabbit skeletal muscle by a protein that is very similar or identical to the G subunit that directs the enzyme to glycogen. *Eur. J. Biochem.* **189,** 243–249.
13. Lavoinne, A., Erikson, E., Maller, J. L., Price, D. J., Avruch, J., and Cohen, P. (1991) Purification and characterization of the insulin-stimulated protein kinase from rabbit skeletal muscle; close similarity to S6 kinase II. *Eur. J. Biochem.* **199,** 723–728.
14. Hubbard, M. J. and Cohen, P.(1989) Regulation of protein phosphatase-1G from rabbit skeletal muscle. 2. Catalytic subunit translocation is a mechanism for reversible inhibition of activity toward glycogen-bound substrates. *Eur. J. Biochem.* **186,** 711–716.

15. MacKintosh, C., Campbell, D. G., Hiraga, A., and Cohen, P. (1988) Phosphorylation of the glycogen-binding subunit of protein phosphatase-1G in response to adrenalin. *FEBS Lett.* **234,** 189–194.
16. Dent, P., Campbell, D. G., Hubbard, M. J., and Cohen, P. (1989) Multisite phosphorylation of the glycogen-binding subunit of protein phosphatse-1G by cAMP-dependent protein kinase and glycogen synthase kinase-3. *FEBS Lett.* **248,** 76–79.
17. Srinivasan, M. and Begum, N. (1994) Regulation of Protein phosphatase-1 and 2A activities by insulin during myogenesis in rat skeletal muscle cells in culture. *J. Biol. Chem.* **269,** 12,514–12,520.
18. Begum, N. (1994) Phenylarsine oxide inhibits insulin-stimulated protein-phosphatase-1 activity and Glut-4 translocation. *Am. J. Physiol.* **260,** 2646–2652.
19. Srinivasan, M. and Begum, N. (1994) Stimulation of protein phosphatase-1 activity by phorbol esters: Evaluation of the regulatory role of protein kinase C in insulin action. *J. Biol. Chem.* **269,** 16,662–16,667.
20. Begum, N. (1995) Stimulation of Protein Phosphatase-1 by insulin. Evaluation of the role of MAP kinase cascade. *J. Biol. Chem.* **270,** 709–714.
21. Begum, N., and Ragolia, L. (1996) Effect of tumor necrosis factor- on insulin action in cultured rat skeletal muscle cells. *Endocrinology* **137,** 2441–2445.
22. Rodbell, M. (1964) Metabolism of isolated fat cells. *J. Biol. Chem.* **239,** 375–380.
23. Fischer, E. H. and Krebs, E. G. (1958) Purification of phosphorylase b and phosphorylase kinase. *Methods Enzymol.* **5,** 369.

13

Analysis of the Isoforms of Protein Phosphatase 1 (PP1) With Polyclonal Peptide Antibodies

Margherita Tognarini and Emma Villa-Moruzzi

1. Introduction

Various isoforms of the catalytic subunit of Protein Phosphatase-1 (Ser/Thr phosphatase of type-1, PP1) were discovered by cDNA cloning *(1)*. PP1α, PP1γ1 and PP1δ (also called β) are found in virtually all mammalian cells, whereas PP1γ2 is testis-specific *(2,3)*. Although PP1α, PP1γ1, and PP1δ are products of different genes, they are about 90% identical *(2)* and migrate as a single protein band on SDS-electrophoresis. For instance, the 37 kDa PP1 purified from rabbit muscle is a mixture of the three isoforms *(4,5)*. The only sequence differences among the isoforms are in the C-termini *(2)*. Consequently, only antibodies to peptides derived from these regions are isoform-specific *(5,6)*. Based on the involvement of PP1 in many different regulatory pathways, differential subcellular localizations for the isoforms may be envisaged. These studies require isoform-specific antibodies, such as those produced and purified using the procedures described herein. The authors also show how the antibodies can be used to distinguish the PP1 isoforms in immunoblot, immunoprecipitate, two-dimensional electrophoresis, and immunofluorescence.

2. Materials

1. Synthetic peptides used to raise antibodies. For PP1γ1 and PP1δ: commercial custom-made, 70% pure peptides (Chiron Mimotopes, Australia). For PP1α: crude preparation, described previously *(7)*.
2. KLH (Keyhole Limpet Hemocyanin, Sigma, St. Louis, MO).
3. EDAC: water-soluble carbodiimide coupling reagent (Bio-Rad, Richmond, CA).
4. PB: 3 mM H$_3$PO$_4$, pH 5.0. Store at 4°C.
5. PBS: 1.96 mM KH$_2$PO$_4$, 8.05 mM Na$_2$HPO$_4$, pH 7.4, and 0.9% NaCl. Store at 4°C.
6. Complete and incomplete Freund's adjuvant (Sigma).

From: *Methods in Molecular Biology, Vol. 93: Protein Phosphatase Protocols*
Edited by: J. W. Ludlow © Humana Press Inc., Totowa, NJ

2.2. Antibody Detection by ELISA

1. Purified recombinant PP1 isoforms (gift from Dr. E. Y. C. Lee, Miami, *3*), Store at −20°C, in the presence of 50% glycerol to prevent freezing.
2. AgB: 0.2 M Na_2CO_3 and 0.2 M $NaHCO_3$, pH 9.6. Store at 4°C.
3. 96-well Microtest flexible plate (Falcon, Los Angeles, CA).
4. WB: 0.9% NaCl and 0.05% Tween-20 (Sigma). Store at 4°C.
5. OVA-PBS: 1% ovalbumin (Sigma) in PBS. Store at −20°C.
6. Antirabbit IgG conjugated to Horseradish peroxidase (Sigma), freshly diluted in PBS to 0.025 U/mL.
7. ABTS-B: 4 mg/mL ABTS peroxidase substrate (2,2′-azino-di[3-ethyl-benzthi-azoline-6-sulfonic acid], Bio Rad) in 50 mM sodium citrate, pH 5.2 and 1.5 µL/mL H_2O_2, freshly prepared.
8. Minireadr II (Dynatech, Chantilly, VA), used to detect the ELISA optical density.

2.3. Antibody Purification

1. Affi-Gel 10 (Bio-Rad). Store at −20°C.
2. NB: 0.1 M $NaHCO_3$, pH 8.6. Store at 4°C.
3. EB: 0.1 M ethanolamine, pH 8.0. Store at 4°C.
4. GB: 50 mM glycine, pH 2.2. Store at 4°C.
5. NP: 1 M Na_2HPO_4. Store at 4°C.
6. Centricon-10 (Amicon).
7. NZ: 2% stock sodium azide, used diluted 1:100. Store at 4°C.

2.4. Immunoblotting

1. Laemmli buffer: 0.1 M Tris-HCl, pH 6.8, 2% SDS, 20% glycerol, 0.01% bromophenol blue, and 2% 2-mercaptoethanol. Store at −20°C. Use diluted 1:2.
2. Polyacrylamide-SDS gel stock solutions: 30% acrylamide/0.8% BIS; 1.5 M Tris-HCl, pH 8.8; 0.5 M Tris-HCl, pH 6.8; 10% SDS, 10% ammonium persulfate, and TEMED (commercial solution) (Bio-Rad). Store at 4°C, SDS at room temperature.
3. Separating gel: 10% acrylamide/0.267% Bis, 0.375 M TRIS-HCl, pH 8.8, 0.2% sodium dodecyl sulfate (SDS), 0.016% ammonium persulfate and 0.167% TEMED. Prepare fresh and use to pour 9 × 5 cm and 0.75-mm thick slab gels.
4. Spacer gel: 5% acrylamide/0.134% Bis, 0.125 M Tris-HCl, pH 6.8, 0.1% SDS, 0.034% ammonium persulfate, and 0.167% TEMED. Prepare fresh and use to pour 1-cm high and 0.75-mm thick spacer gel, with 15 wells. Load samples of up to 18–20 µL.
5. Running buffer: 3 g/L Tris, 14.4 g/L glycine, and 0.1% SDS. Prepared from a 10X stock solution, stored at 4°C.
6. Electrophoresis: run on a Mini-Protean II cell (Bio-Rad).
7. Transblot: run in a Mighty Small Trasphor Tank Transfer Unit (Hoefer).
8. TBB: 4.5 g/L Tris, 21.5 g/L glycine, and 133 mL/L methanol. Store at 4°C. May be reused for up to 10–15 times.
9. Immobilon-P (Millipore).

10. Methanol.
11. TBS: 50 mM Tris-HCl, pH 7.4, and 200 mM NaCl. Store at 4°C.
12. TBS-T: 0.5% Tween-20 (Sigma) in TBS. Store at 4°C.
13. TBS-TM: TBS-T containing 1.5% dry skimmed milk. Store at 4°C, with NZ. Do not reuse.
14. Isoform-specific hyperimmune serum was freshly diluted 1:1000 in TBS.
15. Protein A-peroxidase (Sigma): prepare 65 U/mL stock solution in PBS. Store at −20°C in 10 μL aliquots. Use freshly diluted 1:1000 in TBS.
16. ECL (Amersham): enhanced chemiluminescence solutions, used according to the manufacturer's instructions. Store at 4°C.
17. Hyperfilm ECL (Amersham, Arlington Heights, IL), used to detect chemiluminescence in a dark room.
18. Kodak LX24 developer and A24 rapid fixer: diluted 1:5 in water and used in a dark room.

2.5. Immunoprecipitation

1. Protein A-Sepharose (Sigma) resuspended in 1 vol of PBS per vol of packed resin. Store at 4°C, with NZ.
2. TNE-T: 50 mM Tris-HCl, pH 7.5, 250 mM NaCl, 5 mM EDTA, 0.1% Triton X-100. Store at 4°C and add 20 μg/mL PMSF and 20 μg/mL benzamidine, final concentrations, before use.
3. AB: 25 mM imidazole, pH 7.4, and 50 mM 2-mercaptoethanol (1:300 from commercial stock). Store at 4°C.
4. TNE-TP: TNE-T containing 40 μg/mL PMSF, 40 μg/mL benzamidine, 40 μg/mL TPCK (*N*-tosyl-L-phenylalanine chloromethyl ketone), 4 μg/mL leupeptin and 15 mM 2-mercaptoethanol, all added before use.
5. [^{32}P]phosphorylase a (2–4 × 10^5 cpm/pmol) diluted in AB to 3 mg/mL. Store phosphorylase a crystals *(4)* at 4°C and dilute before use.
6. T: 10% TCA. Store at 4°C.
7. BSA: 25 mg/mL BSA. Store at −20°C.
8. OptiPhase HiSaphe scintillant (Wallac).

2.6. Two-Dimensional Electrophroesis

1. Isoelectrofocusing (first dimension), in a Mini-Protean II 2-D cell with glass capillary tubes (Bio-Rad). The tubes are 7.5 cm long and with an inner diameter of less than 1 mm.
2. First dimension sample buffer: 4.75 M urea, 1% Triton X-100, 2.5% 2-mercaptoethanol, 0.4% Pharmalite (Pharmacia, pH 4.0–6.5), 0.4% Pharmalite (pH 5.0–8.0), 0.4% Pharmalite (pH 3.0–10.0). Store at −70°C. This is used to resuspend the dried immunoprecipitates. For liquid samples, prepare a twofold concentrated sample buffer and dilute 1:2 with sample.
3. First dimesion stock solutions. 30% acrylamide/1.62% Bis, 25% Triton X-100 (Boehringer Mannheim, Germany), Pharmalite commercial solutions, 10% ammonium persulfate and TEMED. Store at 4°C, Triton X-100 at room temperature.

4. First dimesion monomer solution: 4% acrylaminde/0.216% Bis, 9.2 M urea, 0.1% Triton X-100, 2.5% Pharmalite (pH 4.0–6.5), 2.5% Pharmalite (pH 5.0–8.0), 2.5% Pharmalite (pH 3.0–10.0), 0.01% ammonium persulfate, and 0.1% TEMED. Make fresh and use to cast 7-cm long gels.

5. First dimension sample overlay buffer: 9 M urea, 0.4% Pharmalite (pH 4.0–6.5), 0.4% Pharmalite (pH 5.0–8.0), 0.2% Pharmalite (pH 3.0–10.0), and 0.005% Bromophenol blue. Store at –70°C.

6. Upper buffer: 20 mM NaOH. Prepare fresh and degass before use.

7. Lower buffer: 0.1 M H$_3$PO$_4$. Prepare fresh and degass before use.

8. Second dimension SDS sample buffer: 0.0625 M Tris-HCl, pH 6.8, 2% SDS, 5% 2-mercaptoethanol, 10% glycerol, and 0.0012% bromophenol blue. Store at –20°C.

9. Stripping solution: 5 mM phosphate buffer, pH 7.4, 2% SDS, and 2 mM 2-mercaptoethanol. Store at room temperature and add 2-mercaptoethanol before use.

2.7. Immunofluorescence

1. 24-well cell culture plate (Costar, Cambridge, MA).

2. PS-PBS: 3% paraformaldehyde and 2% sucrose in PBS. Store at 4°C.

3. HSNM-T: 20 mM HEPES, pH 7.4, 300 mM sucrose, 50 mM NaCl, 3 mM MgCl$_2$, and 0.5% Triton X-100. Store at 4°C.

4. PBS-BSA-3: 3% BSA in PBS. Store at –20°C.

5. PBS-BSA-0.2: 0.2% BSA in PBS. Store at –20°C.

6. Affinity-purified isoform-specific antibody solution: 5 µg Ig/mL freshly diluted in PBS-BSA-0.2.

7. Rhodamine-tagged swine antirabbit IgGs (DAKO): 10 µg/mL freshly diluted in PBS-BSA-0.2.

8. Fluorescein-tagged phalloidin (Sigma).

9. Mowiol: 20% Mowiol 4-88 (Hoechst, Germany) in PBS. Store at 4°C.

3. Methods

3.1. Production of the Three Isoform-Specific Peptide Antibodies

For the anti-PP1γ1 and anti-PP1δ antibodies, the peptides reproduced the 13 and 12 C-terminal residues of the rat PP1 sequences, respectively *(2,6)*. The sequences: TPPRGMITKQAKK for PP1γ1 and TPPRTANPPKKR for PP1δ. For the anti-PP1α antibodies, the 16-residue peptide reproduced the sequence between the aminoacids 294 and 309, with the following sequence: QILKPADKNKGKYGOL *(7)*. The remaining procedures are the same for all peptides.

1. A 30 mg/mL peptide solution was prepared in PB and 8–9 mg of solubilized peptide was mixed with 5–6 mg of KLH and 9 mg of EDAC.

2. After rotation for 5 h at room temperature and overnight at 4°C, the sample is dialyzed against 500 mL of PBS at 4°C for 2 h (*see* **Note 1**).

3. Approximately 1 mL is recovered and stored at 4°C for several weeks.
4. For immunization, 0.2 mL of the KLH-peptide solution is mixed for few minutes with 0.35 mL of complete Freund's adjuvant, using two 1-mL glass syringes with luer lock, connected to each other through one steel needle carrying two female luer locks, soldered one at each end of the needle (*see* **Note 2**).

   After vigorous mixing for a few minutes, the emulsion obtained is injected subcutaneously in several spots in the neck and back areas of an adult female rabbit. The subsequent injections are performed with material prepared by the same procedure, but using incomplete Freund's adjuvant.
6. The bleedings are performed 7 d after injecting (*see* **Note 3**).
7. Few drops of blood (to test positivity) or larger amounts of blood (for antibody collection) are obtained from the marginal vein of the ear.
8. The injection and bleeding schedule is as follows.

Day 0:	first injection.
Day 10:	second injection.
Day 40:	third injection.
Day 47:	first bleeding.
Day 60:	fourth injection.
Day 67:	second bleeding.

   The subsequent injections (up to 8–10) are performed every 2–3 wk.

3.2. Antibody Detection by ELISA

1. The presence of the antibodies in the sera obtained from each bleeding (by centrifuging the clotted blood at 12,000*g* for 5 min) is tested by ELISA (Enzyme Linked Immunosorbent Assay). Purified recombinant PP1 isoforms are used as antigens.
2. 0.1 mL of 4 μg/mL solution of the appropriate antigen in AgB was used/well of a 96-well flexible plate.
3. Vigorous mixing for 1 h on an orbital shaker is followed by one 0.2 mL wash with WB.
4. The free sites on the plate are then saturated with 0.2 mL OVA-PBS for 2 h, followed by one 0.2 mL wash in WB.
5. 0.1 mL of serum, using several dilutions in PBS, between 1:100 and 1:10,000, is applied for 1 h, followed by three 0.2 mL washes in WB.
6. 0.1 mL of antirabbit IgG conjugated to Horseradish peroxidase is then applied for 1 h, followed by three 0.2 mL washes in WB and by incubation with the substrate-containing solution ABTS-B.
7. The changes in optical density are then detected.
8. All the procedures are performed at room temperature, with vigorous shanking during the incubations. Controls, prepared with preimmune serum are run in parallel with the immune serum.

3.3. Affinity Purification of the Antibodies

1. The sera, obtained from various bleedings and stored at –20°C, are pooled, heat-treated at 56°C for 30 min, and centrifuged at 40,000*g* for 30 min.

2. One vol of saturated ammonium sulfate is added slowly during 20 min on ice with stirring (*see* **Note 4**).
3. After an additional 30 min of stirring, the suspension is centrifuged at 18,000g for 30 min.
4. The pellet is resuspended in the smallest possible vol of PBS, with the help of a Potter homogenizer, dialyzed overnight against 1 L of PBS, and against 1 L of PBS for another 4 h.
5. All the procedures after heat-treatment are performed at 4°C.
6. The samples are then divided into aliquots and stored at -20°C or used immediately for the affinity purification.
7. For each column, 2–3 mL of packed Affi-Gel 10 are washed three times in 2 vols of cold water and centrifuged at 800g for 5 min after each wash.
8. This is followed by one wash in 2 vols of NB (*see* **Note 5**).
9. The packed resin is then mixed with 1 vol of NB containing 10–12 mg of peptide.
10. The slurry is mixed for 2–3 h at 4°C, added with 0.2 mL of EB, and packed on a 5-mL syringe plugged with glass wool.
11. The column is washed with 10 vol of PBS, 10 vol of GB, and 10 vol of PBS, in sequence, at 20 mL/h and stored at 4°C in PBS containing NZ.
12. Up to 5 mL of the antibody solution obtained from the ammonium sulfate precipitation is applied to the column and recirculated on the column for at least 10 times by connecting the end of the column to the top.
13. The column is then washed in PBS until no protein is coming off the column.
14. Bound protien is eluted with GB. The flow rate is 20 mL/h and 0.65 mL fractions are collected.
15. In order to neutralize the elution fractions, 0.08 mL of NP is added to the empty collecting tubes. Each fraction is mixed immediately at the end of collection and the pH is checked and, in case, adjusted to pH 7.0–7.4 with NP.
16. The absence of the antibodies from the wash fractions and their presence in the elution fractions is confirmed by ELISA, performed as described above in **Subheading 3.2.**, after diluting the fractions 1:250 in PBS.
17. The affinity-chromatography is performed at room temperature.
18. The fractions containing the antibodies are pooled, concentrated with Centricon-10 at 4°C, divided into 0.05 mL aliquots, and stored at –20°C. The antibodies, added with NZ, are also stable at 4°C, for several months.

3.4. Use of the Antibodies in Immunoblot

1. The specificity of the antibodies in immunoblot is tested using recombinant PP1 isoforms and PP1 purified from skeletal muscle (**Fig. 1A**).
2. The proteins are boiled for 2 min in 18 µL of Laemmli buffer and applied to a 10% polyacrylamide-SDS gel.
3. The electrophoresis, run at 200 vol for 45 min, is followed by transfer to Immobilon-P which is first soaked for 1 min in methanol, washed twice for 10 min in distilled water, and incubated for 20 min in TBB.
4. During this time, the gel is also equilibrated in TBB for 20 min.

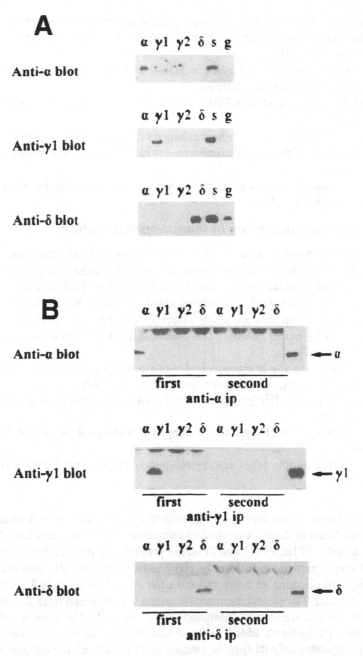

Fig. 1. Specificity of the anti-PP1 isoforms antibodies. (A) 70 ng of purified recombinant PP1α, PP1γ1, PP1γ2 (a testis-specific PP1 isoform) and PP1δ, or 140 ng of PP1-inhibitor 2 complex purified from muscle soluble fraction (s) or 70 ng of PP1 catalytic subunit purified from muscle glycogen particles (g) were detected in immunoblot, using the indicated antibodies. Each antibody recognizes only its specific antigen, and no

5. Transfer is at 200 mA for 2 h, keeping the temperature around 10°C.
6. At the end of transfer, the free sites on the membrane are blocked by incubating for 1 h in TBS-TM.
7. This is followed by two washes in TBS-T and incubation for 1 h with diluted hyperimmune serum (*see* **Note 6**).
8. After three more washes in TBS-T, the membrane is incubated with protein-A-peroxidase.
9. This is followed by five washes, enhanced chemiluminescence (ECL) reaction and detection of the chemiluminescence with ECL films and Kodak developing solutions.
10. All the incubations and washes (7–10 min each) are performed at room temperature with orbital shaking.

3.5. Use of the Antibodies in Immunoprecipitation

1. 5 µL of hyperimmune serum, or 2–3 µg of affinity purified antibodies, and 40 µL of protein A-Sepharose are used to immunoprecipitate the PP1 isoform (**Fig. 1B**).
2. Each recombinant isoform is diluted to 0.3–0.6 mL in TNE-T at 4°C.
3. The tubes are either shaken or rotated upside-down for 90 min at 4°C.
4. The immune complexes are washed three times with 1 mL of TNE-T buffer and either used for immunoblotting (*see* **Subheading 3.4.**) or to assay the PP1 activity (*see* **Note 7**).
5. For PP1 activity, the immune complexes are washed once more in AB without 2-mercaptoethanol, dried and resuspended in 15 µL AB.
6. 25 µL of 3 mg/mL [^{32}P]phosphorylase a ($2–4 \times 10^5$ cpm/pmol) is added to start the reaction.
7. After 20 min at 37°C, the reaction is stopped by adding 0.4 mL of cold T and 50 µL of BSA.
8. After vortexing, the tubes are centrifuged at 12,000g for 3 min at room temperature.

(Figure 1, *continued from previous page*) relevant cross-reactivity is detected among the isoforms. None of the three antibodies recognizes PP1γ2. In the soluble PP1 purified from muscle, PP1α, PP1γ1 and PP1δ are present, whereas the PP1 purified from muscle glycogen particles is PP1δ only. (**B**) The recombinant PP1 isoforms (as in A) were immunoprecipitated with the anti-α or the anti-γ1 or the anti-δ antibodies, as indicated (first ip). The second immunoprecipitate was prepared from the supernatant of the first immunoprecipitate. Immunoblot was as in A. The arrows indicate the recombinant PP1 isoforms added to the electrophoresis as positive controls. Each antibody precipitates only its specific antigen and virtually all the antigen is precipitated in the first ip. Reprinted from *Int. J. Biochem. Cell Biol.*, vol. 28, E. Villa-Moruzzi, F. Puntoni and O. Marin, "Activation of Protein Phosphatase-1 isoforms and Glycogen Synthase Kinase-3ß in Muscle from mdx Mice", pp. 13-22, Copyright 1996, with kind permission from Elsevier Science Ltd, The Boulevard, Langford Lane, Kidlington OX5 1GB, UK.

9. 0.25 mL of the supernatant is counted in a β counter after mixing with 1.2 mL of scintillant.

10. The activity is calculated from the $[^{32}P]H_3PO_4$ released *(4)*.

3.6. Immunodetection of the PP1 Isoforms by Two-Dimensional Electrophoresis

1. For two-dimensional gel electrophoresis, extracts are prepared from HeLa or NIH-3T3 cell (**Fig. 2**).

2. Subconfluent cells are washed twice with PBS, scraped in PBS using a rubber policeman and centrifuged at 2000*g* for 1.5 min.

3. The pellet is washed in 1 mL PBS and resuspended in 10 vol of TNE-TP.

4. After 20 min on ice with occasional vortexing, the suspension is centrifuged at 12,000*g* for 5 min and the supernatant obtained is defined as cell extract.

5. All the procedures are performed at 4°C.

6. A mixture of the three affinity-purified isoform-specific antibodies (2–3 μg of each isoform) and protein A-Sepharose are used to immunoprecipitate the three PP1 isoforms from cell extracts (*see* **Subheading 3.5.**).

7. The washed immune complexes are resuspended in 30 μL of first dimension sample buffer.

8. Following warming at 30°C for 5 min and centrifugation at 2000*g* for 2 min at room temperature, the supernatant is loaded onto the isoelectrofocusing gel, prepared with the first dimension monomer solution into glass capillary tubes (*see* **Notes 9** and **10**).

9. After polymerization and inserting the tubes into the apparatus, the tube gels are equilibrated with 5 μL of first dimension sample buffer for 15 min.

10. Each tube is filled with degassed upper buffer using a syringe before filling the upper chamber.

11. This is followed by filling the lower chamber with degassed lower buffer and pre-electrophoresing at 200 V for 10 min, 300 V for 15 min, and 400 V for 15 min.

12. The upper and lower buffers are then discarded and fresh degassed buffers are applied.

13. 25 μL of the sample (freshly prepared) is applied to each tube gel and overlaid with 30 μL of first dimension sample overlay buffer (*see* **Notes 11–13**).

14. Isoelectrofocusing is run at 500 V for 10 min, followed by 750 V for 3 h and 30 min.

15. After being removed from the capillary tubes, the gels are soaked in SDS sample buffer for 10 min, and overlaid onto the second dimension slab gel, which is a 10% SDS-polyacrylamide gel, 1-mm thick, prepared as in **Subheading 3.4.**) and casted using a two-dimension comb (i.e., with one tooth only) and a 5-mm high spacer gel (*see* **Note 14**).

16. The second dimension electrophoresis is followed by immunoblotting (*see* **Subheading 3.4.**).

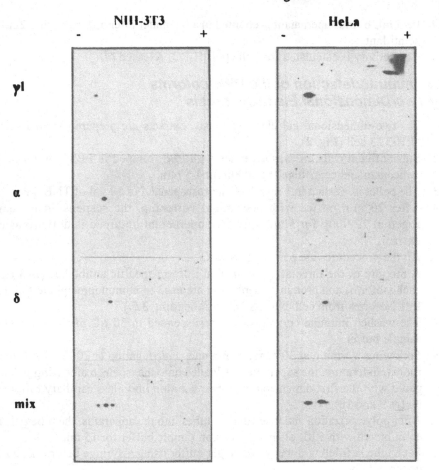

Fig. 2. Separation of the PP1 isoforms from cell extracts by two-dimensional gel electrophoresis. 3.2 mg of HeLa and 2.6 mg of NIH-3T3 cell extracts were used to immunoprecipitate the PP1 isoforms with a mixture of the anti-α, anti-γ1 and anti-δ affinity purified antibodies. The first dimension (isoelectrofocusing) was run from cathode (-) to anode (+). The second dimension (electrophoresis) and immunoblot were as in **Fig. 1**. The HeLa cells blot was probed with the anti-δ, anti-γ1, anti-α antibodies (as indicated) and with the mixture of the three antibodies (mix), in sequence. The NIH-3T3 cells blot was probed only with the three separated antibodies. The picture of the mix was obtained by superimposing the films from the three anti-isoform blots. After each probing, the antibody-protein A-peroxidase complexes were removed as described in the text. In the case of both cell types each antibody recognizes a specific spot.

17. After probing for one isoform, the antibodies are removed by treating the membrane with stripping buffer for 30 min at 60°C, followed by two washes in TBS

for 5 min, incubation for 30 min in TBS-TM, and two more washes with TBS-T
(*see* **Note 15**).

18. Probing with a second antibody is then performed (*see* **Subheading 3.4.**).
19. Up to four probings in sequence can be performed.

3.7. Immunofluorescence Detection of the PP1 Isoforms

1. HeLa cells are grown till subconfluent *(8)* on 12-mm diameter round glass coverslips, positioned at the bottom of a 24-well cell culture plate (*see* **Note 16**).
2. The coverslip-attached cells are washed three times with 1 mL PBS and fixed for
5 min with 0.5 mL PS-PBS.
3. After fixation, the coverslips are washed three more times in PBS-BSA-0.2.
(Unless specifically noted, all these procedures were performed at room temperature and the volumes are intended per well.)
4. The cells are permeabilized for 3 min at 4°C with 1 mL HSNM-T, followed by
three washes in PBS-BSA-0.2.
5. The permeabilization step is followed by an incubation of about 15 min in 1 mL
PBS-BSA-3 at 37°C.
6. 10 μL of diluted affinity-purified antibody is layered on each coverslip and incubation for 30 min at 37°C is carried out in a moist chamber.
7. This incubation is followed by three washes with 1 mL PBS-BSA-0.2 and 15 min
incubation at 37°C with 0.5 mL PBS-BSA-3.
8. After thorough aspiration of the buffer, 10 μL of rhodamine-tagged swine
antirabbit IgGs is layered on each coverslip.
9. To this second antibody solution, 10 μL of fluorescein-tagged phalloidin solution may be added to reach a final concentration of 200 mM. (This optional
labeling is intended to show the F-actin cytoskeleton, and to facilitate the identification of cells in the microscope and monitor their shape. This incubation is
carried out for 30 min at 37°C in a moist chamber.)
10. The coverslips are then washed three times in PBSA-BSA-0.2 and reincubated
for 15 min at 37°C in PBS-BSA-3.
11. Before mounting, each coverslip is soaked once in distilled water to remove
excess salt and protein.
12. Mounting is done by removing individual coverslips from the well bottom with
watch-maker tweezers and inverting them (cells are put in contact with the mounting medium) on standard microscopy slides with a drop of Mowiol (*see* **Note 17**).
13. Slide-mounted coverslips are dried at room temperature in the dark.
14. Slides are observed using a Zeiss Axiphot fluorescence microscope with a Zeiss
plan-apochromatic 63X, NA 1.4, objective.
15. Pictures are taken with Kodak T-Max 400 ISO films exposed at 1000 ISO and
developed in T-Max developer at 1600 ISO (**Fig. 3**).

4. Notes

1. After PBS dialysis, the KLH-bound peptide solution may be turbid and some
precipitation may occur during storage at 4°C. This does not damage the subsequent use as antigen to be injected.

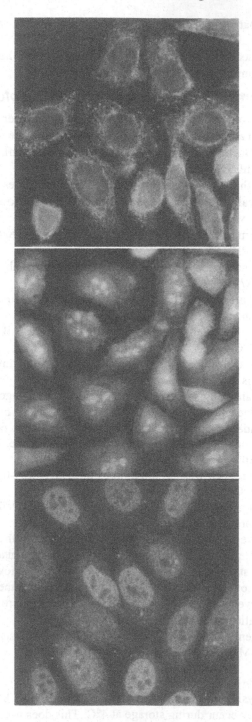

Fig. 3

2. The syringe needle with two female luer locks is built as follows. One luer lock was the original one from the needle. The second is from a second needle from which the needle had been cut off at the insertion with the lock. The second luer lock is soldered at the free end of the first needle.

3. Bleeding is facilitated by wiping the edge of the ear with toluene. At the end of collection, the toluene has to be carefully removed by washing with 70% ethanol, in order to stop bleeding.

4. The saturated ammonium sulfate is prepared and stored at 4°C for 1–2 d. After discarding the crystals at the bottom of the bottle, concentrated NH_4OH is added progressively. The pH is checked from time to time after diluting the solution 1:10 in water. NH_4OH is added until the solution is brought to neutrality.

5. The NB solution is prepared the day before and the pH is rechecked before using, since it tends to increase upon storage.

6. In order to minimize the consumption of antibodies, incubation with the diluted serum is performed on a rocking shaker and the solution can be reused up to five times. Storage is at 4°C, in the presence of NZ.

7. PMSF is added shortly before using the buffer, by diluting a stock solution of 2% PMSF in isopropanol. The stock solution is stored at 4°C and brought to room temperature 10–15 min before use, since at 4°C PMSF may precipitate.

8. Preparation of the first dimension sample buffer. In order to dissolve urea, the solution is warmed in a water bath, to a maximum of 40°C. For storage, it is divided into 0.5-mL aliquots.

9. Preparation of the first dimension monomer solution. The solution is warmed in a water bath to 40°C to dissolve urea, and degassed for 15 min before adding ammonium persulfate and TEMED.

10. Preparation of the first dimension gel. Twenty to thirty capillaries are cast at one time with monomer solution by inserting them into a casting tube. This is sealed with parafilm at the bottom and kept perfectly vertical on a stand fitted with a metal clamp. Glass rods are also inserted if the capillary tubes are not sufficent to fill the casting tube, in order to prevent any movement of the capillaries and to keep them vertical. The degassed monomer solution (with catalysts added) is delivered with a syringe so that the capillaries are filled from the bottom. Trapped air bubbles are removed by tapping the casting tube. After polymerization, the parafilm is removed and the capillary tubes are extracted from the casting tube by pushing the batch of the capillaries from the bottom of the casting tube. Air

Fig. 3. (*opposite page*) Detection of the PP1 isoforms in HeLa cells by immunofluorescence. The topography of the PP1 isoforms is shown by immunohystochemistry employing the three isoform-specific antibodies. PP1α (upper panel) is predominantly located in the cytoplasmic compartment and is presumably associated with vesicles or granules belonging to the endoplasmic reticulum. PP1γ1 (middle panel) and PP1δ (lower panel) display a major nuclear location, but their subnuclear distribution seems to be complementary, in view of their association with different chromatin areas.

bubbles should not be introduced into the bottom of the tubes at this point. After polymerization, the bottom of the capillary tubes are rinsed with lower buffer and inserted into the apparatus.

11. Loading of the sample onto the first dimension. After pre-electrophoresis, the upper buffer is removed with a syringe. Freshly degassed upper buffer is added and the bubbles are removed with a syringe. The sample is loaded underlaying it with a syringe sliding it along the capillary wall.

12. Preparation of the first dimension overlay buffer. In order to dissolve urea, the solution is warmed in a water bath to 40°C. For storage, it is divided into 0.5 mL-aliquots.

13. The upper and lower buffers were freshly prepared and degassed for 30 min before each run.

14. Removal of the gel from the capillary tubes. Once the tube is removed from the apparatus and disconnected from the connecting pieces (specific for each apparatus), some pressure is carefully applied to the top of the gel using a 1 mL syringe, filled with second dimension running buffer. The gel is extruded onto a piece of parafilm wetted with second dimension SDS sample buffer, straightened with a wet spatula, and allowed to sit for 10 min to equilibrate. The gel is then carefully slid between the glass plates onto the top of the slab gel for the second-dimension electrophoresis. If the second dimension is not run immediately, the gel tubes are put into 15-mL tubes filled with second dimension SDS sample buffer and may be stored at 4°C for several days.

15. Stripping of the Immobilon-P after immunoblot is performed by inserting the membrane in a 50-mL plastic tube, with the side of the membrane carrying the proteins facing the inside of the tube, and the opposite side adhering to the tube walls. After filling with stripping solution, the tube is rotated in a water bath.

16. Proper washing and sterilization of the coverslips is critical. This is obtained by soaking the glass coverslips overnight in 1 N HCl, washing them for 1 d in running tap water (or longer, until the pH is back around neutrality), and finally washing them 10 times in distilled water, once for 30 min in 96% ethanol and once for 30 min in acetone. The coverslips are then dried on a paper napkin and placed on aluminium foil in a glass Petri dish. Several layers of aluminium foil and coverslips can be fitted in the same Petri dish. This is then sterilized for 2 h at 200°C.

17. Storage of the fixed cells for up to 4 wk is achieved by keeping them at 4°C in PBS added with NZ. Slides may be kept at 4°C in the dark for several weeks without any significant loss of fluorescence intensity, provided the pH of Mowiol is carefully adjusted to 7.4.

Acknowledgments

The financial support of A.I.R.C. (Milan, I) and Telethon-Italy to the project is gratefully acknowledged. The authors thank Dr. P. C. Marchisio (Milan, I) for performing the immunofluorescence experiment and Dr. E. Y. C. Lee (Miami, FL) for supplying the recombinant PP1 isoforms.

References

1. Peruski, L. F., Wadzinski, B. E., and Johnson, G. L. (1993) Analysis of the multiplicity, structure and function of protein serine/threonine phosphatases. *Adv. Prot. Phosphatases* **7,** 9–30.
2. Sasaki, K., Shima, H., Kitagawa, Y., Irino, S., Sugimura, T., and Nagao, M. (1990) Identification of members of the protein phosphatase 1 gene family in the rat and enhanced expression of protein phosphatase 1α gene in rat hepatocellular carcinomas. *Jpn. J. Cancer Res.* **81,** 1272–1280.
3. Zhang, Z., Bai, G., Shima, M., Zhao, S., Nagao, M., and Lee E. Y. C. (1993) Expression and characterization of rat protein phosphatases-1α, -1gamma1, -1gamma2, and -1δ. *Arch. Biochem. Biophys.* **303,** 402–406.
4. Villa-Moruzzi, E., Ballou, L. M., and Fischer, E. H. (1984) Phosphorylase phosphatase. Interconversion of active and inactive forms. *J. Biol. Chem.* **259,** 5857–5863.
5. Villa-Moruzzi, E., Puntoni, F., and Marin, O. (1996) Activation of protein phosphatase-1 isoforms and glycogen synthase kinase-β in muscle from mdx mice. *Int. J. Biochem. Cell Biol.* **28,** 13–22.
6. Shima, H., Hatano, Y., Chun, Y.-S., Sugimura, T., Zhang, Z., Lee, E. Y. C., and Nagao, M. (1993) Identification of PP1 catalytic subunit isotypes PP1γ1, PP1δ and PP1α in various rat tissues. *Biochem. Biophys. Res. Commun.* **192,** 1289–1296.
7. Villa-Moruzzi, E., Dalla Zonca, P., and Crabb, J. W. (1991) Phosphorylation of the catalytic subunit of type-1 protein phosphatase by the v-abl tyrosine kinase. *FEBS Lett.* **293,** 67–71.
8. Villa-Moruzzi, E. and Puntoni, F. (1996) Phosphorylation of phosphatase-1 in cells expressing v-src. *Biochem. Biophys. Res. Commun.* **219,** 863–867.
9. O'Farrell, P. H. (1975) High resolution two-dimensional electrophoresis of proteins. *J. Biol. Chem.* **250,** 4007–4021.

14

Expression of Mouse Protein Phosphatase 2C in *Eschericia coli* and COS 7 Cells

Takayasu Kobayashi, Kazuyuki Kusuda, Motoko Ohnishi, Naoki Chida, and Shinri Tamura

1. Introduction

Expression of recombinant proteins using plasmid vectors in various kinds of cells has been widely used for the following three purposes:

1. To obtain a large amount of a purified specific protein in a relatively short time,
2. To investigate the phenotypes of the cells caused by a high expression of a protein whose physiological functions are not fully known, and
3. To determine whether posttranslational modification of proteins demonstrated in vitro is also observed in vivo.

For the first purpose, the expression system in *E. coli* cells is commonly used, and for the second and third purposes, the expression systems in various eukaryotic cells such as yeast, insect and mammalian have been developed. A number of useful promoters have been reported in each of these cell systems.

The authors have been taking advantage of these expression systems in order to investigate the molecular properties and physiological functions of protein phosphatase 2C (PP2C), one of the four major protein serine/threonine phosphatases of mammalian cells *(1,2)*, To date, the presence of six distinct isoforms of mammalian PP2C (PP2Cα, β1, β-2, β-3, β-4, and β-5) has been shown *(2)*.

In this chapter, the authors describe the procedure for expression of a fusion protein of glutathione S-transferase and mouse PP2Cα in *E. coli* cells. In addition, the authors also describe a procedure for the transient expression of HA-tagged mouse PP2Cα in COS 7 cells.

From: *Methods in Molecular Biology, Vol. 93: Protein Phosphatase Protocols*
Edited by: J. W. Ludlow © Humana Press Inc., Totowa, NJ

2. Materials

1. LB medium: 1% (w/v) tryptone, 0.5% (w/v) yeast extract, and 0.5 % (w/v) NaCl.
2. Ampicillin stock solution (250X): 2.5% (w/v), stored at –25°C.
3. pGEX-2T vector (Pharmacia, Uppsala, Sweden).
4. pKMC-PP2Cα (mouse PP2Cα cDNA *[4]* ligated into plasmid pKMC *[1]*).
5. 100 m*M* isopropyl-l-thio-β-D-galactoside (IPTG): sterilized by 0.22 μm filter, stored at –25°C.
6. Phosphate-buffered saline (PBS): ice cold.
7. 10% (v/v) Triton X- 100: autoclaved, stored at room temperature.
8. Glutathione agarose beads (Sigma, St. Louis, MO): Before use, swell the lyophilized powder in water (200 mL/g) for 2 h. Pack the gel into a mini-column and wash with at least 10 bed-volumes of PBS and store as 50% (v/v) suspension at 4°C.
9. Washing buffer: 50 m*M* Tris-HCl, pH 7.5, and 150 m*M* NaCl.
10. Thrombin cleavage buffer: 50 m*M* Tris-HCl, pH 7.5, 150 m*M* NaCl, and 2.5 m*M* CaCl$_2$.
11. Thrombin (Sigma): dissolved in water (100 μg/mL), stored at 4°C.
12. Mini-column (Evergreen Scientific, Los Angeles, CA, Cat. No 3383).
13. *E. coli* (DH5α).
14. Dulbeccos minimum essential medium (DMEM) containing 10% (v/v) fetal calf serum.
15. DMEM containing 10% (v/v) NuSerum (Collaborative Research, Bedford, MA).
16. PBS.
17. DEAE-dextran/chloroquine solution: 10 mg/mL (Pharmacia) and 2.5 m*M* chloroquine (Sigma) in PBS, not sterilized.
18. 10% (v/v) dimethyl sulfoxide (DMSO) in PBS.
19. pKMC-PP2Cα (mouse PP2Ca cDNA[4] ligated into plasmid pKMC[1]).
20. pGEM7fz(+) (Promega, Madison, WI).
21. pEUK-C- 1 vector (Clontech, Palo Alto, CA).
22. Plasmid purification kit (Qiagen, Chatsworth, CA).
23. TNE buffer: 10 m*M* Tris-HCl, pH 7.5, 1% (v/v) Triton X-100, 150 m*M* NaCl, and 1 m*M* EDTA.
24. Anti-HA antibody (Boehringer Mannheim, Mannheim, Germany).
25. HA peptide (YPYDVPDYA) (Boehringer Mannheim).
26. Protein A-agarose beads (Sigma): Equilibrated with TNE buffer and stored as 50% (v/v) suspension at 4°C.
27. TE buffer: 10 m*M* Tris-HCl, pH 7.5, and 1 m*M* EDTA.
28. SDS-sample buffer: 62.5 m*M* Tris-HCl, pH 6.8, 2% (w/v) SDS, 10% (v/v) glycerol, and 5% (v/v) 2-mercaptoethanol.

3. Methods

3.1. Expression of GST-PP2Cα Fusion Protein in E. coli Cells (3)

1. Cut out the *Nco*I-*Eco*RI fragment from pKMC-PP2Cα and make the *Nco*I site blunt end by Klenow fragments. Subclone the resultant DNA fragments into a multicloning site of the pGEX 2T vector in the correct reading frame. Transform *E. coli* (DH5α) cells with the plasmid.

2. Streak out the transformants onto a LB/ampicillin plate. Incubate the plate for 18 h at 30 CC (*see* **Note 1**).

3. Inoculate one colony into 30 mL LB/ampicillin medium in 100-mL flask and grow for 16 h at 30°C in a shaking incubator (60 strokes/min). Collect the cells by centrifugation (1000*g* for 15 min).

4. Resuspend the cells in 1 mL fresh LB/ampicillin (1X) medium. Inoculate them into 300 mL of the LB/ampicillin medium in 1-L flask and incubate for 1 h at 30°C in the shaking incubator (120 strokes/min). Add a 0.3 mL aliquot of 100 m*M* IPTG to the medium and incubate for another 5 h.

5. Collect the cells by centrifugation (1000*g* for 20 min) and wash them twice with 10 mL ice cold PBS. Suspend the cells in 3 mL ice cold PBS containing 1 m*M* PMSF and keep the cell-containing tubes on ice for 10 min.

6. Lyse the cells using a sonicator with a 6-mm diameter probe (*see* **Note 2**). Add 10% (v/v) Triton X-100 to 1% final concentration and centrifuge at 10,000*g* for 10 min at 4°C.

7. Add to the supernatant 200 µL of 50% (v/v) suspension of glutathione agarose beads, and mix gently by rotary shaker at 4°C for 15 min.

8. Transfer the suspension of the beads into a mini-column. Wash the beads with 12 mL PBS containing 1% (v/v) Triton X-100, then with 4 mL washing buffer, and finally with 4 mL thrombin cleavage buffer.

9. Cut the mini-column at the middle portion using a red-hot razor blade, and put the lower half of the plastic column into a 1.5-mL disposable tube (*see* **Note 3**).

10. Put the column into a new 1.5-mL tube after a centrifugation at 200*g* for 10 s. Add 300 µL of thrombin cleavage buffer containing thrombin (10 µg/mL) and incubate for 1 h at 25°C with occasional shaking.

11. Collect the supernatant containing the purified PP2Ca after a centrifugation at 200*g* for 10 s and store it at −70°C before use.

3.2. Expression of HA-PP2Cα in COS 7 Cells (3)

1. Prepare the HA-tagged PP2Cα expression vector as follows. Digest the plasmid pEUK-C- 1 with EcoRI (*see* **Note 1**). Fill the resultant gap with Klenow fragments and relegate it with T4 DNA ligase. The product is termed pEUK-C-w/oE. Cut out the DNA fragment containing the multicloning sites (*Xba*I, *Xho*I, *Eco*RI, *Kpn*I, *Sma*I, *Asu*II, *Cla*I, *Hin*dIII, and *Bam*HI) from the plasmid pGEM7fz(+) with *Xba*I and *Bam*HI and insert it into pEUK-C-w/oE previously double digested by *Xba*I and *Bam*HI. The resultant plasmid is termed pEUK-C-7fz.

2. Insert an oligonucleotide encoding the epitope. sequence (YPYDVPDYA) in frame into the 5' end portion of the open reading frame of mouse PP2Cα cDNA using PCR as follows. Perform first PCR using primer- 1, 5'-ATG TAT CCA TAT GAT GTT CCA GAT TAT GCT GGA GCA TTT TTA GAC AAG-3', primer-2, 5'-CCC CTG CAT TCT GAA TTC-3' and the PP2Cα cDNA as the template. Perform second PCR using primer-3, 5'-GCT CTA GAG GCA CGA TGT ATC CAT ATG ATG TT-3', primer-2 and the first PCR product as the template. Double digest the second PCR product with *Xba*I and *Eco*RI and insert

the resultant DNA fragment into the expression vector pEUK-C-7fz previously double digested by *Xba*I and *Eco*RI. Cut out the remaining 3' portion of mouse PP2Cα cDNA with *Eco*RI from the PP2Ca cDNA *(4)* and introduce it into the *Eco*RI site of the expression vector c containing the *Xba*I/*Eco*RI fragment of the second PCR product.

3. Purify the plasmid by QIAGEN plasmid purification kit.
4. Seed 5×10^5 COS 7 cells in a 10-cm dish with 10 mL of DMEM containing 10% (v/v) NuSerum (*see* **Note 5**).
5. Add 0.4 mL DEAE-dextran/chloroquine solution to 10 mL of DMEM containing 10% (v/v) NuSerum and mix well. Add 10 μg of the expression plasmid to this mixture (*see* **Note 6**).
6. Aspirate the medium from the dish and wash the cells with 5 mL of DMEM containing 10% (v/v) NuSerum. Add 5 mL of the DEAE-dextran/chloroquine/ plasmid DNA solution, and incubate for 1 h.
7. Aspirate the DEAE-dextran/chloroquine/plasmid DNA solution from the dish and add 5 mL 10% (v/v) DMSO in PBS. Incubate the cells for 2 min at 25°C.
8. Aspirate the DMSO solution and wash the cells with 5 mL PBS.
9. Add 10 mL DMEM containing 10% (v/v) FC'S to the dish and culture the cells overnight at 37°C in a CO_2 incubator.
10. Replate one of the two halves of the transfected COS 7 cells into a new 10-cm dish and culture the cells overnight at 37°C in a CO_2 incubator. Discard the other half.
11. Wash the cells with 10 mL PBS. Add 1 mL TNE buffer to lyse the cells. Transfer the lysate into a 1.5-mL disposable plastic tube.
12. Centrifuge the tube at 12,000*g* for 10 min at 4°C.
13. Collect the supernatant and add to it 2 μL anti-HA antibody (1 mg/mL). Incubate the solution for 1 h at 4°C.
14. Add 20 μL of 50% (v/v) suspension of protein A-agarose beads to the mixture. Mix gently by a rotary shaker for 1 h at 4°C. Centrifuge at 1000*g* for 1 min at 4°C and discard the supernatant.
15. Wash the agarose-beads 3 times with 1 mL TNE buffer and once with 1 mL TE buffer.
16. Add 25 μL SDS sample buffer and incubate for 2 min at 100°C for SDS-polyacrylamide gel electrophoresis. Alternatively, elute HA-PP2Cα with 25 μ*M* HA peptide solution for the assay of protein phosphatase activity.

4. Notes

1. *E. coli* cells must be grown at 30°C or below.
2. Too severe sonication may result in a loss of binding of GST-PP2Cα protein to glutathinone beads.
3. The mini-column must be fitted to a 1.5 mL disposable tube. We usually use mini-column purchased from Evergreen.
4. The expression vector for COS 7 cells must have both an SV40-derived origin of replication and eukaryotic transcription regulatory elements such as enhancers and a promoter.

5. COS 7 cells grow rapidly and require passage every 3 d: typically, a confluent plate of cells is split 1 to 10.
6. DNA should be added to the medium containing DEAE dextran/chloroquine. Do not mix the DNA with DEAE dextran/chloroquine solution directly. Otherwise, the DNA and DEAE dextran will form large precipitates.

References

1. Terasawa, T., Kobayashi, T., Murakami, T., Ohnishi, M., Kato, S., Tanaka, O., Kondo, H., Yamamoto, H., Takeuchi, T., and Tamura, S. (1993) Molecular cloning of a novel isotype of Mg^{2+}-dependent protein phosphatase β (type 2Cβ) enriched in brain and heart. *Arch. Biochem. Biophys.* **307,** 342–349.
2. Kato, S., Yerasawa, T., Kobayashi, T., Ohnishi, M., Sasahara, Y., Kusuda, K., Yanagawa, Y., Hiraga, A., Matsui, Y., and Tamura, S. (1995) Molecular cloning and expression of Mouse Mg^{2+}-dependent protein phosphatase β-4 (type 2Cβ-4). *Arch. Biochem. Biophys.* **318,** 387–393.
3. Ausubel, F. M., Brent, R., Kingston, R. E., Moore, D. D., Seidman, J. G., Smith, J. A., and Struhl, K. (1991) *Current Protocols in Molecular Biology,* Wiley, New York.
4. Kato, S., Kobayashi, T., Terasawa, T., Ohnishi, M., Sasahara, Y., Kanamaru, R., and Tamura, S. (1994) The cDNA sequence encoding mouse Mg^{2+}-dependent protein phosphatase α. *Gene* **145,** 311,312.

References

15

Expression of Functional Protein Phosphatase 1 Catalytic Subunit in *E. coli*

Mariam Dohadwala and Norbert Berndt

1. Introduction

Ten years ago, the purification of mammalian protein phosphatase 1 (PP1) was done routinely in only a handful of laboratories around the world. It typically involved the sacrifice of five rabbits and a cumbersome procedure that lasted 5 or 6 d, not to mention the requirement for heavy equipment, such as large-capacity centrifuges and tissue blenders. If all went well, you were rewarded with 1 or 2 mg of a close to homogeneously pure PP1 preparation. As in so many other instances, the classical purification of PP1 from its natural source has been supplanted by the recombinant DNA approach. This usually involves several steps, i.e., cloning of the gene or cDNA, subcloning the cDNA into a suitable vector, expressing the cDNA in a suitable host, and finally, purifying the recombinant protein.

In this chapter, we will describe the methodology that has been developed and optimized in our laboratory over the last few years to produce purified recombinant PP1 using an *Escherichia coli* expression system. The first two steps, although necessary for the approach taken here, will be described only in passing, since the methods involved are not unique to PP1 and are covered in-depth in many laboratory manuals on molecular biology. Rather, we will focus on the steps taken to optimize the system, that is, finding the conditions for both optimal expression and extraction. Second, we will describe in detail the purification of PP1, which consists of up to three chromatographic steps, affinity chromatography followed by gel filtration and ion-exchange chromatography. Although we have exclusively dealt with one isozyme, PP1α, this procedure can easily be adopted for the other known isozymes of PP1 as well as engi-

From: *Methods in Molecular Biology, Vol. 93: Protein Phosphatase Protocols*
Edited by: J. W. Ludlow © Humana Press Inc., Totowa, NJ

neered forms, such as the phosphorylation site knockout mutant PP1αT320A *(1)*. In a companion article, we will discuss two applications of recombinant PP1 (Chapter 6).

The technique developed in our laboratory can yield spontaneously active, nearly pure PP1 in 2 d and has been used to identify a site in PP1α phosphorylated by cyclin-dependent kinases *(1)* and—after scaling up—to solve the crystal structure of PP1α *(2)*.

2. Materials

2.1. Nonstandard Laboratory Equipment

1. An orbital shaker fitted with a plate that can safely accommodate up to four 4-L flasks. We use the model "Bigger Bill" manufactured by ThermoLyne, Dubuque, IA.
2. A sonicator or ultrasonic cell disrupter equipped with a 0.5-in. probe. We are using the VirSonic 300 Cell Disrupter by Virtis, New York, NY.
3. A 2.6 × 70 cm glass chromatography column (Pharmacia, Piscataway, NJ) to be used for the gel-filtration step and a 1.5 x 12 cm Econo-Pac polypropylene column (available in quantities of 50 from Bio-Rad, Hercules, CA) to be used for the affinity chromatography step (*see* **Note 1**).
4. Centriprep-10 devices (Amicon, Beverly, MA) to concentrate dilute PP1 solutions.
5. A scintillation counter to determine the PP1 activity.

2.2. Reagents and Solutions

All chemicals are from Sigma (St. Louis, MO) if not otherwise indicated.

2.2.1. Expression of PP1 in E. coli

1. Suitable plasmid harboring the recombinant phosphatase cDNA: To shorten the distance between the bacterial *tac* promoter and the ATG start codon, we have prepared an *NarI-Hind*III fragment of rabbit PP1α cDNA *(3)* and subcloned it into the *Bam*HI site of the *E. coli* expression vector pDR540 (Pharmacia). The construction of the constitutively active mutant, PP1αT320A, has been described previously *(1)*. The recombinant plasmid was then used to transform *E. coli* DH5α.
2. LB /amp medium: Mix 10 g bacto-tryptone (Difco, Detroit, MI), 5 g bacto-yeast extract (Difco), and 10 g solid NaCl with 950 mL of H_2O, adjust the pH to 7.4 with 1 N NaOH, add H_2O to a final volume of 1 L, and autoclave to sterilize. Just before use, add ampicillin (IBI, New Haven, CT) to a final concentration of 50 µg/mL.
3. Stock solution of 1 M $MnCl_2$.
4. 100 mM IPTG stock solution (Boehringer-Mannheim, Indianapolis, IN).
5. 100 mM phenylmethylsulfonyl fluoride (PMSF) (Pierce, Rockford, IL) in isopropanol; stable at room temperature for up to 1 yr.
6. 1 M benzamidine; stable at room temperature for up to 1 yr.
7. 2 mg/mL Aprotinin; stable at a pH of 7.0–8.0 and –20°C for 6 mo.
8. 0.5 mg/mL leupeptin; stable at –20°C for 6 mo.
9. 1 mg/mL pepstatin A in methanol; stable at –20°C for 6 mo.

10. Lysis buffer: 50 mM Tris-HCl, pH 7.5, containing 0.1 mM EGTA, 10% glycerol (Aldrich, Milwaukee, WI). Store at 4°C for up to 3 mo. Just before use, add: $\frac{1}{1000}$ vol of 2-mercaptoethanol and the above protease inhibitors (*see* **items 5–9**). The final concentrations will then be 0.1% (v/v) 2-mercaptoethanol, 0.1 mM PMSF, 1 mM benzamidine, 2 μg/mL aprotinin, 0.5 μg/mL leupeptin, and 1 μg/mL pepstatin A.

2.2.2. Purification of Recombinant PP1

1. Buffer A: 20 mM Tris-HCl, pH 7.5, containing 0.1 mM EDTA, 0.1 mM EGTA, 10% (v/v) glycerol, and freshly added 0.1% (v/v) 2-mercaptoethanol. For the column chromatographic steps, also add 1 mM MnCl$_2$, 1 mM benzamidine, and 0.1 mM PMSF.
2. Chromatography media: Heparin Sepharose CL-6B, Sephacryl S-100 HR, and HiTrap Q (Pharmacia).

2.2.3. Assay of PP1 Activity

Although phosphatase assay kits are now commercially available through Gibco-BRL (Gaithersburg, MD), we prefer to make up our own solutions (*see* **Note 2**).

1. Solution B: 50 mM Tris-HCl, pH 7.0. Store this buffer solution at 4°C for up to 1 y. Just before an experiment, freshly add 2-mercaptoethanol to 0.1% (v/v).
2. 1 mg/mL bovine serum albumin (BSA) in solution B. Store at 4°C for 1 mo.
3. 75 mM caffeine in solution B. Store at room temperature for up to 6 mo.
4. 10% (w/v) trichloroacetic acid (TCA).
5. Phosphoprotein substrate: As a substrate, we routinely use ^{32}P-labeled phosphorylase-*a* that we prepare by incubating 10 mg/mL phosphorylase-*b* with 0.1 mg/mL phosphorylase kinase for 1.5 h at 30°C in a reaction buffer consisting of 50 mM Tris-HCl, pH 8.2, 50 mM glycerol 1-phosphate, 0.1 mM CaCl$_2$, 10 mM Mg acetate, and 0.2 mM [γ-^{32}P]ATP (ICN, Irvine, CA). Detailed descriptions of this procedure can be found in refs. *4* and *5*, and are not the focus of this chapter. We routinely store ^{32}P-phosphorylase-*a* crystals in the refrigerator at a concentration of $\geq$ 15 mg/mL in a solution consisting of 50 mM Tris-HCl, pH 7.0, 0.1 mM EGTA, 0.3% (v/v) 2-mercaptoethanol, and 250 mM NaCl (*see* **Note 3**).

3. Methods

3.1. Production of Recombinant PP1 in E. coli Bacteria

Maintain a stock of recombinant bacteria in 20% (v/v) sterile glycerol frozen at −70°C. Several days before use, streak the glycerol stock onto an LB/amp agar plate, and incubate at 37°C up to 24 h. Single colonies of *E. coli* strain DH5α harboring pDR540-PP1α will then form. This plate can be kept sealed at 4°C for several weeks.

3.1.1. Induction of Gene Expression

1. In the afternoon, inoculate 25–100 mL LB/amp media with a single colony containing the recombinant plasmid and grow overnight at 37°C on an orbital shaker at 300 rpm.

2. Inoculate between 0.5 and 4.0 L (in one 2-L or four 4-L flasks) of LB/amp containing 1 mM MnCl$_2$ with 10–80 mL of the overnight culture (OD$_{600}$ ≤0.1), and grow at 37°C until the OD$_{600}$ is 0.6–0.7 under vigorous shaking (≥300 rpm). This should take about 3–4 h (*see* **Note 4**).
3. At this point, add IPTG to a final concentration of 1 mM, and grow for 18–20 h at 37°C. Record the OD$_{600}$ of a 1:10 dilution of the *E. coli* culture, which should be ≥0.350 (*see* **Note 5**).
4. Harvest the bacteria by centrifugation at 4000g for 25 min. Freeze the bacterial pellet at -20°C until further use.

3.1.2. Bacterial Lysis

1. Make sure the Virtis VirSonic 300 Cell Disrupter is ready to use. Mount the 0.5-in. probe with a flat tip onto a laboratory stand. This probe is suitable for volumes between 10 and 50 mL. The optimal position of the Power setting dial depends on the individual application. In this case, set the dial to 5. Before using the instrument in your experiment, make sure that at the given power setting the instrument is tuned correctly (*see* **Note 6**).
2. If frozen, thaw the bacteria for 15 min on ice. Suspend the bacterial pellet in ice-cold lysis buffer. The final volume should correspond to $\frac{1}{20}$ of the original culture. Transfer the bacterial suspension into a glass beaker (at least twice the size of the volume), and place into an ice bucket.
3. Lower the probe into the suspension with the tip immersed approx 1 cm. Sonicate the bacteria six times for 30 s (with 15-s breaks) making sure that the sample is properly cooled at all times. Centrifuge the lysate for 20 min at 20,000g.
4. Immediately assay for PP1 activity in the clarified lysate (*see* **Note 7**).

3.2. PP1 Assay

1. Dilute samples containing PP1 in solution B containing 1 mg/mL BSA (*see* **Note 8**).
2. Preincubate for up to 10 min at 30°C in labeled microfuge tubes: 10 μL solution B with or without 3 mM MnCl$_2$ and 10 μL diluted PP1 (*see* **Note 9**).
3. Start reaction by adding 10 μL of ^{32}P-phosphorylase-a. Mix well by pipeting up and down, and continue to incubate at 30°C for 10–15 min. Change pipet tips after each sample.
4. Stop reaction by adding 100 μL 10% TCA, and place tubes on ice for a few minutes.
5. Spin tubes in a microfuge for 3 min at 16,000g (maximum speed in the Eppendorf Model).
6. Transfer 100 μL of the supernatant into fresh tubes. Be careful not to touch the protein pellet. Add 1 mL of scintillant. Mix well and count radioactivity. Include the following controls containing solution A + 1 mg/mL BSA instead of PP1: totals (samples that have not been centrifuged) to calculate the specific radioactivity of the substrate protein and blanks to calculate the spontaneous release of inorganic phosphate from the substrate protein.
7. Calculate the PP1 activity by the formula:

$$(\text{Net cpm} \times 1.3 \times 100 \times d)/(SRA \times t \times 1000) = \text{u/mL} \qquad (1)$$

where d is dilution of PP before assay, SRA is specific radioactivity of ^{32}P-phosphorylase-a in cpm/pmol, t is incubation time in min, and unit is nmol/min (*see* **Note 10**).

3.3. Purification of Recombinant PP1

3.3.1. Affinity Chromatography on Heparin Sepharose CL-6B

1. Load the clear lysate onto a heparin Sepharose CL-6B column (1.5 inner diameter × 9 cm = 16 mL bed volume) equilibrated with buffer A (*see* **Note 11**). Adjust the flow rate to 1 mL/min.
2. Wash the column with 5 bed volumes or 80 mL buffer A containing 150 mM NaCl. The OD$_{280}$ trace recording the eluted protein should have returned to the baseline by then.
3. Elute PP1 with five to six bed volumes or 80–96 mL of a gradient from 150–600 mM NaCl. Collect 1.5-mL fractions, and assay for PP1 activity (*see* **Note 12**).
4. Pool the active fractions, and measure their volume. Centrifuge the PP1-containing fractions in Amicon's Centriprep-10 devices at 4°C and 1500g, thereby concentrating them to ≤2 mL. Although this step should not take more than a few hours, it can safely be done overnight, since PP1 is quite stable from now on.

3.3.2. Gel Filtration on Sephacryl S-100 HR

1. Apply the concentrated solution obtained in the previous step to a Sephacryl S-100 HR column (2.6 inner diameter × 66 cm = 350 mL bed volume) equilibrated with buffer A + 150 mM NaCl, and develop the column at a flow rate of 1 mL/min.
2. Direct the void volume $v_0 \approx 130$ mL into the waste, and then collect 3-mL fractions.
3. Assay the fractions for PP1 activity (*see* **Note 13**). For many purposes, the PP1 protein will be sufficiently pure after completion of this step. If this is so, proceed directly to **Subheading 3.3.3., step 3**.

3.3.3. Ion-Exchange Chromatography on Hi-Trap Q

1. Apply the pooled fractions from the previous step to two prepacked 1-mL HiTrap Q columns equilibrated with buffer A (*see* **Note 14**). Maintain a flow rate of 1 mL/min. After loading the sample, wash the column with 10 mL buffer A.
2. Elute bound proteins with 25 mL of a linear gradient from 0 to 700 mM NaCl in buffer A, and collect 1-mL fractions (*see* **Note 15**).
3. Pool the active fractions, and concentrate them as described (**Subheading 3.3.1., step 4**) to ≤1 mL (*see* **Note 16**). (Optional step: Dialyze against buffer A overnight to remove excess Mn^{2+}). Add glycerol with a pH of 7.0 to a final concentration of 50% (v/v) glycerol, and store the enzyme preparation at −20°C (*see* **Note 17**).

4. Notes

1. Although hand-blown columns that are operated under gravity flow conditions and attached to a generic fraction collector will suffice in principle, we found that

the use of a programmable, integrated chromatography system helps enormously to control the conditions more carefully and make efficient use of time. In our lab, we use columns fitted with flow adapters and the Econo System from Bio-Rad, which includes a gradient controller, a gradient mixer, and a peristaltic pump for flow control and a UV monitor.

2. For a lab that performs phosphatase assays only occasionally, using a kit may be more economical. Bear in mind that even with a kit, you still have to label phosphorylase-*b* yourself and factor in the cost of radioactive ATP.

3. Determine the number of assays you are going to perform on a given day. Dilute the required amount of ^{32}P-phosphorylase-*a* crystal suspension to 3 mg/mL solution B containing 15 mM caffeine, and incubate at 30°C. The phosphorylase-*a* crystals should dissolve within a few minutes.

4. The flasks used to grow the recombinant bacteria should be filled with culture medium up to one-quarter of their capacity. Provided there is only one orbital shaker available, the upper limit is thus 4 L of bacterial culture. Keep in mind that the available centrifuge capacity should match this figure; otherwise, one would have to perform several runs in succession, and valuable PP1 activity would be lost. Like others before us *(6,7)*, we found that without Mn^{2+} in the medium, there was very little PP1 protein expressed. The presence of Mn^{2+} may be required because PP1 is a metalloenzyme with two divalent cations bound in the active center *(2)*.

5. Previously published procedures to express PP1 in bacteria used a growth temperature of 27°C *(6,7)*. However, the vectors used in these two protocols were different from the one used in our laboratory. Nonetheless, we have compared the expression levels of PP1 at three different growth temperatures, 21, 27, and 37°C. The yield of soluble activity was two- to threefold higher at 37°C than at the other temperatures.

6. For reproducible results, it is crucial that the instrument is properly tuned. This is especially important in situations where the instrument is being shared by several laboratories who might have to use different probes. If not already performed earlier, the instrument should be tuned according to the manufacturer's instructions. In principle, the procedure consists of finding the lowest "percent error" reading for each power setting. Ideally, this procedure should be done only once after each change of the probe.

7. It is imperative that the time spent for lysis is kept at a minimum, since the PP1 activity at this stage is relatively labile. We have lysed the recombinant bacteria in $\frac{1}{5}$, $\frac{1}{10}$, and $\frac{1}{20}$ vol of the original culture. Although there was no difference in recovery of total PP1 activity between these three conditions, using a volume as small as reasonably possible ($\frac{1}{20}$) helps to load the sample quickly onto the first column. It also appears important that the lysate should be kept on ice at all times.

8. The dilution of samples for the PP1 assay is critically important in order to avoid repeating the assay of certain samples: The dephosphorylation reaction proceeds with a linear rate until up to 20% of ^{32}P has been released from phosphorylase-*a*. The dilution of the sample should be high enough to allow for incubation times of

up to 10–15 min. In our experience, we found the following dilution factors ideally suited to the different stages of the purification:

Bacterial lysate:	1000–2000.
Chromatographic steps:	5000–10,000.
Concentrated solutions:	25,000–50,000.

Depending on the chosen incubation time, 40–60 samples can be performed per session, provided subsequent samples are being started every 15 s. The time required to process these is approx 60–75 min. In the interest of minimizing the time, we sometimes forgo duplicate determinations when screening column eluates.

9. In contrast to mammalian cells, *E. coli* bacteria do not usually express phosphorylase phosphatase activity. Thus, all the activity measured in this system is a result of mammalian recombinant PP1. Therefore, there is no need to suppress other phosphatases, such as PP2A, that dephosphorylate phosphorylase-*a*. In other words, there is no need to include either inhibitor-2 or okadaic acid in the phosphatase assay. However, it is useful, but not essential, to measure each sample in the absence and presence of 1 mM Mn^{2+}. In contrast to previous reports where the recombinant PP1 was found to be absolutely dependent on Mn^{2+} for activity *(6,7)*, this system yields a form of PP1 that is spontaneously active in the absence of Mn^{2+}. However, we note that a slight activation of PP1 occurs upon addition of Mn^{2+} to the assay. In our hands, the ± Mn^{2+} activity ratio usually varies between 0.5 and 0.7 (*see also* **Note 17**).

10. The calculation of activity data for dozens of samples can be sped up significantly by using customized worksheet templates created with a spreadsheet program, such as Microsoft Excel. The experimenter simply has to enter the primary data, and the activity is computed automatically. Information about these aides are available from the authors on request.

11. The capacity of a 16-mL heparin column is sufficient to bind all of PP1 from a 4-L *E. coli* culture. The column size, and therefore, the time required to complete this step can be reduced if smaller cultures are used as starting material.

12. PP1 activity peaks at around 400 mM NaCl.

13. When this protocol is followed, the peak of PP1 activity can be expected in fraction 13–15.

14. If necessary, scaling up or down at this stage is very convenient, since prepacked HiTrap Q columns can be arranged in succession.

15. PP1 activity peaks at around 500 mM NaCl.

16. In summary, this preparation can conveniently be performed in about 2 d. The progress of the purification is summarized in **Table 1**. From 1 L of culture, one should comfortably expect about 1–2 mg of PP1 protein. **Figure 1** shows the purity of PP1 after each of the chromatographic steps.

17. For storage, it is crucial that PP1 is kept in vials that have a low protein binding capacity. Polypropylene is most appropriate. Second, since the buffering capacity of buffer A (20 mM Tris-HCl) is relatively low, it is crucial that the glycerol added to the final preparation is in the neutral pH range. Finally, as already

Table 1
Purification of recombinant PP1α Catalytic Subunit from 1 L of *E. coli* Culture

Purification step	Protein, mg	PP1 activity, U	Specific activity, U/mg	Purification, x-fold	Yield, %
Cell lysate	187.6	102,650	547.2	1	100
Heparin Sepharose	23.8	86,020	3,614	6.6	83.8
Sephacryl S-100 HR	4.48	69,298	15,468	28.3	67.5
HiTrap Q	1.62	52,736	32,553	59.5	51.4

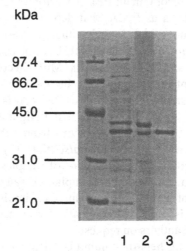

Fig. 1. Purification of recombinant PP1 catalytic subunit. Aliquots of samples obtained after each chromatographic step were separated by SDS gel electrophoresis and stained with Coomassie blue R. In addition to the mole-wt standards, the lanes represent the following purification stages: 1—Heparin Sepharose CL-6B, 2—Sephacryl S-100 HR, 3—HiTrap Q.

pointed out, recombinant PP1 produced in this system is not dependent on Mn^{2+}, based on the following two results: (a) even after diluting the enzyme 50,000-fold, the $\pm Mn^{2+}$ activity ratio is still approx 0.6, and (b) dialyzing the preparation against a buffer lacking Mn^{2+} does not alter this ratio. If kept under these conditions, the enzyme is stable for about 6 mo. Thereafter, the enzyme gradually, yet slowly loses its activity although the protein remains intact as judged by SDS gel electrophoresis and Western blotting with antibodies specific for the C-terminus of PP1α *(8)*. It may be possible to restore the enzyme's activity by adding divalent cations (Co^{2+} or Mn^{2+}) back to the preparation *(9)*.

Acknowledgments

Work in our laboratory was supported in part by a CHLA Research Institute Fellowship (to M. D.) and grants (to N. B.) from the Tobacco-related Disease Research Program of the State of California (3KT-0193) and the National Institutes of Health (1R01-CA54167).

References

1. Dohadwala, M., Da Cruz e Silva, E. F., Hall, F. L., Williams, R. T., Carbonaro-Hall, D. A., Nairn, A. C., et al. (1994) Phosphorylation and inactivation of protein phosphatase 1 by cyclin-dependent kinases. *Proc. Natl. Acad. Sci. USA* **91**, 6408–6412.
2. Goldberg, J., Huang, H.-B., Kwon, Y.-G., Greengard, P., Nairn, A. C., and Kuriyan, J. (1995) Three-dimensional structure of the catalytic subunit of protein serine/threonine phosphatase-1. *Nature* **376**, 745–753.
3. Berndt, N. and Cohen, P. T. W. (1990) Renaturation of protein phosphatase 1 expressed at high levels in insect cells using a baculovirus vector. *Eur. J. Biochem.* **190**, 291–297.
4. Shenolikar, S. and Ingebritsen, T. S. (1984) Protein (serine and threonine) phosphate phosphatases. *Methods Enzymol.* **107**, 102–129.
5. MacKintosh, C. (1993) Assay and purification of protein (serine/threonine) phosphatases, in *Protein Phosphorylation—A Practical Approach*, 1st ed. (Hardie, D.G., ed.), IRL, Oxford, England, pp. 197–230.
6. Zhang, Z., Bai, G., Deans-Zirattu, S., Browner, M. F., and Lee, E. Y. C. (1992) Expression of the catalytic subunit of phosphorylase phosphatase (protein phosphatase-1) in *Escherichia coli*. *J. Biol. Chem.* **267**, 1484–1490.
7. Alessi, D. R., Street, A. J., Cohen, P., and Cohen, P. T. W. (1993) Inhibitor-2 functions like a chaperone to fold three expressed isoforms of mammalian protein phosphatase-1 into a conformation with the specificity and regulatory properties of the native enzyme. *Eur. J. Biochem.* **213**, 1055–1066.
8. Runnegar, M. T., Berndt, N., Kong, S., Lee, E. Y. C., and Zhang, L. (1995) In vivo and *in vitro* binding of microcystin to protein phosphatases 1 and 2A. *Biochem. Biophys. Res. Commun.* **216**, 162–169.
9. Chu, Y., Lee, E. Y. C., and Schlender, K. K. (1996) Activation of protein phosphatase 1: Formation of a metalloenzyme. *J. Biol. Chem.* **271**, 2574–2577.

16

Protein Phosphatase 2A and Protein Phosphatase X Genes in *Arabidopsis thaliana*

Gemma Pujol, Albert Ferrer, and Joaquin Ariño

1. Introduction

The presence of Ser/Thr protein phosphatase activities in higher plants was recognized some years ago *(1–3)*. The molecular cloning of the cDNAs encoding the catalytic subunits of several of these phosphatases made clear two points: First, that plant Ser/Thr phosphatases were very similar in sequence to those found in other organisms. Second, that they constitute an extremely complex gene family. For instance, at least six genes encoding proteins highly related to the catalytic subunit of PP1 have been identified in *Arabidopsis thaliana (4,5)*, whereas only four appear to exist in mammals and only one in other organisms, such as budding yeast. The author's laboratory has proven the existence in *Arabidopsis thaliana* of at least five isoforms of the catalytic subunit of PP2A *(6,7)*, whereas two forms have been identified in mammals *(8,9)* and only one in other organisms *(10–12)*. On the basis of their amino acid sequence, *Arabidopsis thaliana* PP2A isoforms can be classified into two subfamilies. The first subfamily is composed of PP2A-1, PP2A-2, and PP2A-5 and the second is composed of PP2A-3 and PP2A-4 *(4,5,13)*. The sequence identity among the members of each family is very high (97–98%), whereas members of different subfamilies are about 80% identical. Plant PP2A isoforms are strongly related to the PP2A proteins from other organisms (about 80% identical their mammalian counterparts). A protein phosphatase related to type 2A phosphatases, protein phosphatase X (PPX), was initially identified in rabbit *(14)*, although it has also been found in insects. The author's group has cloned two different isoforms of PPX from *Arabidopsis thaliana (15)*, demonstrating that this phosphatase is present in very different types of eukaryotic

From: *Methods in Molecular Biology, Vol. 93: Protein Phosphatase Protocols*
Edited by: J. W. Ludlow © Humana Press Inc., Totowa, NJ

cells. Both isoforms are very similar (93% at the amino acid level), and PPX is also an extremely conserved protein (the plant protein is 83% identical to mammalian PPX). A phosphatase related to type 2C (type 2C proteins are largely unrelated in sequence to type 1, 2A, or 2B) has also been reported in Arabidopsis thaliana *(16,17)*. However, no type 2B activity has been reported nor has a type 2B-related gene been cloned in plants.

It is known that type 1 and 2A enzymes are oligomers composed of a catalytic subunit and one or more regulatory subunits. The same is true for plants. In fact, homologs to the 65 and 55 kDa regulatory subunits of PP2A have been recently identified in Arabidopsis thaliana or pea *(18–21)*. If it is considered that, in mammalian cells, the number of isoforms of the regulatory polypeptides can be very high (*see* **ref. 22** and references contained therein) and assume at least a similar complexity for plants, the emerging picture is that Ser/Thr phosphatases in plants constitute a family of a very high complexity.

To elucidate the biological role of all these proteins, specific DNA probes and antibodies are very useful tools. In this chapter protocols are provided for the simultaneous amplification from genomic DNA of probes corresponding to isoforms of PP2A and PPX, as well as for the expression of PPX in *E. coli* to generate polyclonal antibodies.

2. Materials

2.1. Isolation of PP2A and PPX Probes from A. thaliana Genomic DNA by Polymerase Chain Reaction

1. 10X reaction buffer: 100 mM Tris-HCl, pH 8.4, 0.5 M potassium chloride, 1 mg/mL gelatin.
2. dNTP solution (2 mM each). Store at –20°C.
3. 10 mM magnesium chloride solution.
4. 5 M solution of the following oligonucleotides: MP-1s (GGG*AATTC*GA(CG)(CAT)T(GC)(CT)T(AGC)TGGTC(ATG)GA(TC)CC and MP-3a (CC*AAGCTT*(CT)(CA)(GA) CA(GA)TA(GA)TT(GT)GG(N)GC (*see* **Note 1**).
   Italicized sequences correspond to *Eco*RI and *Hin*dIII restriction sites added to facilitate cloning.
5. *Taq* DNA polymerase (*see* **Note 1**).
6. Automated thermal cycler, Eppendorf tubes, and mineral oil (Sigma, St. Louis, MO).
7. Standard reagents and equipment for DNA electrophoresis.

2.2. Expression of PPX in E. coli for Antibody Production

1. *E. coli* strains: NM522, *supE thi-1* Δ(*lac-proAB*) Δ(*mcrB-hsdSM*)[5](r$_K$⁻m$_K$⁻) [F' *proAB lacI^{q}Z*ΔM15], and BL21(DE3), *hsd*S *gal ompT* λDE3 Δ*nin5 imm21*.
2. Ampicillin stock solution: 50 mg/mL in H$_2$O. Aliquot and store at –20°C.
3. Isopropyl-β-D-thiogalactopyranoside (IPTG) stock solution: 100 mM in H$_2$O. Aliquot and store at –20°C.

4. Sonication buffer: 50 mM Tris-HCl, pH 7.5, 1 mM MgCl$_2$, and 1 mM dithiothreitol. Store at 4°C.

5. Washing buffer: 50 mM Tris-HCl, pH 7.5, 10 mM ethylene diamine tetraacetic acid (EDTA), 1 mM dithiothreitol and 0.5% (v/v) Triton X-100. Store at 4°C.

6. Solubilization buffer: 50 mM Tris-HCl, pH 8.0, and 6 M urea. Make immediately before use.

7. 2X SDS-PAGE sample buffer: 125 mM Tris-HCl, pH 6.8, 4% (w/v) SDS, 10% (v/v) β-mercaptoethanol, and 20% (v/v) glycerol. Store at −20°C.

8. Elution buffer: 25 mM Tris base, 192 mM glycine, and 0.1% SDS. Store at 4°C.

9. Soaking buffer: 2.5 mM Tris base, 19.2 mM glycine, and 0.1% SDS. Store at 4°C.

10. Dialysis tubing (Sigma): Tubing of 12,000 mol wt cutoff (flat width 25 mm; diameter 16 mm) should be washed once in 50% ethanol for 60 min, twice in 10 mM sodium bicarbonate for 10 min, and once in 4 mM EDTA for 60 min. Rinse tubing with distilled water and store in 0.02% NaN$_3$ at 4°C.

11. 10X Phosphate-buffered saline (PBS): 1.5 M NaCl and 100 mM NaH$_2$PO$_4$, pH 7.5. Store at 4°C.

12. Saturated ammonium sulfate solution: 66.2 g (NH$_4$)$_2$SO$_4$, 100 mL 10X PBS made up to 1 L with distilled water. Store at 4°C.

13. Extraction buffer: 50 mM Tris-HCl, pH 8.0, 10 mM NaCl, 1% (w/v) SDS, 5% (v/v) β-mercaptoethanol, 10 μg/mL aprotinin, 1 μg/mL E64 (Sigma), 0.5 μg/mL leupeptin, 1 μg/mL pepstatin, and 0.1 mM phenylmethylsulphonyl fluoride (PMSF). Add β-mercaptoethanol and protease inhibitors immediately before use (*see* **Note 2**).

14. LB Medium: 1% (w/v) tryptone, 1% (w/v) NaCl, 0.5% (w/v) yeast extract.

3. Methods

3.1. Isolation of PP2A and PPX Probes from A. thaliana *Genomic DNA by Polymerase Chain Reaction*

The following protocol allows the simultaneous amplification of at least one isoform of PP2A (PP2A-5) and one of PPX (PPX-1) as bands of about 220 bp in length (*see* **Note 3**).

1. Set up a 100 μL PCR reaction containing 0.15 μg of *A. thaliana* genomic DNA (*see* **Note 3**), 10 mM Tris-HCl (pH 8.4), 50 mM potassium chloride, 0.1 mg/mL gelatin, 0.2 mM concentration of each dNTP, 1 mM magnesium chloride, 0.5 μM concentration of each oligonucleotide, and 2.5 U of *Taq* polymerase.

2. Perform the DNA amplification (30 cycles) under the following conditions: 94°C for 105 s, 45°C for 2 min, and 72°C for 2 min (*see* **Note 4**). Analyze 10 μL of the reaction on 2% agarose gels (*see* **Note 5**) for presence of 200–220 bp amplification products.

3. Extract the remaining PCR reaction with phenol/chloroform, precipitate with ethanol and digest with *Eco*RI and *Hind*III (*see* **Note 1**). Run the reaction in an agarose gel and recover the DNA from the gel.

4. Ligate into *Eco*RI and *Hind*III-cleaved pUC18 or equivalent vector. Sequence the insert to ascertain the nature of the cloned DNA. In our hands, most inserts correspond to fragments AP-1 (PP2A-5) or AP-2 (PPX-1) *(6)*.

5. The aforementioned fragments can be used for a number of purposes, including library screening and Northern and Southern blotting. It is important to note that, because the high level of identity between PPX-1 and PPX-2, very high stringency conditions must be used to specifically detect one of the isoforms. For instance, when AP-2 is used in Southern blot experiments, both isoforms are still detected even when filters are washed under conditions usually considered as rather stringent (68°C in 0.2X SSC, 0.5% SDS). To detect specific isoforms, filters must be washed at 68°C in 0.1X SSC, 0.5% SDS *(15)*. Similar caution against unwanted cross-hybridization must be taken in the case of PP2A, particularly within members of the same subfamily *(6)*.

3.2. Expression of PPX in E. coli for Antibody Production

3.2.1. Cloning of a PPX-1 cDNA Fragment into the Expression Vector pRSET-B

1. Digest clone pcEP124 (*see* **Note 6**) with *Eco*RI to release a 843-bp cDNA fragment. Separate this fragment from the plasmid by agarose gel electrophoresis and purify it using your preferred technique.
2. Digest plasmid pRSET-B (Invitrogen, San Diego, CA) with *Eco*RI (*see* **Note 7**). Confirm the success of the digestion by agarose gel electrophoresis and dephosphorylate with calf intestinal phosphatase to prevent religation of the vector. Extract the digest once with an equal volume of phenol-chloroform to inactivate the phosphatase and collect the DNA by ethanol precipitation.
3. Ligate approx 100 ng of the vector DNA and 30 ng of the cDNA fragment. Set up in parallel a control ligation lacking the cDNA fragment to confirm the success of the dephosphorylation reaction.
4. Transform competent NM522 *E. coli* cells with the ligation mix and select the transformed colonies on LB plates supplemented with 100 µg/mL ampicillin. Confirm the presence of the recombinant plasmid and the orientation of the insert by digesting plasmid DNA isolated from different colonies with *Eco*RV and *Hin*dIII. Clones containing the insert in the correct orientation yield two DNA fragments of 546 and 3209 bp.
5. Select the appropriate recombinant clones and confirm that the plasmid contains the cDNA sequence in the correct reading frame by sequencing the sense strand using the T7 primer.
6. For expression studies, transform competent BL21(DE3) *E. coli* cells (*see* **Note 8**) with the selected recombinant plasmid.

3.2.2. Expression and Solubilization of Recombinant PPX-1 Protein

1. Inoculate 3 mL of LB medium supplemented with 100 µg/mL ampicillin with a freshly isolated colony of BL21(DE3) cells harboring the expression plasmid. Grow overnight at 37ºC with shaking.
2. On the following day, inoculate 100 mL of LB medium supplemented with 200 µg/mL ampicillin with 1 mL of the overnight culture and incubate at 37ºC with shaking. Monitor the growth at 30-min intervals by measuring optical den-

sity at 600 nm. When an optical density value of 0.7–0.8 is reached, cells are ready to induce the expression of the recombinant protein.

3. To induce the expression add IPTG to a final concentration of 0.4 m*M* and incubate at 37ºC with shaking for 4–5 h. Chill the flask rapidly on ice.

4. Harvest cells by centrifugation at 7000*g* for 5 min at 4ºC. Discard the supernatant and thoroughly resuspend cell pellets in 20 mL of ice-cold sonication buffer. Divide cells in 5-mL aliquots.

5. Disrupt cells of 5-mL aliquots by sonication at 12–15 microns for 2.5 min at 30 s intervals in an ultrasonic disintegrator (MSE, 60W) while being chilled in a –10ºC bath.

6. Centrifuge at 15,000*g* for 15 min to pellet the inclusion bodies, which contain the insoluble recombinant PPX-1 protein. Remove the supernatant carefully and keep pellets on ice. Pellets may be stored frozen at –20ºC at this point.

7. Thoroughly resuspend each pellet in 5 mL of washing buffer and incubate 15–20 min on ice. Centrifuge at 15,000*g* for 15 min and discard the supernatant. Keep pellets containing the inclusion bodies on ice.

8. To solubilize the recombinant PPX-1 protein, resuspend each pellet in 5 mL of solubilization buffer and incubate on ice for 30 min. Centrifuge at 15,000*g* for 15 min. Remove the supernatant to a clean centrifuge tube and centrifuge again at 15,000*g* for 15 min. Transfer the supernatant from the second centrifugation to a clean flask, carefully avoiding any of the pellet, and place on ice. Divide the supernatant into aliquots (1 mL) and either process immediately or store at –20ºC (*see* **Note 9**).

3.2.3. Preparative SDS-PAGE

1. To purify the recombinant PPX-1 protein the authors use preparative denaturing 12.5% polyacrylamide slab gel electrophoresis (SDS-PAGE) performed by the method of Laemmli *(23)*. For preparative SDS-PAGE they routinely use a BIO-RAD PROTEAN II xi Cell with 16 cm × 16 cm × 1.5 mm gels, which are cast with 15 sample wells (each 6 mm × 17 mm × 1.5 mm and 2.5 mm apart).

2. Add 1 mL of 2X SDS-PAGE sample buffer to 1 mL of the supernatant from **step 8** (**Subheading 3.3.**), heat at 90–100ºC for 5 min and load about 130 µL of sample per well. Perform electrophoresis at 15ºC with water cooling at a constant current of 15 mA in the stacking gel (5% polyacrylamide) and 30 mA in the separating gel (12.5% polyacrylamide). Run gels until the tracking dye reaches the front (5–5 1/2 h).

3. After electrophoresis, visualize the recombinant PPX-1 protein by precipitation of the SDS-PPX-1 complex in the presence of potassium ions. To this end, wash the gel in several changes of distilled water and carefully place the gel in a shallow tray containing cold 0.25 *M* KCl. Cover the tray and shake gently at 4ºC only as long as is necessary to identify the desired band (30–60 min), which should be visible as a pale white band against a dark background. No destaining is required.

3.2.4. Electroelution of Recombinant PPX-1 Protein

1. Excise the band corresponding to recombinant PPX using a scalpel and soak in distilled water for 10–15 min. Drain off water and either process immediately or

store at –20ºC in a sealed plastic tube. Stain the remainder of the gel with Coomassie blue to check the accuracy of excision.

2. Place gel slice in a Petri dish and cover with 10–20 mL of distilled water. Dice the gel into approx 1.5 mm cubes (do not mash). Remove water by suction and/or blotting with filter paper. Soak 5 min in 10–20 mL elution buffer and then remove buffer by suction and/or blotting.

3. For electroelution of the protein the author uses the ISCO model 1750 electrophoretic sample concentrator. Transfer the gel pieces to the larger well (sample well) of the electrophoretic elution cell, which has been fitted with the appropriate molecular weight cutoff dialysis disks (*see* **Note 10**). Carefully fill the cell with soaking buffer. Place the elution cell(s) into the elution tank and fill the tank with elution buffer in both sides. Connect electrodes (cathode near sample well) and electroelute (set the current at 10 mA/elution cell) for about 8 h. The author routinely uses 2 elution cells to electroelute the protein from one preparative gel (*see* **Subheading 3.4.**).

4. The protein will have migrated to the sample collection well. Carefully remove the buffer, which is not in the conical part of the sample collection well. Being very careful not to damage the membranes and using a pipet with a disposable plastic tip, mix the protein solution in the collection well (0.20–0.25 mL) and transfer this solution to a 1.5-mL Eppendorf tube. Rinse the well with 50 µL of fresh buffer and add this to the protein collected. Store the protein sample at –20ºC.

5. Transfer the solution containing the electroeluted protein to dialysis tubing (molecular weight cutoff value of 12,000) that has previously been soaked in PBS. Dialyze the protein sample against 2 L of cold PBS. Change the buffer 2 times at 2-h intervals. Determine the protein concentration by the method of Bradford *(24)* (*see* **Note 11**) and check the purity of the recombinant PPX-1 protein by SDS-PAGE followed by Coomassie blue staining (*see* **Note 12**).

3.2.5. Production of Antibodies Against PPX

1. Antibodies against Arabidopsis PPX are raised into male New Zealand white rabbits. A 2 kg rabbit is injected with 1–1.5 mL of protein antigen solution containing 0.1 mg of recombinant PPX-1 in PBS after it has been emulsified with an equal volume of complete Freund's adjuvant (Difco, Detroit, MI) (*see* **Note 13**).

2. Give two additional 0.1-mg booster injections at 4-wk intervals, as on d 0 with the exception that incomplete Freund's adjuvant (Difco) is used instead of complete Freund's adjuvant.

3. Collect venous blood from the ear (use 30-mL Corex tubes) 14 d after the last injection. If more antiserum is required the rabbit may be given another injection of antigen, exactly as described in step 2, and bleed 14 d later.

4. Allow blood to clot for 1 h at 37ºC. Separate clot from the sides of the tube using a Pasteur pipet and left overnight at 4ºC to allow it to contract. On the following day, centrifuge at 10,000g for 10 min at 4ºC and collect the supernatant.

5. To obtain a crude antibody preparation, mix slowly a volume of serum (V) with 0.6V of saturated ammonium sulfate solution and stir gently for 30 min at 4ºC. The precipitate is spun down at 10,000g for 15 min.

6. Wash the pellet with a 50% of saturation ammonium sulfate solution made up in PBS. Repeat washing (3–4 times) until the precipitate is white. The final precipitate is dissolved in 0.5V of PBS.
7. Dialyze the crude antibody solution against 2 L of cold PBS. Change the buffer two times at 2-h intervals. Centrifuge the dialyzed antibody solution at 10,000g for 10 min. Divide the supernatant in 0.5 mL aliquots and store at –20ºC (*see* **Note 14**).

3.2.6. Immunodetection of PPX in A. thaliana *Protein Extracts*

3.2.6.1. PREPARATION OF PROTEIN EXTRACTS

1. Collect *A. thaliana* seedlings (light-grown 6-d-old), divide in aliquots of 1 g and freeze in liquid nitrogen. Plant material may be processed immediately or stored at –80ºC until used.
2. Grind 1 g of plant material to a fine powder in a mortar precooled in liquid nitrogen. Transfer to an Eppendorf tube and lyophilize. Do not allow sample to thaw at any time.
3. To delipidize, add 1 mL of an acetone:hexane mixture (59:41) to the lyophilized powder (10–15 mg) and shake vigorously. Centrifuge at 10,000g for 5 min at room temperature and discard the supernatant. Extract sample twice more with 1 mL of acetone:hexane and once with 1 mL of acetone.
4. Carefully remove acetone and allow the remaining of the acetone to evaporate at room temperature (60 min).
5. Resuspend the dried pellet in 0.5 mL of extraction buffer and shake gently for 1 h 30 min at room temperature. Centrifuge at 10,000g for 5 min and transfer the supernatant to a clean tube, carefully avoiding any of the pellet.
6. Determine protein concentration by the method of Bradford *(24)* (usually 0.5–1.0 mg/mL). Apply appropriate aliquots of the supernatant (usually 10–20 µL) to the SDS-PAGE gel.

3.2.6.2. IMMUNOBLOTTING

1. Transfer proteins by electroblotting from the resulting gel onto ECL-nitrocellulose membranes (Amersham, Arlington Heights, IL) by the method of Towbin *(25)*.
2. Block nonspecific binding to the blot with a 10% solution of non fat dried milk in PBS overnight. On the following day rinse the blot with PBS.
3. Incubate with anti-PPX antibody (usually 20 mL of a 1:1000 dilution in PBS + 5% bovine serum albumin) for 1 h at room temperature with shaking.
4. Wash the blot twice for 15 min with 50 mL of PBS.
5. For detection, the author routinely uses the chemiluminescent ECL Western blotting system from Amersham. Incubate with a secondary antibody enzyme conjugate (usually 20 mL of a 1:25,000 dilution in PBS + 5% bovine serum albumin) for 1 h at room temperature with shaking.
6. Wash the blot twice for 15 min with 50 mL of PBS, incubate with the appropriate substrate according to the manufacturer's recommendations, and expose the blot to autoradiography film.

4. Notes

1. The oligonucleotides described here correspond to conserved sequences in PP1, PP2A, and PP-2B located near the C-terminus of type 1 and 2A proteins (*see* **ref. 26** for sequence comparison). They have been successfully used for amplification of phosphatase DNA fragments from a large variety of sources (from human to yeast, see also **refs. 27–29**). Although the reagents and conditions described here correspond exactly to those successfully used a few years ago in the author's laboratory, since then reagents and equipments have evolutioned. Therefore, the following points should be taken into account. For instance, the extra restriction enzyme sites included in oligonucleotides MPs and MPas were designed years ago to allow easy cloning when blunt-end cloning of PCR fragments was a difficult task. Currently, they can be eliminated since new types of thermostable polymerase (or mixtures of them) allow efficient blunt-end cloning. Expand™ (from Boehringer Mannheim) performs well under conditions similar to those used for standard Taq DNA polymerase. Alternatively, the use of standard Taq DNA polymerase can be combined with specific vectors carrying 3'-T overhangs at the insertion site (as pGEM®-T, from Promega, Madison, WI). Of course, if the extra restriction sites are not included in the oligonucleotide, **steps 3** and **4** in **Subheading 3.1.** must be modified accordingly. In fact, direct cloning of the PCR fragments is preferable, since incubation with restriction enzymes might result in internal cleavage of some PCR fragments.

2. Stock solutions of protease inhibitors: Aprotinin 1 mg/mL in H_2O (store at −20ºC). E64 3 mg/mL in H_2O (store at −20ºC). Leupeptin 10 mM in H_2O (prepared fresh, do not store). Pepstatin 3 mg/mL in methanol (store at −20ºC). PMSF 100 mM in isopropanol (store at room temperature protected from light).

3. Since genomic DNA is used in this protocol, in some cases the presence of introns prevents the identification of certain amplification bands as putative phosphatases. This is probably the case for PP2A-3 and PP2A-4 since, in these genes, the exons containing the sequences corresponding to MPs and MPas are separated by two introns *(30)*. In the case of PPX-2 (as in PPX-1) both oligonucleotides hybridize to sequences within the same exon (Pujol et al., unpublished). The recovery of PPX-1, but not of PPX-2, sequences can be explained by the lower level of identity between MP-1s and the PPX-2 sequence. In any case, MP-1s and MP-3as must be considered as "multipurpose" oligonucleotides, suitable for phosphatase amplification in different organisms. For instance, the author has recently amplified type 1 and 2A fragments from alfalfa cDNA (Vissi et al. unpublished). However, if the amplification of one or more specific type 2A or type X phosphatases from *Arabidopsis thaliana* is desired, the proposed oligonucleotides should be modified accordingly the published sequences to fit the specific needs.

4. The step at 72°C can be probably reduced to 60–90 s if a fast automated thermal cycler is used.

5. Alternatively, low molecular weight bands can be nicely resolved using a 4% NuSieve® 3:1 agarose or the low melting NuSieve GTG agarose gels (FMC BioProducts, Rockland, ME).

6. Clone pcEP124 contains a cDNA encoding the *Arabidopsis* catalytic subunit of protein phosphatase X (PPX-1) *(15)*. The cDNA is cloned into the *Eco*RI site of plasmid pBluescript (Stratagene), and contains three nucleotides of 5'-untranslated region, a 915-bp open reading frame ending with a UAA stop codon, and a 342-bp 3'-untranslated region.

7. Plasmid pRSET-B is a prokaryotic expression vector designed to produce recombinant proteins fused to a short leader peptide from the bacteriophage T7 gene 10 protein. Expression of the DNA sequences cloned in this vector is driven by the bacteriophage T7 promoter.

8. Strain BL21(DE3) has the advantage that, as a B strain, it is deficient in the *lon* protease, and it also lacks the *ompT* outer membrane protease that can degrade proteins during purification. Thus, the expressed proteins might be expected to be more stable in BL21(DE3) strain than in host strains that contain these proteases. Bacteriophage DE3 is a λ derivative that has the immunity region of phage 21 and carries a DNA fragment containing the *lacI* gene, the *lacUV5* promoter, the beginning of the *lacZ* gene and the gene for T7 RNA polymerase. Once a DE3 lysogen is formed, the only promoter known to direct transcription of the T7 RNA polymerase gene is the lacUV5 promoter, which is inducible by IPTG. Addition of 0.4 m*M* IPTG to a growing culture of BL21(DE3) lysogen induces T7 RNA polymerase, which in turn transcribes the target DNA in the plasmid.

9. To confirm the success of both the expression and the solubilization of the recombinant PPX-1 protein, analyze samples from each step by SDS-PAGE. After Coomassie blue staining of the gel, a prominent protein band migrating with an apparent molecular mass of 36.5 kDa should be observed. The recombinant PPX-1 protein contains a leader peptide of 49 amino acid residues fused in frame to the N-terminal 280 amino acid residues of the *Arabidopsis* PPX-1. This represents a truncated version of the catalytic subunit of PPX-1 lacking 25 amino acid residues from the C-terminal end.

10. Use dialysis tubing with a molecular weight cutoff value of 12,000. Cut the tubing into approx 15-cm lengths, soak in elution buffer for 1 h and rinse with water. Lay the tubing on a paper towel and cut disks using a cork borer (No. 8 for the small cell and no. 11 for the large). Discard the disks in contact with the towel, wash the upper disks with water and store them in elution buffer supplemented with 0.02% NaN$_3$ at 4°C.

11. Dialyze in a single dialysis bag the protein electroeluted from four preparative gels (2–2.5 mL). The authors usually recover an average of 150 μg of recombinant PPX-1 protein from each preparative SDS-PAGE.

12. The electroeluted protein should migrate as a single band. However, some minor protein bands of lower molecular mass, resulting from proteolytic degradation of the PPX-1 protein, have been be detected in some experiments.

13. To emulsify complete Freund's adjuvant with the protein solution, the constituents are pumped in and out of a 3-mL glass hypodermic syringe fitted with a 19-gage needle until a thick white emulsion is formed. The stability of the emulsion can be tested by dropping it onto water. Drops should float on the sur-

face of the water and not disperse. The emulsion is injected subcutaneously into the back of the rabbit (0.2–0.3 mL of emulsion at each site). It is a safe policy to immunize at least two rabbits in parallel.

14. Antibody solutions should not be repeatedly frozen or thawed, as this can lead to aggregation of the immunoglobulins.

References

1. Mackintosh, C. and Cohen, P. (1989) Identification of high levels of type 1 and type 2A protein phosphatases in higher plants. *Biochem. J.* **262,** 335–339.
2. Mackintosh, C., Coggins, J., and Cohen, P. (1990) Plant protein phosphatases. Subcellular distribution, detection of protein phosphatase 2C and identification of protein phosphatase 2A as the major quinate dehydrogenase phosphatase. *Biochem J.* **273,** 733–738.
3. Jagiello. I., Donella-Deana, A., Szczegielniak, J., Pinna, L. A., and Muszynska, G. (1992) Identification of protein phosphatase activities in maize seedlings. *Biochem. Biophys. Acta* **1134,** 129–136.
4. Smith, R. D. and Walker, J. C.(1993) Expression of multiple type 1 phosphoprotein phosphatases in *Arabidopsis thaliana*. *Plant Mol. Biol.* **21,** 307–316.
5. Arundhati, A., Feiler, H., Traas, J., Zhang, H., Lunness, P. A., and Doonan, J. H. (1995) A novel *Arabidopsis* type 1 protein phosphatase is highly expressed in male and female tissues and functionally complements a conditional cell cycle mutant of *Aspergillus*. *Plant J.* **7,** 823–834.
6. Arino, J., Perez-Callejon, E., Cunillera, N., Camps, M., Posas, F., and Ferrer, A. (1993) Protein phosphatases in higher plants: multiplicity of type 2A phosphatases in Arabidopsis thaliana. *Plant Mol. Biol.* **21,** 475–485.
7. Casamayor, A., Perez-Callejon, E., Pujol, G., Arino, J., and Ferrer, A. (1994) Molecular characterization of a fourth isoform of the catalytic subunit of protein phosphatase 2A from Arabidopsis thaliana. *Plant Mol. Biol.* **26,** 523–528.
8. Stone, S. R., Hoofstenge, J., and Hemmings, B. A. (1987) Molecular cloning of cDNAs encoding two isoforms of the catalytic subunit of protein phosphatase 2A. *Biochemistry* **26,** 7215–7220.
9. Ariño, J., Woon, C. W., Brautigan, D. L., Miller, T. B., and Johnson, G. L. (1988) Human liver phosphatase 2A: cDNA and amino acid sequence of two catalytic subunits isotypes. *Proc. Natl. Acad. Sci. USA* **85,** 4252–4256.
10. Erondu, N. E. and Donelson, J. E. (1991) Characterization of trypanosome protein phosphatase 1 and 2A catalytic subunits. *Mol. Biochem. Parasitol.* **49,** 303–314
11. Orgad, S., Brewis, N. D., Alphey, L., Axton, J. M., Dudai, Y., and Cohen, P. T. W. (1990) The structure of protein phosphatase 2A is as highly conserved as that of protein phosphatase 1. *FEBS Lett.* **275,** 44–48.
12. Cormier, P., Osborne, H. B., Bassez, T., Poulhe, R., Bellé, R., and Mulner-Lorillon, O. (1991) Protein phosphatase 2A from Xenopus oocytes: characterization during meiotic cell division. *FEBS Lett.* **295,** 185–188.
13. Stamey, R. T. and Rundle, S. J. (1995) Characterization of a novel isoform of a type 2A serine/threonine protein phosphatase from Arabidopsis thaliana. *Plant Physiol.* **110,** 335.

14. Brewis, N. D., Street, A. J., Prescott, A. R., and Cohen, P. T. (1993) PPX, a novel protein serine/threonine phosphatase localized to centrosomes. *EMBO J.* **12**, 987–996.
15. Perez-Callejon, E., Casamayor, A., Pujol, G., Clua, E., Ferrer, A., and Arino, J. (1993) Identification and molecular cloning of two homologues of protein phosphatase X from Arabidopsis thaliana. *Plant Mol. Biol.* **23**, 1177–1185.
16. Leung, J., Bouvier-Durand, M., Morris, P-C., Guerrier, D., Chefdor, F., and Giraudat, J. (1994) Arabidopsis ABA response gene ABI1: Features of a calcium-modulated protein phosphatase. *Science* **264**, 1448–1452.
17. Meyer, K., Leube, M. P., and Grill, E. (1994) A protein phosphatase 2C involved in ABA signal transduction in Arabidopsis thaliana. *Science* **264**, 1452–1455.
18. Evans, I. M., Fawcett, T., Boulter, D., and Fordham-Skelton, A. P. (1994) A homologue of the 65 kDa regulatory subunit of protein phosphatase 2A in early pea (Pisum sativum L.) embryos. *Plant Mol. Biol.* **24**, 689–695.
19. Slabas, A. R., Fordham-Skelton, A. P, Fletcher, D., Martinez-Rivas, J. M., Swinhoe, R., Croy, R. R., and Evans, I. M. (1994) Characterization of cDNA and genomic clones encoding homologues of the 65 kDa regulatory subunit of protein phosphatase 2A in *Arabidopsis thaliana*. *Plant Mol. Biol.* **26**, 1125–1138.
20. Rundle, S. J., Harding, A. J., Corum, J. W., and O'Neill, M. (1995). Characterization of a cDNA encoding the 55 kDa B regulatory subunit of *Arabidopsis* protein phosphatase 2A. *Plant Mol. Biol.* **28**, 257–266.
21. Corum, J. W., Harding, A. J., Stamey, R. T., and Rundle, S. J. (1996) Characterization of DNA sequences encoding a novel isoform of the 55 kDa B regulatory subunit of the type 2A serine/threonine phosphatase of *Arabidopsis thaliana*. *Plant Mol. Biol.* **31**, 419–427.
22. Csortos, C., Zolnierowicz, S., Bakó, E., Durbin, S. D., and DePaoli-Roach, A. A. (1996) High complexity in the expression of the B' subunit of protein phosphatase 2A. Evidence for the existence of at least seven novel isoforms. *J. Biol. Chem.* **271**, 2578–2588.
23. Laemmli, U. K. (1970) Cleavage of structural proteins during the assembly of the head of bacteriophage T4. *Nature* **227**, 680–685.
24. Bradford, M. (1976) A rapid and sensitive method for quantitation of microgram quantities of protein utilizing the principle of protein-dye binding. *Anal. Biochem.* **72**, 248–254.
25. Towbin, H., Staehelin, T., and Gordon, J. (1979) Electrophoretic transfer of proteins from polyacrylamide gels to nitrocellulose sheets: procedure and some applications. *Proc. Nat. Acad. Sci. USA* **76**, 4350–4354.
26. Jenny, T. F., Gerloff, D. L., Cohen, M. A., and Benner, S. A. (1995) Predicted secondary and supersecondary structure for the Serine/Threonine-specific protein phosphatase family. *Proteins: Structure, Fuctions and Genetics* **21**, 1–10.
27. Wadzinsky, B. E., Heasley, L. E., and Johnson, G. L. (1990) Multiplicity of protein serine-threonine phosphatases in PC12 pheochromocytoma and FTO-2B hepatoma cells. *J. Biol. Chem.* **265**, 21,504–21,508.
28. Posas, F., Clotet, J., Muns, M. T., Corominas, J., Casamayor, A., and Ariño, J. (1993) The gene PPG encodes a novel yeast protein phosphatase involved in glycogen accumulation. *J. Biol. Chem.* **268**, 1349–1354.

29. Posas, F., Casamayor, A., Morral, N., and Ariño, J. (1992) Molecular cloning and analysis of a yeast protein phosphatase with an unusual amino-terminal region. *J. Biol. Chem.* **267,** 11,734–11,740.
30. Pérez-Callejón, E., Casamayor, A., Pujol, G, Camps, M., Ferrer, A., and Ariño, J. (1998) Molecular cloning and characterization of two phophatase 2A catalytic subunit genes from *Arabidopsis thaliana. Gene* (in press).

17

Separation of Protein Phosphatase Type 2C Isozymes by Chromatography on Blue Sepharose

Susanne Klumpp and Dagmar Selke

1. Introduction

Unlike the other serine/threonine protein phosphatases (for reviews *see* **refs.** *1–3*), type 2C (PPM) is a monomeric enzyme and its activity depends on the presence of Mg^{2+} *(4)*. Two major isoforms of protein phosphatase type 2C (PP2C) were isolated from various vertebrate organisms and tissues: PP2Cα and PP2Cβ (formerly called $PP2C_1$ and $PP2C_2$) with a molecular mass of 43–48 kDa *(5,6)*. The isoforms cannot be distinguished by substrate specificity. In addition, neither a specific activator nor an inhibitor is known.

Ion exchange chromatography on Mono Q is an established method to separate PP2Cα and PP2Cβ. It works fast, reproducibly, and, is very efficient for the PP2C proteins from rabbit skeletal muscle and liver *(5)*. The versatility of this method is based on the differences in the net charge of the proteins, e.g., the isoelectric point (IP) of PP2Cα from rat liver is 4.98, whereas that of PP2Cβ from the same tissue source is 4.62. Consequently, the proteins can be separated by chromatography on Mono Q. In contrast to those values for the isozymes from rat liver, the IP values of PP2C isozymes form bovine retina do not differ significantly (PP2Cα: 4.87, PP2Cβ: 4.82). Therefore, separation on Mono Q is not applicable in general. As an alternative, chromatography on Blue Sepharose in the presence of Mg^{2+} proved a novel technique to separate PP2Cα and PP2Cβ proteins.

In Blue Sepharose, the dye Cibachron Blue F3G-A is immobilized to an agarose-based matrix *(7)*. The concentration of the coupled dye is 7–8 μmole/mL drained gel *(7)*. In general, proteins with affinity to Blue Sepharose can be subdivided into two main categories: Enzymes requiring adenylyl-containing

From: *Methods in Molecular Biology, Vol. 93: Protein Phosphatase Protocols*
Edited by: J. W. Ludlow © Humana Press Inc., Totowa, NJ

cofactors (e.g., kinases, dehydrogenases) that are eluted by NAD⁺ or NADP⁺; and nonenzymic proteins (e.g., albumin, interferons), which are eluted by increasing salt concentration. It is known from X-ray crystallography that Cibachron Blue binds to the nucleotide-binding pocket of horse liver alcohol dehydrogenase.

This chapter describes the experimental details to separate bovine retinal PP2Cα and PP2Cβ isozymes by chromatography on Blue Sepharose. The authors decided to study the interaction of PP2C with Cibachron Blue mainly for two reasons: The general knowledge of Mg^{2+} being important for nucleotide-binding and the observation that many nucleotide-binding proteins do have affinity to Blue Sepharose; and PP2C strictly requires Mg^{2+} for activity, suggesting the presence of at least one Mg^{2+}-binding site within the polypeptide chain.

2. Materials

2.1. Chemicals

1. Blue Sepharose HiTrap (Pharmacia, Uppsala, Sweden).
2. The supply of the other chemicals is not considered critical.

2.2. Equipment

1. FPLC equipped with a LCC-500 programmer, two P-500 dual piston pumps, an MV-7 automated injection valve, solvent mixer, prefilter, sample loops, UV-monitor, and recorder (Pharmacia).
2. Vertical electrophoresis apparatus, sodium dodecyl sulfate polyacrylamide gel elctrophoresis (SDS-PAGE) minigels, and power supply.
3. Liquid scintillation counter.

2.3. Solutions

The pH of buffers was adjusted at room temperature (*see* **Notes 1–3**).

1. Column buffer A: 20 m*M* Tris-HCl, pH 7.0, 40 m*M* NaCl, 0.1 m*M* EDTA, 5% glycerol (v/v), 0.1% β-mercaptoethanol (v/v), 0.02% NaN₃ (w/v).
2. Column buffer B: Column buffer A plus 3 m*M* $MgCl_2$.
3. Regeneration buffer I: 0.1 *M* Tris-HCl, pH 8.5, 0.5 *M* NaCl.
4. Regeneration buffer II: 0.1 *M* sodium acetate, pH 4.5, 0.5 *M* NaCl.
5. Regeneration buffer III: 6 *M* urea.
6. Storage solution: 20% ethanol.
7. PP2C: e.g., after gel filtration in 20 m*M* Tris-HCl, pH 7.0, 40 m*M* NaCl, 1 m*M* EDTA, 0.1% β-mercaptoethanol, 0.02% NaN₃.

3. Methods

3.1. Blue Sepharose Chromatography (see Notes 4–8)

1. Prepare a 1-mL column (HR 5/5) with Blue Sepharose or use the prepacked ready-to-go HiTrap Blue column.

2. Wash pump A with column buffer A; pump B with column buffer B.
3. Equilibrate column with at least 10 column volumes of column buffer B at a flow rate of 0.75 mL/min.
4. After bulb has been switched on for some time, adjust the baseline (OD_{280}) and wait for steadiness.
   Program FPLC: e.g., program #1 for the actual column run and program #2 as subcommands for the application procedure.

#1)	0 mL	B = 100		#2)	0 mL	hold
		portset 6.1				valve 1.2
		mL/min = 0.75				clear data
		cm/mL = 0.3				monitor 1
		call #2)				level % 5
	5 mL	B = 100				mL/mark 3
		B = 70				integrate 1
	10 mL	B = 70				
	70 mL	B = 0			5 mL	valve 1.1
	75 mL	B = 0				

5. Inject protein (e.g., in this case, #2: 5 mL) and start program #1 (*see* **Notes 9–11**).
6. Collect fractions of 1 mL and monitor OD_{280}.
7. Continue eluting until OD_{280} readings have returned to baseline value or at least remain constant.
8. May assay fractions immediately for PP2C activity as outlined in detail *(4)*.
9. May assay fractions by denaturing SDS-PAGE (preferentially minigel size 89 × 100 × 0.5 mm), stain with Coomassie Blue or $AgNO_3$. For silver-staining the method of Blum et al. is recommended *(8)*.
10. Wash column with 10 volumes of regeneration buffer I, followed by 5 mL of water (*see* **Note 12**).
11. Wash column with 10 volumes of regeneration buffer II, followed by 5 mL of water (*see* **Note 13**).
12. Wash column with 5 volumes of regeneration buffer III. Continue washing with 10 mL water.
13. Equilibrate column in 20% ethanol and store at 4–6°C (*see* **Note 14**).

3.2. Elution Profile of PP2C

1. Blue Sepharose proved a useful and novel technique to separate PP2C isozymes (*see* **Note 15**).
2. The proteins were retained by the resin in the presence of Mg^{2+}.
3. Differential elution of native PP2Cα and PP2Cβ was achieved by gradually reducing the concentration of the divalent cation (*see* **Note 16**).
4. A first peak of PP2C activity eluted from Blue Sepharose at 1.5 m*M* $MgCl_2$ (**Fig. 1A**). Silver-stained SDS-PAGE showed a protein of 43 kDa (**Fig. 1B**). Detection (presence and intensity) of this protein correlated with PP2C activity in the corresponding fractions (*see* **Note 17**).
5. A second peak of PP2C activity eluted at 0.5 m*M* $MgCl_2$ (**Fig. 1A**). Silver-stained SDS-PAGE revealed the presence of a 45 kDa protein (**Fig. 1B**).

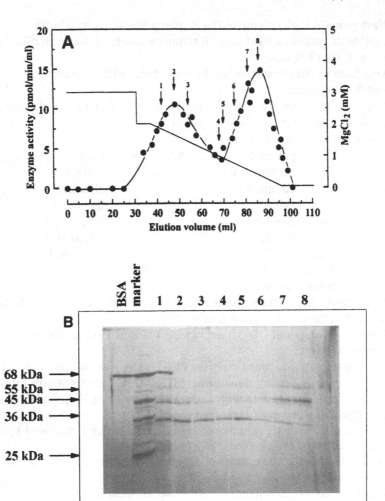

Fig. 1. Affinity chromatography of bovine retinal PP2C isozymes on HiTrap Blue
(A), and analysis of the Blue Sepharose fractions by 12.5% SDS-PAGE **(B)**. After
binding of PP2C (3 mM Mg^{2+}) and washing (2 mM Mg2+) the column (1 mL) was
developed with a linear gradient (60 mL; 2 mM Mg^{2+}–0 mM Mg^{2+}). The flow rate was
0.75 mL/min and fractions of 1 mL were collected. Numbering in (A) refers to the
fractions analyzed by silver-stained minigels in (B). Molecular mass markers (0.1 μg,
respectively) were bovine serum albumin (68 kDa), glutamate dehydrogenase
(55 kDa), ovalbumin (45 kDa), glyceraldehyde-phosphate dehydrogenase (36 kDa),
and triosephosphate isomerase (27 kDa).

6. A 36 kDa protein was also retained by Blue Sepharose (**Fig. 1B**). This protein,
however, did not show phosphatase activity using either casein or phosphorylase

a as a substrate. Yet, most of this protein was found in the flow through fractions together with many other contaminants (data not shown).

4. Notes

1. Use at least double distilled water for FPLC-solutions.
2. Have 0.02% NaN_3 present in all solutions in order to prevent microbiological contamination.
3. All solutions that are to be passed through FPLC columns must be prefiltered (0.22 µm; cellulose acetate).
4. All operations should be carried out at 4–6°C: Have centrifuge, tubes, and rotor chilled; FPLC columns, valve, and superloop should be within a cold room or refrigerator; keep protein samples always on ice and solutions in a refrigerator. Do not forget to chill the tubes in the fraction collector as well.
5. Remove other ser/thr-protein phosphatases from PP2C prior to chromatography on Blue Sepharose. Both, PP1 and PP2A have affinity to Cibachron Blue and will interfere with PP2C isozyme separation. PP1 and PP2A are efficiently removed from PP2C by chromatography on Heparin Sepharose (PP1) and gel filtration (PP2A).
6. It is recommended to have PP2C partially purified and enriched prior to separation of the isozymes via chromatography on Blue Sepharose. In this case (**Fig. 1**) PP2C from a soluble extract of bovine retinae has been chromatographed on DEAE Sephacel, Heparin Sepharose, Phenyl Sepharose, and Superdex 75. Dealing with retinal tissue, the addition of protease inhibitors is necessary only for the very first step (homogenization in the presence of 0.1 mM phenylmethylsulfonyl fluoride, 1 mM benzamidine).
7. Make sure that PP2C solution to be applied to Blue Sepharose does not contain high salt (maximally 50 mM).
8. If there is low amount of protein available and you have to operate at maximal sensitivity of absorbance range, then performance of a ′test run′ is recommended: Do not apply protein, but run the column with buffers, speed, sensitivity as if sample would be present. You may well discover peaks or baseline shifts that cannot be explained, but have to be considered as intrinsic background. A typical artificial air-bubble peak, in contrast, is recognized by a sudden and maximal increase in absorbance.
9. Spin protein sample prior to loading (10,000*g*, 10 min, 6°C).
10. As little as 15 µg total protein still is sufficient to give reasonable recovery of PP2C isozymes and to yield nice separation. On the other hand, maximal binding capacity is 15 mg human serum albumin per mL of Blue Sepharose.
11. Make sure that the valve positions are correct. Otherwise you might inject the sample into waste rather than loading as intended.
12. For regeneration of Blue Sepharose after the column run: Use the 10 mL superloop to apply the various buffers. This saves time (pump washing) and buffer (pump volume).
13. Note that rinsing the column with water is essential in between the regeneration steps in order to avoid contact of acid with alkaline buffers. If this happens, the

pressure will increase dramatically. In case you forgot the water step, a reduction in flow rate is recommended.

14. Regeneration with urea (or other chaotropic agents) is appropriate to get rid of lipoproteins that tightly bind to Blue Sepharose.
15. A 60 mL gradient for a 1 mL column seems fairly oversized. The standard protocol of 5–10 column volumes, however, did not yield satisfying results. Successful separation of PP2C isozymes strongly depends on an extremely flat gradient.
16. Elution of PP2C is also possible by increasing salt concentration (NaCl). This, however, results in unsatisfactory preparations since many proteins are eluted unspecifically.
17. Binding of PP2C to Blue Sepharose required a minimum of 3 mM Mg^{2+}. The optimal concentration of Mg^{2+}, however, depends on the amount of protein applied. At the beginning, to make sure there is sufficient Mg^{2+} around for PP2C to bind to Blue Sepharose, it is recommended to operate with 10 mM Mg^{2+}.

Acknowledgments

This work was supported by the Deutsche Forschungsgemeinschaft (Kl 601/3-3).

Note Added in Proof

In the meantime, antisera specific for PP2Ca and PP2Cd, respectfully, did confirm separation of PP2C isozymes by Blue Sepharose (Western blotting, data not shown).

References

1. Ingebritsen, T. S. and Cohen, P. (1983) Protein Phosphatases: properties and role in cellular regulation. *Science* **221,** 331–337.
2. Cohen, P. (1989) The structure and regulation of protein phosphatases. *Annu. Rev. Biochem.* **58,** 453–508.
3. Wera, S. and Hemmings, B. A. (1995) Serine/threonine protein phosphatases. *Biochem. J.* **311,** 17–29.
4. McGowan, C. H. and Cohen, P. (1988) Protein phosphatase 2C from rabbit skeletal muscle and liver: a Mg^{2+}-dependent enzyme. *Meth. Enzymol.* **159,** 416–426.
5. McGowan, C. H. and Cohen, P. (1987) Identification of two isozymes of protein phosphatase 2C in both rabbit skeletal muscle and liver. *Eur. J. Biochem.* **166,** 713–722.
6. McGowan, C. H., Campbell, D. G., and Cohen, P. (1987) Primary structure analysis proves that protein phosphatase 2C1 and 2C2 are isozymes. *Biochim. Biophys. Acta* **930,** 279–282.
7. Pharmacia Biotech: Instruction manual ´Blue Sepharose,´ flyer ´HiTrap Blue,´ and brochure ´Affinity Chromatography.´ Pharmacia Biotech AB, Uppsala, Sweden.
8. Blum, H., Beier, H., and Gross, H. J. (1987) Improved silver staining of plant proteins, RNA and DNA in polyacrylamide gels. *Electrophoresis* **8,** 93–99.

18

Chromatographic Isolation
of PP2A from *Limulus* Lateral Eyes

Conventional and Small Scale Methods

Samuel C. Edwards, Travis B. Van Dyke, Timothy H. Van Dyke, and David L. Brautigan

1. Introduction

A major component of the biochemical mechanisms involved in the photoresponse in both invertebrate and vertebrate organisms is the phosphorylation and dephosphorylation of specific protein intermediates whose phosphorylation state is transitory in response to light, darkness, and other stimuli (*see* **ref.** *1* for review), and results from changes in the relative activities of specific protein kinases and phosphatases. Of the known protein serine\threonine protein phosphatases, type 2A (PP2A) dephosphorylates many of the identified phosphoproteins involved in the light-dependent processes in photoreceptors from both vertebrates and invertebrate organisms. For example, in vertebrates rod photoreceptors, dephosphorylation of rhodopsin by PP2A *(2–5)* is necessary for the restoration of increased light sensitivity (dark adaptation).

Photoreceptors from the ventral and lateral eyes of the horseshoe crab, *Limulus polyphemus*, have been used extensively to elucidate the basic biochemical and physiological mechanisms involved in the photoresponse (*see* **ref.** *6* for review). The authors have evidence that in *Limulus* photoreceptors, the activities of PP2A, PP1, and PP2B (calcineurin) are all regulated by light *(7,8)*. Moreover, PP2A dephosphorylates arrestin and reduces its affinity for membranes of *Limulus* photoreceptor cells *(9)*. In eyes from *Limulus*, and also *Drosophila* and *Calliphora*, arrestin is phosphorylated by a calcium/calmodulin-dependent protein kinase II upon illumination

From: *Methods in Molecular Biology, Vol. 93: Protein Phosphatase Protocols*
Edited by: J. W. Ludlow © Humana Press Inc., Totowa, NJ

(10–12). It is believed that the phosphorylation of arrestin leads to the formation of an inactive arrestin/rhodopsin complex (light-adapted). Therefore, it is likely that PP2A is involved in regulating the photo-response in these photoreceptors.

PP2A has been isolated from a number of tissues in the form of many different high molecular weight complexes *(13–15)*. The basic holoenzyme in mammalian tissue consists of a heterotrimer that includes the catalytic subunit (37–38 kDa), a conserved A subunit (60–63 kDa), and a family of diverse B subunits (B, B', or B" at 55 kDa, 54 kDa, or 74 kDa, respectively). The B subunit seems to be responsible for giving the PP2A holoenzymes their specificity toward certain substrates *(15–19)*. Therefore, the cell-specific expression of different PP2A heterotrimers containing different B subunits may occur in different cells or within the same cell, and consequently, may target PP2A for specific cellular functions.

Posttranslational modifications can modulate PP2A activity *(20)*. Phosphorylation of the catalytic and A subunits by a specific autophosphorylation-activated protein kinase produced an 80% decrease in PP2A activity *(21)*. Phosphorylation of tyrosine 307 at the C terminus of the catalytic subunit of PP2A by tyrosine kinases caused up to a 90% decrease in activity *(22)*. These phosphorylations might also affect the association of the subunits in the holoenzyme.

As a first step in understanding the role PP2A in invertebrate visual processes and the mechanism by which light may regulate its activity, the authors have used the following conventional chromatographic methods, i.e., heparin Sepharose affinity chromatography and anion exchange chromatography on Mono Q Sepharose, to isolate four fractions from the *Limulus* lateral eye soluble fraction that contain PP2A-like activity. They have also developed a small-scale method to prepare PP2A-enriched fractions from lateral eyes. Using this method, PP2A from the eyes of one or two animals can be absorbed and eluted from an anion exchange matrix, DEAE cellulose, and subsequently an affinity matrix, thiophosphorylase *a* Sepharose. PP2A activity is measured using radiolabeled glycogen phosphorylase *a* as the substrate in the presence of pharmacological agents that modulate PP2A activity. The presence of PP2A in the fractions eluted from DEAE cellulose were monitored by Western blot analysis with an antibody directed against a unique and conserved region of the PP2A catalytic subunit.

2. Materials

2.1. Preparation of Limulus Lateral Eye Homogenates

1. Adult horseshoe crabs are purchased from the Marine Biological Laboratory, Woods Hole, MA.

2. *Limulus* ringers (LR, from **ref. 23**): 420 mM NaCl, 11 mM KCl, 11 mM CaCl$_2$, 20 mM MgSO$_4$, 30 mM MgCl$_2$, 5 mM NaHCO$_3$, 5 mM MOPS, pH adjusted to 7.5 with NaOH. For use, add glucose (5 mM final concentration) to the stock LR.

3. *Limulus* organ culture medium (LOCM, from **ref. 24**): 16.7 g NaCl, 0.28 g KCl, 1.41 g MgCl$_2$. 6 H$_2$O, 5.6 g MgSO$_4$·7 H$_2$O, 1.17 g CaCl$_2$·2 H$_2$O, 0.12 g HEPES, 0.115 g TES, 2.97 g glucose, 0.0125 g penicillin-G, 100 mL horse serum (Life Technologies, Gaithersburg, MD), 100 mL of Medium 199 (10X) with Hanks salts and L-glutamine without NaHCO$_3$, 800 mL of H$_2$O, adjusted to pH 7.5 with NaOH. Filter sterilize the solution (0.22 μm filter unit, Nalgene). Store 1-mL aliquots at −20°C. After thawing, insoluble material will be present in the LOCM. This can be removed by centrifugation at 13,000g for 5 min.

4. Homogenizing buffer (HB modified from **ref. 25**): 50 mM MOPS, 0.1 mM Na$_2$EDTA, 1 mM EGTA, pH adjusted to 7.2 with NaOH, and stored at 4°C. For use, add 0.1 mM phenylmethylsulfonyl fluoride (PMSF), 0.1 mM leupeptin, and 0.1 mM dithiothreitol (DTT) (final concentrations) to the stock HB.

2.2. Chromatographic Separation of PP2A from PP1

2.2.1. Conventional Method

1. Buffer A (from **ref. 26**): 20 mM Tris base, 1 mM EGTA, 1 mM benzamidine, 5% (v/v) glycerol, 0.1 M NaCl, which is adjusted to pH 7.0 with HCl. Store stock at 4°C. Add 0.1 % (v/v) β-mercaptoethanol and 0.1 mM PMSF (final concentrations) prior to use.

2. Buffer A containing 0.5 M NaCl.

3. Buffer B: 50 mM Tris base, 0.1 mM PMSF, 10% glycerol, 1 mM benzamidine, 1 mM EDTA. Adjusted the pH to 7.4 with HCl. Maintain the stock at 4°C. Add 0.1% (v/v) β-mercaptoethanol and 0.1 mM PMSF (final concentrations) prior to use.

4. Buffer B containing 0.5 M NaCl.

5. Filtration microconcentrators (Centricon 10, Amicon, Danvers, MA).

6. Heparin Sepharose column (Hitrap, 1 mL; Pharmacia, Uppsala, Sweden).

7. Liquid chromatography system, such as GradiFrac System (Pharmacia) or comparable equipment, with a UV detector (280 nM) and a fraction collector.

8. Biocompatible Rainin Dynamax Binary Gradient HPLC (or equivalent) or FPLC (Pharmacia) system equipped with a 10 mL Superloop (Pharmacia), UV detector, and fraction collector.

9. Mono Q FPLC column (HR 5/5; Pharmacia).

2.2.2. Small Scale Preparation of a Lateral Eye PP2A-Enriched Fraction

1. Buffer C: 20 mM Tris base, 0.1 mM Na$_2$ EDTA, which is adjusted to pH 7.2 with HCl.

2. Buffer D (from **ref. 27**): Add 0.2% β-mercaptoethanol, 1 mM PMSF, 0.1 mM leupeptin, 2 μg/mL aprotinin (final concentrations) to buffer C prior to use.

3. Buffer D containing 0.1, 0.2, or 0.3 M NaCl.

4. DEAE-cellulose: diethylaminoethyl-cellulose (medium mesh, Sigma, St. Louis, MO). Hydrate 1 g in buffer C. Wash several times in buffer C. Maintain at 4°C. Prior to use, equilibrate the resin in buffer D.

5. 0.2 μm filter units (Millipore Ultrafree®-MC).

6. Phosphorylation reaction buffer (from **ref. 28**): 250 mM Tris-HCl, pH 8.2 containing 16.7 mM MgCl$_2$, 1.67 mM ATP-γ-S (Boehringer Mannheim, Germany), 0.83 mM CaCl$_2$, and 1.33 mM β-glycerophosphate.

7. Glycogen phosphorylase b: (Sigma); 100 mg/mL in buffer E: 50 mM β-glycerophosphate, 2 mM Na$_2$ EDTA, 0.1 % β-mercaptoethanol, 50% glycerol.

8. Phosphorylase kinase: (Sigma; 20 mg/mL in buffer E).

9. CNBr-activated Sepharose 4B (Pharmacia Biotech).

10. Thiophosphorylase a-Sepharose:
    a. Thiophosphorylase a (10 mg) is synthesized from glycogen phosphorylase b and ATP-γ-S (Calbiochem or Boehringer Mannheim) using the detailed procedures described by Brautigan and Shriner *(29)* or Cohen et al. *(28)*.
    b. Coupling of thiophosphorylase a to Sepharose:
        i. Suspend the thiophosphorylase a in coupling buffer (0.1 M sodium phosphate [pH 8.3] containing 0.5 M NaCl. Then transfer the protein to a filtration microconcentrator (Centricon 10, Amicon). Reduce the volume of the solution to 400 μL by centrifugation (5000g, 20 min, 20°C). Add coupling buffer to restore the volume to 3 mL.
        ii. Add the thiophosphorylase a in coupling buffer in a polypropylene tube to 1 mL of hydrated CNBr-activated Sepharose 4B (in 1 mM HCl). Couple the thiophosphorylase a to the matrix by gentle mixing overnight at 4°C.
        iii. Remove excess ligand by washing the matrix with 5 ml of coupling buffer.
        iv. Block any remaining active groups on the matrix by transferring the gel to 1 M ethanolamine (pH 8.0) and allowing the suspension to stand for 2 h.
        v. Wash the resin with three cycles of exposure to 0.1 M sodium acetate (pH 4.0) containing 0.5 M NaCl, and then 0.1 M Tris-HCl, pH 8.0, containing 0.5 M NaCl.
        vi. Resuspend the resin in 20 mM Tris HCl (pH 7.2) containing 0.1 mM Na$_2$ EDTA and 1 mM Na azide. Maintain at 4°C. Pre-equilibrate an aliquot in buffer D prior to use.

2.3. Identification of PP2A

2.3.1. Phosphorylase a Protein Phosphatase Activity

1. Glycogen phosphorylase b: (Sigma) 100 mg/mL in buffer E: 50 mM β-glycerophosphate, 2 mM Na$_2$EDTA, 0.1% β-mercaptoethanol, 50% glycerol.

2. Phosphorylase kinase: (Sigma) 20 mg/mL in buffer E.

3. Buffer E (from **ref. 28**): 50 mM Tris-HCl, pH 7.0, 0.1 mM EDTA, 15 mM caffeine, 0.1% (v/v) β-mercaptoethanol.

4. γ-^{32}P-ATP (3000–6000 Ci/ mmole; New England Nuclear).

5. Buffer F (from **ref. 28**): 20 mM imidazole-HCl (pH 7.63) containing 0.1 mM Na$_2$EDTA, 1 mg/mL bovine serum albumin (Fraction V), 0.1% β-mercaptoethanol.

6. Ice-cold 20% (w/v) trichloroacetic acid (20% TCA).
7. Protamine sulfate (Sigma).
8. Okadaic acid (LC Laboratories or other vendors).
9. Inhibitor 2 (recombinant form of the heat stable inhibitor of PP1, New England Biolabs, Beverly, MA).
10. Scintillation cocktail (Ecolume, ICN).

2.3.2. Western Blot Analysis for PP2A Catalytic Subunit

1. Vertical gel electrophoresis equipment (such as Bio-Rad Mini-Protean II), transfer apparatus, and power supplies.
2. Electrophoresis grade chemicals from Bio-Rad or other vendors for sodium dodecylsulfate (SDS) polyacrylamide gel electrophoresis and electrophoretic transfers.
3. 0.15% sodium deoxycholate.
4. 72% trichloroacetic acid.
5. Ice-cold acetone.
6. 2 M N-ethylmaleimide (Sigma) in 95% ethanol.
7. Polyvinylidene fluoride (PVDF) transfer membrane (Immobilon-P, Millipore, Bedford, MA).
8. Rabbit polyclonal antiserum generated using synthetic peptides from the carboxy terminus of the catalytic subunit of mammalian PP2A (kindly provided by Dr. Marc Mumby, University of Texas Southwestern Medical Center, also available from Upstate Biotechnology Inc.).
9. Horseradish peroxidase-conjugated goat antirabbit IgG (Promega, Madison, WI).
10. Enhanced chemiluminescence reagents (ECL, Amersham, Arlingotn Heights, IL).

3. Methods

3.1. Preparation of Limulus *Lateral Eye Homogenates*

1. Quickly decerebrate the animal by making two deep cuts through the soft portion of the exoskeleton at the anterior portion of the mouth to severe the circumesophageal ring. Proceed anteriorly to cut the soft exoskeleton and the underlying tissue along the perimeter at the base of the first pair of walking legs to beyond the ventral eye. Then make another deep cut in the soft exoskeleton (*parallel to the anterior of the animal*) to connect the two other incisions. This section of tissue, which includes the brain and the associated ventral eye, which are both embedded in hepatopancreas, can now be removed and placed into chilled (0°C) LR for use in other experiments.
2. Remove the lateral eyes from the animal using a single edged razor blade. *(Be careful that the cornea remain attached to the eye [see* **Note 1**].*)* Carefully make an oblique incision along the dorsal ridge of the eye. Then remove the eye by making another incision along the ventral surface of the eye. While maintaining the blade parallel to the surface of the exoskeleton, proceed toward the dorsal cut to complete excision of the eye. Place the two eyes into chilled (0°C) LR.

3. Depending on the particular experiment, sufficient material may require that eyes are pooled from several animals. Also, eyes may be placed into LOCM and stored overnight at 4°C in the dark prior to proceeding.

4. Allow the eyes to become light- or dark-adapted by maintaining them for 2 h at room temperature in either ambient light or in complete darkness (*see* **Note 2**).

5 While in the light or the dark (aided by an infrared viewer, Finder Scope, FJW Optical Systems, Inc., Elgin, IL), use fine-tipped forceps (Dumont #5) to remove the eye tissue from the cornea.

6. Homogenize the tissue in HB (20 µL/ mg wet weight) using a motor-driven glass-Teflon homogenizer at 0°C. *The extracellular matrix will not break down.* The homogenates from both light- and dark-adapted eyes are then treated in the same manner (*see* **Note 3**).

7. Centrifuge the crude homogenate at 10,000g for 1 min (4°C) to remove the insoluble screening pigment present in the lateral eye as well as any unbroken cells and cell debris.

8. Centrifuge the resulting supernatant at 100,000g for 1 h to prepare the soluble fraction. Alternatively, centrifuge the 10,000g supernatant at 30 psi (130,000g) for 15 min in a Beckman Airfuge.

3.2. Chromatographic Separation of PP2A from PP1

3.2.1. Conventional Method

1. Equilibrate the soluble fraction with Buffer A using an Amicon Centricon 10 microconcentrator by alternatively concentrating and resuspending the sample in Buffer A. In the microconcentrator, centrifuge the sample at 5000g (4°C) to reduce the volume to 500 µL. Then add buffer A to resuspend the sample in a total volume of 2 mL. Repeat this procedure three times.

Perform **steps 2–5** in a 4°C cold room.

2. Using a liquid chromatography system, pre-equilibrate a 1-mL heparin Sepharose Hitrap column (Pharmacia) with three bed volumes of buffer A (flow rate of 900 µL/min. Monitor the eluent from the column using a UV monitor set at 280 nm (*see* **Note 4**).

3. Throughout the chromatographic procedure collect 300 µL fractions. Slowly inject the sample onto the column.

4. When all of the proteins that do not bind to heparin (including PP2A) have passed through the column, which is determined by return of the UV absorbance of the mobile phase to baseline (0 OD units), change the mobile phase to buffer A containing 0.5 *M* NaCl.

5. Assay the fractions for protein phosphatase Type 1 and Type 2A activities using the assay described in **Subheading 3.3.1.** The PP1 activity should be in the 0.5 *M* NaCl fraction, whereas the PP2A will be in the unbound, flow-through fraction.

Perform the following steps at room temperature using a biocompatible HPLC or FPLC system to which is connected a 10-mL Superloop® (Pharmacia) and a Mono Q FPLC column (HR 5/5; Pharmacia) anion exchange column.

6. Pool the flow-through (0.1 M NaCl) fraction from the heparin Sepharose column. Equilibrate this material with buffer B using the washing and concentration procedures in Amicon 10 concentrators described above in **step 1**.
7. The Mono Q FPLC column (HR 5/5; Pharmacia), connected to the HPLC or FPLC should be pre-equilibrated with buffer B (*see* **Note 5**).
8. Place the pooled sample into the HPLC system through a Rheodyne injector equipped with a 10-mL sample loop (Superloop, Pharmacia). Then inject it onto the column at a flow rate of 0.5 mL/min. Collect the eluent in 1 min (500 μL) fractions. Continuously monitor the UV absorbance of the proteins in the eluent at 280 nm. Collect the eluent in 1 min (500 μL) fractions.
9. Use a linear NaCl concentration gradient (0–500 μM) that develops over 85 min to elute proteins from the column.
10. Assay the fractions for protein phosphatase Type 1 and Type 2A activities using the assay described in **Subheading 3.3.1.**
11. In the fractions that demonstrate protein phosphatase activity, determine the amount of protein present by the Bradford assay *(29)* using commercial reagent (Bio-Rad) and bovine serum albumin (BSA) as the standard.

3.2.2. Small Scale Preparation of a Lateral Eye PP2A-Enriched Fraction

1. Add 200 μL of DEAE cellulose to a 4-mL polypropylene tube containing the lateral eye soluble fraction. Incubate the sample with the resin with gentle rocking overnight at 4°C (*see* **Note 6**).
2. Transfer the mixture to the upper chamber of a 0.2 μm filter unit (Millipore Ultrafree®-MC). Centrifuge the filter unit at 3000g for 1 min (4°C) in a microcentrifuge.
3. The lower reservoir contains the eluent with the unbound proteins.
4. Place the upper reservoir, which contains the retained anion exchange matrix, into another microcentrifuge tube. Resuspend the matrix with 400 μL of Buffer D containing 0.1 M NaCl.
5. Centrifuge the filter unit at 3000g for 1 min (4°C). Again, the lower reservoir contains the proteins that are eluted from the resin with 0.1 M NaCl.
6. Repeat **steps 4** and **5** with Buffer D containing 0.1 M NaCl.
7. Repeat **steps 4** and **5** with Buffer D containing 0.3 M NaCl.
8. The 0.3 M NaCl eluent contains PP2A and is mostly devoid of PP1. Dilute 200 μL of this eluent 1:8 with Buffer D to lower the NaCl concentration to below 0.05 M. Add 200 μL of thiophosphorylase a Sepharose. Incubate the sample with the resin with gentle rocking overnight at 4°C.
9. Repeat **steps 2** and **3**.
10. Repeat **steps 4** and **5** with Buffer D.
11. Repeat **steps 4** and **5** with Buffer D containing 0.3 M NaCl.

3.3. Identification of PP2A

3.3.1. Phosphorylase a Protein Phosphatase Activity

1. Prepare $^{32}PO_4$-labeled phosphorylase a from glycogen phosphorylase b using the detailed procedures described by Brautigan and Shriner *(29)* or Cohen et al. *(28)*.

Maintain the $^{32}PO_4$-labeled phosphorylase a at approx 3 mg/mL in solution E at 4°C (*see* **Note 7**).

2. In 1.5 mL microcentrifuge tubes combine 20 μL of the sample eluent with 20 μL of Buffer F or Buffer F containing a protein phosphatase activator or inhibitor (final reaction concentrations, protamine sulfate, 15 μg/mL; okadaic acid, 10 nM, inhibitor 2, 100 nM). Preincubate this mixture for 5 min at 30°C.

3. Add 20 μL of the $^{32}PO_4$-labeled phosphorylase a (0.31 nmoles, 1-4 × 10^5 cpm/nmole). Allow the reaction to proceed for 10 min at 30°C. (Validate that the reaction remains linear for the 10 min incubation time.)

4. Terminate the reaction by adding 180 μL of ice-cold 20% trichloroacetic acid (TCA) followed by placing each sample on ice for 10 min.

5. Centrifuge each tube at 13,000g for 4 min to separate the precipitated proteins from the free inorganic phosphate ($^{32}P_i$) released by the reaction.

6. Being careful not to disturb the pellet, transfer 200 μL of the supernatant to a scintillation vial and determine the amount of $^{32}P_i$ in the supernatant using liquid scintillation spectroscopy.

7. Knowing the specific activity of the the $^{32}PO_4$-labeled phosphorylase a, calculate the amount of phosphate released/min/mL of eluent. Remember to compensate for taking 83% of the total sample, and to subtract the amount of radioactivity in the supernatant that is present in the absence of any eluent (blank).

8. The PP2A-like protein phosphatase activity is that percentage of the phosphorylase a protein phosphatase (Phos a PP) activity in the eluent that is inhibited by 10 nM okadaic acid or is stimulated by protamine. The PP1-like activity is that percentage of the Phos a PP activity in the eluent that is inhibited by inhibitor 2 and/or protamine.

3.3.2 . Western Blot Analysis for PP2A Catalytic Subunit

Steps 1–6 are used to concentrate the protein in the eluents to help facilitate Western blot analysis.

1. Mix 400 μL of each of the eluents from the eluents with 600 μL of deionized water.

2. Add 100 μL of 0.15% sodium deoxycholate to each sample, and then mix (*see* **Note 8**).

3. Add 100 μL of 72% TCA, and then mix.

4. Centrifuge at 13,000g for 5 min.

5. Decant the supernatant. Then add 1 mL of ice-cold acetone, and mix.

6. Centrifuge at 13,000g for 5 min.

7. Decant the supernatant, and allow the pellets to air dry.

8. Resolubilize the sample proteins by the addition of Laemmli buffer (1970) and sonication (two 10-s bursts, Misonik) (*see* **Note 9**).

9. Add 1/10 vol of 2 M n-ethylmaleimide.

10. Separate proteins by SDS PAGE on a 10% acrylamide resolving gel.

11. Electrophoretically transfer the sample proteins to PVDF membrane.

12. Perform Western blot analysis using a polyclonal antibody directed against the PP2A catalytic subunit, a goat antirabbit IgG conjugated to horseradish peroxidase secondary antibody, and enhanced chemoluminescence detection (ECL) (*see* **Note 10**).

4. Notes

1. Make sure to order adult animals with good "clear" eyes that form a pseudopupil. This insures that the lateral eye tissue will remain attached to the cornea during removal from the animal. The authors have obtained more consistent results with these animals. There is no apparent difference in experiments performed with males or female animals. Be aware, however, that the great number of eggs in the females makes dissections of the brain and ventral eye more tedious. Maintain the animals in good, recirculating, filtered artificial sea water (e.g., Instant Ocean®, Aquarium Systems), at 926 mosm (36 ppt), and 14–18°C. The authors feed the animals on a weekly basis, and allow them to adapt to these conditions for at least 1 wk prior to use. The visual system is modulated by efferent input to the eyes, which is driven by a circadian clock in the brain. It is important to maintain the animals on a controlled light cycle or in natural light conditions.

2. During dissection of the lateral and ventral eyes, it is possible that you will stimulate the efferent innervation to the eyes. The efferent neurotransmitter is octopamine. The octopamine signal transduction cascade involves the activation of a cyclic AMP-dependent protein kinase and the phosphorylation of a least one phosphoprotein, a ninaC-like protein. More reproducible light-dependent effects are obtained on PP2A, PP1, and PP2B activities when the tissues have been incubated in *Limulus* saline for 2 h at room temperature in either ambient room light or complete darkness. Light-stimulated phosphorylation of the proteins in the lateral and ventral eyes are most consistently observed after overnight incubation of the eyes in the dark in LOCM at 4°C. *(31)*.

3. In the methods described, homogenates of freshly dissected eyes were used to prepare the PP2A-enriched fractions. Previously it had been determined that the PP2A-like activities in homogenates prepared from eyes that had been immersed in liquid nitrogen immediately after dissection are not the same *(8)*. There is evidence that the combination of the rapid freezing and homogenization in the presence of reducing agents such as β-mercaptoethanol or dithiothreitol reduces the interaction of the B subunit with the other two subunits (AC) *(32)*.

4. In the traditional chromatographic procedures, most of the *Limulus* lateral eye PP2A activity is present in the flow through fraction from the heparin-Sepharose (**Fig. 1**). Most of the PP1 remains bound, and is eluted with buffer containing 0.5 M NaCl. When the PP2A-enriched fraction is subsequently subjected to anion exchange chromatography on Mono Q Sepharose, four peaks of PP2A-like activity are resolved (**Fig. 2**). When these fractions were analyzed by size exclusion chromatography (Rainin Hydropore 5/5 HPLC column), the apparent molecular weights of the first, second, and third peaks were 100, 158, and 158 kDa. Therefore peak "1" may be the PP2A AC complex, and peaks "2" and "3" may be different forms of the heterotrimeric complex.

5. The anion exchange chromatography on a Mono Q Sepharose column can be performed on either a biocompatible HPLC or FPLC. Proper adapters are available from Pharmacia for attaching the Superloop®, and the Mono Q column to the sample injector (Rheodyne or the like) of a conventional HPLC system. The

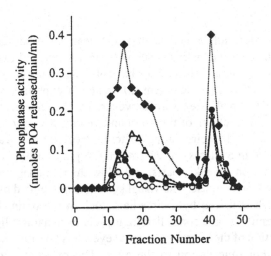

Fig. 1. Most of the PP 2A activity in the *Limulus* lateral eye soluble fraction can separated from PP 1 by heparin Sepharose affinity chromatography (HrTrap, Pharmacia). Proteins in the soluble fraction were loaded onto the column in buffer A. Buffer A was then pumped across the column at a flow rate of 0.9 mL/min. Minute fractions were collected. After fraction 35, the mobile phase was changed to buffer A containing 0.5 M NaCl (arrow). Phosphatase activity in every other fraction was assayed using phosphorylase a with or without protamine (pro; 15 µg/mL), okadaic acid (OKA; 3 nM), or both protamine and okadaic acid. PP 2A activity is the phosphorylase a protein phosphatase activity that is inhibited by 10 nM OKA or can be stimulated by protamine. Control --●--: ; Protamine --◆--; OKA: --○--; Pro & OKA: --△--.

operating pressures must not exceed 1500 psi. The Superloop facilitates the easy application of large volumes of sample to the columns. The only caution is to be sure to bypass flow through the loop after injection of the sample onto the column.

6. The method for small scale preparation of a PP2A-enriched fraction was undertaken to provide a procedure by which the eyes from a single animal or small number of animals can be used in experiments designed to understand the mechanism by which light or other modulators regulate PP2A activity. Confinement of the samples to small filter units are also useful for experiments in which radioisotopes are being employed to examine phosphorylation of PP2A. Using the small scale procedure, the PP2A activity binds to the DEAE cellulose, and is eluted in buffer containing increasing concentrations of NaCl (**Fig. 3**). Most of the PP1 is present in the 0.1 M NaCl fraction. Therefore in subsequent experiments the matrix was washed twice with 0.1 M NaCl, and then the PP2A was eluted in buffer containing 0.3 M NaCl. Affinity chromatography using thiophosphorylase a conjugated to Sepharose was subsequently used to further purify *Limulus* lateral eye PP2A. Some of the PP2A activity does not bind to the matrix; some of the activity is eluted with buffer containing 0.3 M NaCl (**Fig. 4**).

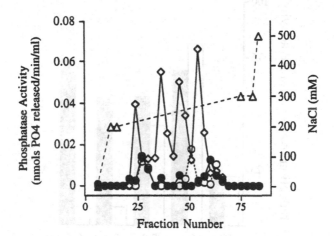

Fig. 2. Separation of *Limulus* lateral eye PP2A by Mono Q anion exchange chromatography. The flow-through fraction from the heparin Sepharose column was equilibrated in buffer B and applied to the Mono Q FPLC column. Sample proteins were eluted from the column using a linear NaCl gradient (0–500 mM). Minute fractions at a flow rate of 0.5 mL/min were assayed for PP2A activity using the phosphorylase a assay with or without protamine (15 µg/mL) or okadaic acid (3 nM). Control: —●—; Protamine —◇—; OKA: --○--; NaCl: --△--.

7. ^{32}P-labeled glycogen phosphorylase a (Phos a) is a traditional phosphoprotein substrate used to monitor PP2A and PP1 activity when assays are performed in the presence of inhibitors and stimulators such as okadaic acid, inhibitor 2, and protamine. Both glycogen phosphorylase b and phosphorylase kinase are available from Sigma. The procedures for their purification from rabbit skeletal muscle are described in **ref. 28**. Other phosphoprotein substrates can also be used. For example, myelin basic protein, which can be phosphorylated by cyclic AMP-dependent protein kinase, is a substrate for both PP1 and PP2A. Another phosphoprotein substrate for the study of PP2A is cardiac myosin light chain phosphorylated by myosin light chain kinase. It is a better substrate for PP2A than PP1 *(33)*. A nonradioactive assay that employs phosphopeptides can be used to assay PP2A activity. Kits are being marketed by Promega and Upstate Biotechnologies. The sensitivity of these assays require about a nmole of P_i to be formed. In assays that employ phosphoprotein or phosphopeptides that are substates for both PP1 and PP2A, the phosphatase activity caused by the particular enzyme can be discriminated using specific modulatory agents. This includes okadaic acid that inhibits PP2A at a lower dose (IC_{50} 0.5 nM) than PP1 (IC_{50} 50 nM). Also PP1, but not PP2A, is inhibited by the heat stable inhibitors, inhibitors 1 (I-1) and 2 (I-2). There are now commercial sources of I-1 and I-2 (Promega), or recombinant I-2 (New England BioLabs). I-2 is somewhat easier to use than I-1, since I-1 is a PP1 inhibitor only after it is phosphorylated by cyclic AMP-dependent protein kinase. Be aware that the effects of protamine are dependent on the

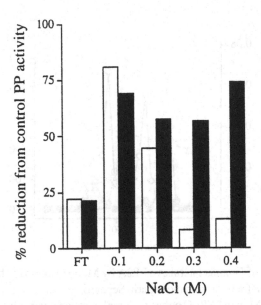

Fig. 3. Partial separation of PP 2A from PP 1 in the *Limulus* lateral eye soluble fraction by a small scale batch method employing DEAE cellulose. The soluble fraction of the homogenate prepared from the lateral eyes of two animals was incubated with 200 µL of DEAE cellulose preequilibrated in buffer D. Centrifugation of the matrix in 0.22 µm filter units was used for the stepwise elution of proteins from the matrix in buffer D (FT) or buffer D containing 0.1–0.4 M NaCl. Fractions were assayed for phosphatase activity using the phosphorylase a assay with or without protamine (15 µg/mL) and inhibitor 2 (I2, 100 nM) or okadaic acid (OKA, 10 nM). Values are expressed as the percent reduction from the activity in each fraction in presence of the phosphatase assay buffer alone (Control). PP 2A activity is the phosphorylase a protein phosphatase activity that is inhibited by 10 nM OKA; PP1 activity is that which is inhibited by I2 and protamine. □ I2 + Protamine, ■ OKA.

phosphoprotein substrate being used *(34)*. Using phosphorylase *a*, protamine inhibits PP1, but stimulates PP2A. Once the two phosphatases are separated by heparin Sepharose, the addition of protamine, which markedly stimulates PP2A activity, does facilitate identification of the PP2A-containing fractions.

8. There are probably relatively small amounts of PP2A catalytic subunit in the eluents of the chromatographic matrices. To facilitate Western blot analysis, the proteins in the eluents from fractions can be concentrated without the need of a carrier protein using a method described for protein precipitation prior to performance of a modified Lowry protein assay *(35)*. In this method sodium deoxycholate (DOC), which is not soluble at acid pH, is added to each sample. Therefore the addition of ice-cold trichloroacetic acid (TCA, final concentration 7.2–10%) precipitates both the sample proteins and the DOC. This produces a significant pellet during centrifugation of the precipitated material. The DOC together with the TCA, but

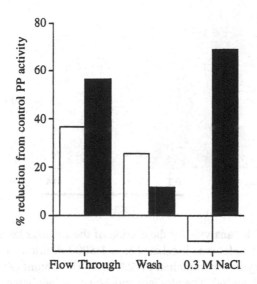

Fig. 4. Small scale affinity chromatography of PP 2A-like activity in the *Limulus* lateral eye soluble fraction. The PP 2A-enriched fraction from DEAE cellulose (0.3 *M* NaCl fraction) was incubated overnight with thiophosphorylase a Sepharose in buffer E. Centrifugation of the matrix in 0.22 μm filter units was used for the elution of proteins from the matrix in Buffer E (Wash) and buffer E containing 0.3 *M* NaCl. Fractions were assayed for phosphatase activity using the phosphorylase a assay with or without protamine and inhibitor 2 (I2) or okadaic acid. Values are expressed as the percent reduction from the activity in each fraction in presence of the phosphatase assay buffer alone (Control). ☐ I2 + Protamine, ■ Okadaic Acid.

   not the precipitated proteins, are both solubilized in ice-cold acetone and are subsequently removed. The precipitated proteins are then solubilized in Laemmli buffer *(36)*.
9. Previous studies have shown that PP2A catalytic subunit (C) is not fully reduced by the normal amount of β-mercaptoethanol present in Laemmli buffer or it reoxidizes into different forms during SDS-PAGE. Consequently, during analysis by SDS-PAGE, the PP2A C does not form a tightly defined band or migrates as multiple bands. N-ethylmaleimide, when added in excess of β-mercaptoethanol, reacts with the sulfhydryl in the protein, and alleviates the problem with the PP2A C.
10. The catalytic subunits of PP1 and PP2A are two of the most conserved proteins that exist in nature. And yet some of the commercial antibodies directed against PP1, and PP2A do not recognize the *Limulus* enzyme. Most of these antibodies are directed against the carboxy terminus, the least conserved region of the enzyme. The authors have been fortunate that Dr. Marc Mumby has kindly provided his antibody directed against the PP2A C subunit and that it indeed recognizes the *Limulus* protein (**Fig. 5**). When choosing which antibodies to use with *Limulus* tissue, be careful. If it is known that the antibodies recognize *Drosophila* PP2A, it is more likely that it will recognize the *Limulus* protein.

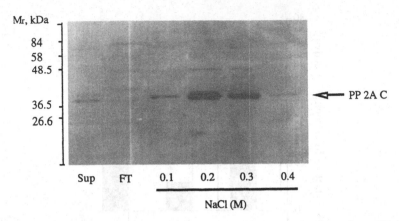

Fig. 5. Western blot analysis for the elution of the *Limulus* lateral eye PP 2A from DEAE cellulose. The blot was incubated in a 1:5000 dilution of a rabbit polyclonal antibody directed against mammalian PP 2A catalytic subunit (PP 2A C; kindly provided by Dr. Marc Mumby). The blot was subsequently incubated with a goat antirabbit polyclonal antibody conjugated to horseradish peroxidase (Promega). PP 2A C in the fractions was visualized by enhanced chemiluminescence (ECL).

References

1. Hargrave, P. A. and Hamm, H. E. (1993) in *Regulation of Cellular Signal Transduction Pathways by Desensitization and Amplification.* (Sibley, D. R., and Houslay, M. D., eds.), Wiley, New York, NY, pp. 1–104.
2. Palczewski, K., Hargrave, P. A., McDowell, J. H.and Ingebritsen, T. S. (1989) The catalytic subunit of phosphatase 2A dephosphorylates phosphoopsin. *Biochemistry* **28,** 415–419.
3. Palczewski, K., McDowell, J. H., Ingebritsen, T. S., and Hargrave, P. A. (1989) Regulation of rhodopsin dephosphorylation by arrestin. *J. Biol. Chem.* **264,** 15,770–15,773.
4. Fowles, C., Akhtar, M., and Cohen, P. (1989) Interplay of phosphorylation and dephosphorylation in vision: protein phosphatases of bovine outer segments. *Biochemistry* **28,** 9385–9391.
5. King, A. J., Andjelkovic, N., Hemmings, B. A., and Akhtar, M. (1994) The phosph-opsin phosphatase from bovine rod outer segments. An insight into the mechanism of stimulation of type 2A protein phosphatase activity. *Eur. J. Biochem.* **225,** 383–394.
6. Battelle, B. -A. (1991) Regulation of retinal functions by octopaminergic efferent neurons in *Limulus,* in *Progress in Retinal Research,* vol. 10, (Osborne, N. and Chader, H., eds.), Pergamon, pp. 333–355.
7. Edwards, S. C., O'Day, P. M., and Herrera, D. C. (1996). Characterization of protein phosphatases type 1 and type 2A in *Limulus* nervous tissue: Their light regulation in the lateral eye and evidence of involvement in the photoresponse. *Visual Neurosci.* **13,** 73–85.

8. Ellis, D. Z. and Edwards, S. C. (1994) Characterization of a calcium/calmodulin-dependent protein phosphatase in the *Limulus* nervous tissue and its light regulation in the lateral eye. *Visual Neurosci.* **11**, 851–860.

9. Van Dyke, T. H.,Tsibris, A., Windelspekt, M., and Edwards, S. C. (1994) Protein Phosphatase Type 2A: Characterization and evidence of responsibility for dephosphorylation of *Limulus* and *Drosophila* arrestin. *Invest. Ophthamol. Vis. Sci. Abstr.* **35**, 2129.

10. Wiebe, E., Wishart, A., Edwards, S., and Battelle, B. (1989) Calcium/calmodulin stimulate phosphorylation of photoreceptor proteins in *Limulus*. *Visual Neurosci.* **3**, 107–118.

11. Calman, B. G., Andrews, A. W., Edwards, S. C., Rissler, H. M., and Battelle, B-A. (1996) Calcium/calmodulin-dependent protein kinase II and arrestin phosphorylation in *Limulus* eyes. *J. Photochem. Photobiol. B* **35**, 33–44.

12. Matsumoto, H., Kurien, B. T., Takagi, Y., Kahn, E. S., Kinumi, T., Komori, N., Yamada, T., Hayashi, F., Isono, K., Pak, W. L., Jackson, K. W., and Tobin, S. L. (1994) Phosrestin I undergoes the earliest light-induced phosphorylation by a calcium/calmodulin-dependent protein kinase in *Drosophila* photoreceptors. *Neuron* **12**, 997–1010.

13. Mumby, M., Russell, K., Garrard, L., and Green, D. (1987) Cardiac contractile protein phosphatases. *J. Biol. Chem.* **262**, 6257–6265.

14. Waelkens, E., Agostinis, P., Goris, J., Merlevde, W. (1987) The polycation-stimulated protein phosphatases: regulation and specificty. *Adv. Enzyme Regulation* **26**, 241–270.

15. Usui, H., Imazu, M., Maeta, K., Tsukamoto, H., Azuma, K., and Takeda, M. (1988) Three distinct forms of type 2A protein phosphatase in human erythrocyte cytosol. *J. Biol. Chem.* **263**, 3752–3761.

16. Pato, M. and Kerc, E. (1986) Limited Proteolytic Digestion and Dissociation of Smooth Muscle Phosphatase 1 Modifies its substrate specificty. *J. Biol. Chem.* **261**, 3770–3774.

17. Scheidtmann, K., Mumby, M., Rundell, K., and Walter, G. (1991) Dephosphorylation of simian virus 40 large-T antigen and p53 protein by protein phosphatase 2A: inhibition by small-T antigen. *Molec. Cell.* **11**, 1996–2003.

18. Yang, S., Lickteig, R., Estes, R., Rundell, K., Walter, G., and Mumby, M. (1991) Control of protein phosphatase 2A by simian virus 40 small-t antigen. *Molec. Cell Biol.* **11**, 1988–1995.

19. Sontag, E., Fedorov, S., Kamibayashi, C., Robbins, D., Cobb, M., and Mumby, M. (1993) The interaction of SV40 small tumor antigen with protein phosphatase 2A stimulates the map kinase pathway and induces cell proliferation. *Cell* **75**, 887–897.

20. Brautigan, D. L. (1995) Flicking the switches: phosphorylation of serine/threonine protein phosphatases. *Semin. Cancer Biol.* **6**, 211–217.

21. Guo, H., Reddy, S. A. G. and Damuni, Z. (1993) Purification and characterization of an autophosphorylation-activated protein serine threonine kinase that phosphorylates and inactivates protein phosphatase 2A. *J. Biol. Chem.* **268**, 11,193–11,198.

22. Chen, J., Martin, B., and Brautigan (1992) Regulation of protein serine-threonine phospatase type-2A by tyrosine phosphorylation. *Science* **257,** 1261–1264.
23. Warren, M. and Pierce, S., (1982) Two cell volume regulatory systems in the *Limulus* myocardium: an interaction of ions and quaternary ammonium compounds. *Biol. Bull.* **163,** 504–516.
24. Kass, L. and Renninger, G. H. (1988) Circadian change in function of *Limulus* photoreceptors. *Visual Neurosc.* **1,** 3–11.
25. Novak-Hofer, I., Lemos, J. R., Villerman, M., and Levitan, I. B. (1985) Calcium and cyclic nucleotide-dependent protein kinases and their substrates in *Aplysia* nervous system. *J. Biol. Chem.* **269,** 10,283–10,287.
26. Orgad, S., Dudai, Y., and Cohen, P. (1987) The protein phosphatases of *Drosophila melanogaster* and their inhibitors. *Eur. J. Biochem.* **164,** 31–38.
27. Zolnierowicz, S., Csortos, C., Bondor, J., Verin, A., Mumby, M. C., DePauli-Roach, A. (1994) Diversity in the regulatory B-subunits of protein phosphatase 2A: identification of a novel isoform highly expressed in brain. *Biochemistry* **33,** 11,858–11,867.
28. Cohen, P., Alemany, S., Hemmings, B. A., Resink, T. H., Stralfors, P., and Lim Tung, H. Y. (1988) Protein phosphatase-1 and protein phosphtase-2A from rabbit skeletal muscle. *Methods Enzymol.* **159,** 391–409.
29. Brautigan, D. L.and Shriner, C. L. (1988) Methods to distinguish various types of protein phosphatase activity. *Methods Enzymol.* **159,** 329–347.
30. Bradford, M. M. (1976) A rapid and sensitive method for the quantification of microgram quantities of protein utilizing the principle of protein dye binding. *Anal. Biochem.* **72,** 248–254.
31. Edwards, S. C., Wishart, A. C., Wiebe, E. M., and Battelle, B.-A. (1989) Light-regulated proteins in the *Limulus* ventral photoreceptor. *Visual Neurosc.* **3,** 95–105.
32. Kamibayashi, C., Estes, R., Slaughter, C., and Mumby, M. (1991) Subunit interactions control protein phosphatase 2A. *J. Biol. Chem.* **266,** 13,251–13,260.
33. Kamibayashi, C., Estes, R., Lickteig, R. L., Yang, S. I., Craft, C. and Mumby, M. C. (1994) Comparison of heterotrimeric protein phosphatase 2A containing different B subunits. *J. Biol. Chem.* **269,** 20,139–20,148.
34. Pelech, S. and Cohen, P. (1985) The protein phosphatases involved in cellular regulation. 1. Modulation of protein phosphatases-1 and 2A by histone H1, protamine, and heparin. *Eur. J. Biochem.* **148,** 245–251.
35. Peterson, G. L. (1977) A simplification of the protein assay method of Lowry et al. which is more applicable. *Anal. Biochem.* **83,** 346–356.
36. Laemmli, U. (1970) Cleavage of structural proteins during assembly of the bacteriophage T4. *Nature* **277,** 680–685.

19

Purification and Assay
of the Ptc1/Tpd1 Protein Phosphatase 2C
from the Yeast *Saccharomyces cerevisiae*

Matthew K. Robinson and Eric M. Phizicky

1. Introduction

One species of protein phosphatase 2C (PP2C) in the yeast *S. cerevisiae* is encoded by the gene PTC1/TPD1 *(1,2)*. This gene encodes a protein that is highly conserved in all eukaryotes. It is 38% identical to the rat protein over the entire sequence, with identity reaching up to 80% in distinct regions. TPD1 was shown to encode protein phosphatase 2C activity based on two lines of evidence: first, Ptc1/Tpd1 protein expressed in *E. coli* exhibits readily detectable Mg^{2+} or Mn^{2+} dependent protein phosphatase activity with ^{32}P-labeled casein as a substrate. Second, this activity does not require Ca^{2+} and is resistant to okadaic acid at concentrations capable of inhibiting all the other main families of protein phosphatases in eukaryotic organisms. These are the primary distinguishing enzymatic characteristics of mammalian PP2C *(3)*. Yeast has at least two other PP2C species since extracts made from cells deleted for PTC1/TPD1 exhibit substantial PP2C activity *(1,2)*. Putative genes for this activity have been cloned *(1)*.

Although Ptc1/Tpd1 protein comprises a fraction of total PP2C activity extracted from cells, deletion of PTC1/TPD1 results in a distinct set of phenotypes, implying that Ptc1/Tpd1 protein must play a specific role in the cell for which the predominant protein phosphatase 2C cannot readily compensate. This genetic analysis has shown that Ptc1/Tpd1 protein is involved, either directly or indirectly, in regulating a wide variety of cellular processes including: growth at high temperature, cell separation, small carbon source metabolism, tRNA splicing, and at least two different MAP kinase pathways that in

From: *Methods in Molecular Biology, Vol. 93: Protein Phosphatase Protocols*
Edited by: J. W. Ludlow © Humana Press Inc., Totowa, NJ

part regulate response to environmental stresses *(1,2,4)*. Although genetic studies have implicated Ptc1/Tpd1 protein in a wide variety of cellular processes, there is as yet no information about the molecular basis for the control of protein activity or the nature of its substrates.

By studying the biochemical activity of Ptc1/Tpd1 protein the authors hope to gain a better understanding of its regulation and its recognition of substrates. It may also lend insight into the differences between the species of PP2C in yeast and how they are functioning in vivo. To begin to study biochemical activity, the authors describe a method to purify Ptc1/Tpd1 protein to near homogeneity from yeast and an assay for its protein phosphatase activity with a synthetic peptide substrate.

2. Materials

Unless explicitly stated general chemicals are BAKER ANALYZED A. C. S. Reagents, manufactured by the J. T. Baker Company (Phillipsburg, NJ).

1. Yeast strain: Relevant genotype ura3⁻ GAL⁺ (example of such a strain is DBY 747 Mat *a his3-Δ 1 leu2,3-112 ura3-52 trp1-289* atcc # 44774) American Type Culture Collection 12301 Parklawn Drive, Rockville, MD.
2. pEG(KT): GST fusion protein vector for expression in yeast *(5)*. The plasmid encodes a thrombin cleavage site between the GST and Ptc1/Tpd1 moieties of the fusion protein. This facilitates release of the Ptc1/Tpd1 portion of the fusion protein from the GST portion. The entire PTC1/TPD1 open reading frame was inserted into the polylinker to form an in-frame translational fusion.
3. YPGAL media = 1% yeast extract, 2% bactopeptone, 2% galactose.
4. -Ura raffinose media = 0.67% Difco yeast nitrogen base without amino acids, 2% raffinose and supplemented with amino acids and nucleotides as needed. For the above strain histidine, leucine and tryptophan must be added to a final concentration 30 mg/L. (Components purchased from either Sigma, St. Louis, MO or Difco, Detroit MI).
5. Yeast extract buffer: 50 mM Tris-HCl, pH 7.5 (25°C) , 5 mM DTT (Boehringer Mannheim, cat. no. 100 034), 1 mM EDTA, 500 mM NaCl, 1 mM PMSF, 1 μg/ mL pepstatin, 2 μg/mL leupeptin. Store at 4°C. Add DTT and protease inhibitors just before use.
6. Extract dilution buffer: Identical to yeast extract buffer except it contains no NaCl.
7. Glass beads: 0.5 mm glass beads. (BioSpec Products, Inc., Bartlesville, OK).
8. Phenylmethylsulfonylflouride (PMSF): Stock solution is 17.4 mg/mL (w/v) in DMSO made just before use. (purchased from Sigma).
9. Leupeptin: 10 mg/mL in distilled water. Store at −20°C in small aliquots to limit the number of freeze thaws. (purchased from Sigma).
10. Pepstatin A: 1 mg/mL in 100% ethanol. Store at -20°C. (Purchased from Sigma).
11. Glutathione Sepharose 4B: Purchased as a preswollen resin in 20% ethanol. Store at 4°C until use. (Pharmacia Biotech, cat. no. 17-0756-01).

12. GST wash buffer: 50 mM Tris-HCl, pH 7.5, 5 mM DTT, 1 mM EDTA, 100 mM NaCl, 1 mM PMSF, 1 μg/mL pepstatin, 2 μg/mL leupeptin. Store at 4°C. Add DTT and protease inhibitors just before use.

13. Human Thrombin: Thrombin is purchased as a lyophilized solid and resuspended in water to reconstitute its buffer (50 mM sodium citrate pH 6.5, 0.2 M NaCl, 0.1% PEG-8000). It is aliquoted into single use aliquots, quick frozen, and stored at −70°C for long term storage. Immediately before use thrombin is thawed at 37°C.

14. Thrombin buffer: 50 mM Tris-HCl, pH 8.0, 10 mM NaCl, 2.5 mM CaCl$_2$, 0.1% β-mercaptoethanol. Store at 4°C. Add β-mercaptoethanol just before use. (Sigma, cat. no. M-6250).

15. Dialysis buffer: 20 mM Tris-HCl pH 7.5, 5 mM DTT, 0.5 mM EDTA, 50 mM NaCl, 50% glycerol (w/v). Buffer must be chilled to 4°C before use.

16. Dialysis tubing: 10 mm dialysis tubing with mol. weight cutoff = 6–8,000 Daltons (Spectrum Medical Industries, Inc., Houston, TX).

17. Substrate Peptide: The peptide RRATVA, with a C-terminal carboxyl group rather than the amide form, was purchased crude (94% pure by mass spectrometry) and not further purified. Peptide is stored as a lyophilized solid at −20°C until use, when it is resuspended in distilled water at a concentration of 25 mg/mL (37 mM). (Quality Controlled Biochemical, Inc., Hopkinton, MA).

18. Protein Kinase A: 100 U/μL in 10 mM Tris-HCl, pH 7.0. The kinase is aliquoted and stored at −20°C. Kinase is thawed on ice before use and refrozen on dry ice/isopropanol after use. Multiple freeze thaws do not diminish the activity of the kinase. (Sigma, cat. no. P-5511; 1 U will transfer 1.0 pmol of phosphate from [γ-^{32}P] ATP to hydrolyzed and partially dephosphorylated casein per minute at pH 6.5 at 30°C).

19. Adenosine 3′: 5′ cyclic monophosphate (cAMP): 100 μM stock solution in 10 mM Tris-HCl, pH 7.5. Store at −20°C. (Sigma, cat. # A-4137)

20. [γ-^{32}P] ATP: Specific activity of radionucleotide is 3000 Ci/mmol and a concentration of 10 mCi/mL. Radionucleotide is stored at −20°C until use. (Dupont NEN, cat. no. NEG-002A).

21. Peptide column buffer: 50 mM Tris-HCl, pH 7.5, 0.1 mM EGTA, 30% acetic acid.

22. Peptide purification resin: Analytical grade (mesh size 100–200) Bio-Rad AG1-8X column resin with acetate counterion. Store at 4°C. Equilibrate to 50 mM Tris-HCl, pH 7.5, 0.1 mM EGTA, 30% acetic acid. (Biorad Laboratories, Hercules, CA, cat. no. 140-1443).

23. Disposable columns: disposable polypropylene columns (Pierce, Rockford, IL, cat. no. 29922).

3. Methods

3.1. Purification of GST-TPD1 Protein from Yeast

1. A starter culture of yeast cells containing the GST-TPD1 expression plasmid are grown to a density of 6 × 10^6 cells/mL in -Ura raffinose media (nonrepressing conditions, *see* **Note 1**). One liter cultures of YP galactose media (inducing conditions) are inoculated with the starter culture to a density of 4 × 10^5 cells/mL. Cells are grown at 30°C in an air shaker (200 rpm) until they reach a density of 2.4 × 10^7 cells/mL (~6 generations, *see* **Note 3**).

2. Yeast cells are harvested by centrifugation at 4000*g* for 5 min and resuspended in 3 mL of yeast extract buffer. Cells are lysed by shearing with 0.5 mm glass beads for five, 20 s pulses (incubate on ice for 1 min between pulses to keep extract chilled). Extract is removed from glass beads by decanting or by passage through a plastic tube that has 25 gage holes in the bottom, beads are washed with 1 mL of yeast extract buffer, and cell debris is removed by centrifugation at 14,000*g* for 10 min. Extract is aliquoted, quick frozen on dry ice/isopropanol, and stored at –70°C until needed.

3. 2.5 mL (50 mg/mL, *see* **Note 5**) extract is diluted to 100 m*M* NaCl by the addition of 4 vol of extract dilution buffer and incubated with 0.5 mL glutathione sepharose that is equilibrated in GST wash buffer. Extract is allowed to incubate with resin for 30 min at 4°C with gentle mixing.

4. Pellet resin by spinning at 1500*g* for 1 min in centrifuge. Wash five times with 1-mL aliquots of GST wash buffer to remove nonspecifically bound proteins (*see* **Note 6**). Wash one time with 1 mL of thrombin buffer to equilibrate for thrombin cleavage. Save 50 µL sample as a prethrombin treated control, pulse spin, and remove supernatant.

5. Add 1 mL of thrombin buffer and 20 U of thrombin, incubate for 5 min at room temperature with gentle mixing. Pulse spin and save the supernatant. Wash the resin with 750 µL of thrombin buffer, pulse spin, and combine the supernatant with the initial eluant. Repeat the wash step but keep this fraction separate (*see* **Note 8**). Save 50 µL of the pellet as a postthrombin control to show that proteolytic cleavage occurred.

6. Pipet the eluted protein fractions into dialysis tubing and dialyze against 1 L of dialysis buffer for >5 h. Remove samples from dialysis tubing and store at –20°C (*see* **Note 9**). Save multiple aliquots of the dialysis buffer, which will be required when assaying purified protein for activity.

7. If desired, quantitate the purification by analysis on SDS-PAGE and compare to known standards. Load equal percentage of all fractions, thereby being able to quantitate percent yield of the thrombin elution by directly comparing band intensities after silver staining the gel (*see* **Note 10**).

3.2. Phosphorylation of Peptide

Peptide can be phosphorylated at different specific activities depending upon the situation in which it is to be used (*see* **Note 11**). A reaction for the phosphorylation of peptide at 4×10^3 dpm/pmol is detailed in **Table 1** (*see* **Note 12**).

Incubate reaction at 30°C for 6 h. Peptide can be frozen at –20°C at this stage or it can be immediately purified as detailed below.

3.3. Purification of Peptide

1. 0.5 mL peptide purification resin is equilibrated to 50 m*M* Tris-HCl, pH 7.5, 0.1 m*M* EGTA, 30% acetic acid in a disposable polypropylene column by passing 20 column volumes of peptide column buffer through resin. Column is allowed to settle and buffer is run through the column until almost no head is left.

2. Peptide phosphorylation reaction is brought to 30% glacial acetic acid (v/v), mixed gently with the column resin by stirring with glass Pasteur pipet, and allowed to

Table 1
Phosphorylation of Peptide with Protein Kinase A

Peptide phosphorylation reaction	Stock solutions	4×10^3 cpm/pmol
100 mM Tris-HCl, pH 7.0	1 M	10 µL
12 mM MgCl$_2$	1 M	1.2 µL
100 µM ATP (unlabeled)	10 mM	1.0 µL
20 µCi [γ-^{32}P] ATP (3000 Ci/mmol, 10 mCi/mL)		2.0 µL
1 µM 3':5' cAMP	100 µM	1.0 µL
100 U Protein kinase A	10 U/µL	10 µL
3 mM Peptide	37 mM, 25 mg/mL	8.0 µL
Distilled water		84 µL
Total volume		100 µL

resettle for 30 min. Column is started and a 0.5 mL fraction is collected while keeping the head size low. Two other fractions are collected in the same manner. At this point, the column can be disposed of because unincorporated ATP is all that remains in the column. The majority of the peptide is recovered in the first fraction.

3. Lyophilize the peptide under vacuum to concentrate it and to remove acetic acid. Add 0.5 mL distilled water to resuspend peptide, measure pH, and lyophilize again. Repeat until pH of suspension is >6.5 (requires ~1.5 mL).

4. Resuspend the peptide in distilled water at an appropriate concentration and count to determine the percent incorporation.

3.4. Protein Phosphatase Assay

1. Tpd1 protein is assayed for protein phosphatase activity with the ^{32}P-labeled peptide in a 10 µL reaction containing 50 mM Tris-HCl, pH 7.5, 20 mM MgCl$_2$ (*see* **Note 13**). The reaction is initiated by the addition of substrate at varying concentrations and allowed to proceed for 10 min at 30°C.

2. The reaction is stopped by the addition of 10 µL 10% Trichloroacetic acid. Phosphate released in the reaction is then converted to a phosphomolybdic complex followed by organic extraction of the complex. This is done by adding, in order, 3.5 µL 5 M H$_2$SO$_4$, 30 µL of isobutyl alcohol/toluene (1:1; v/v), and 3.5 µL of 10% ammonium molybdate. The reaction is vortexed for 10 s and then centrifuged in a microfuge at 12,000g for 2 min. A 15 µL aliquot of the top layer is then added to scintillation fluid and counted.

3. Based on the specific activity of the peptide substrate, counts per minute (cpm) can be converted to moles of phosphate released and plotted to determine the Km and Vmax of the protein phosphatase with the peptide substrate.

4. Tpd1 protein has a Km of 10 µM and a Vmax of 91 fmols phosphate released/minute for the phosphorylated peptide RRATVA (**Fig. 1**). This is within an order of magnitude of the Km determined for the mammalian enzyme.

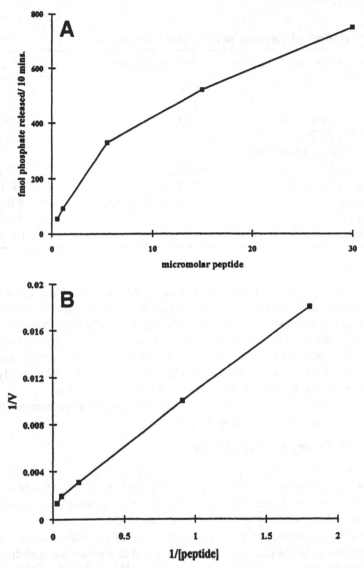

Fig. 1. Purified Ptc1/Tpd1 protein was incubated for 10 min at 30°C with increasing concentrations of the ^{32}P labeled peptide (RRATVA), as indicated. Reactions were stopped by the addition of trichloroacetic acid to 5% (w/v) and released inorganic phosphate was quantitated as described above to determine the kinetic parameters. (**A**) Michaelis-Menten plot. (**B**) Lineweaver-Burk plot.

4. Notes

1. GST-TPD1 can function as the sole source of PTC1/TPD1 in the cell even when expressed at low levels. Therefore, fusion to GST does not disrupt Ptc1/Tpd1 protein acitivity.

2. A_{600} = 1.0 corresponds to 2×10^7 cells/mL for yeast. Optical density reading should be measured between 0.1–0.7 to be significant.

3. Expect the generation time for yeast in -Ura raffinose media to be approx 6 h, and in YP galactose media to be roughly 4 h.

4. PMSF is very unstable in aqueous solutions. A second aliquot should be added after cells have been lysed and extract removed from the glass beads.

5. Expected concentration for these extracts is between 30–50 mg/mL as measured by Bradford assay.

6. At the time of first wash it is easier to move resin into microfuge tubes. Subsequent spins are for 20 s at top speed in microfuge.

7. GST-TPD1 can be bound to glutathione sepharose at NaCl concentrations in excess of 500 m*M*. This can facilitate purification because less nonspecific proteins are retained. However equilibration to thrombin cleavage conditions requires more washing steps to lower the NaCl concentration.

8. If extracts are not precleared well, a lipid layer will form on the glutathione Sepharose pellets. This interferes with washing of nonspecific proteins, and should be avoided as much as possible. If a lipid layer is observed on pellets, removal of the layer can be accomplished through shorter pulse spins.

9. After dialysis, storage of protein at −20°C should not result freezing of the sample.

10. Protein purified by this procedure routinely results in nearly homogeneous preparations of protein as judged by silver staining, with the only major contaminant being the thrombin used to cleave the fusion protein.

11. Sequence of peptide and method for phosphorylation are essentially as described by Deana, et al. *(6)*.

12. Using the outlined method, the authors have been able to label peptides to a wide range of specific activities.

13. Unphosphorylated peptide has no effect on mammalian enzyme activity over a range of 3–300 µ*M* in reactions *(6)*.

References

1. Maeda, T., Tsai, A. Y., and Saito, H. (1993) Mutatins in a protein tyrosine phosphatase gene (PTP2) and a protein serine/threonine phosphatase gene (PTC1) cause a synthetic growth defect in *Saccharomyces cerevisiae*. *Mol. Cell. Biol.* **13**, 5408–5417.

2. Robinson, M. K., van Zyl, W. H., Phizicky, E. M., and Broach, J. R. (1994). TPD1 of *Saccharomyces cerevisiae* encodes a protein phosphatase 2C-like activity implicated in tRNA splicing and cell separation. *Mol. Cell. Biol.* **14**, 3634–3645.

3. Cohen, P. (1989). The structure and regulation of protein phosphatases. *Annu. Rev. Biochem.* **58**, 453–508.

4. Huang, K. N. and Symington, L. A. (1995) Suppressors of a *Saccharomyces cerevisiae pkc1* mutation identify alleles of the phosphatase gene PTC1 an of a novel gene encoding a putative basic leucine zipper protein. *Genetics* **141**, 1275–1285.

5. Mitchell, D. A., Marshal, T. K., and Deschenes, R. J. (1993). Vectors for the inducible overexpression of glutathione S-transferase fusion proteins in yeast. *Yeast* **9,** 715–723.
6. Deana, A. D., Mac Gowan, C. H., Cohen, P., Marchiori, F., Meyer, H. E., and Pinna, L. A. (1990) An investigation of the substrate specifictiy of protein phosphatase 2C using synthteic peptide substrates; comparison with protein phosphatase 2A. *Biochem. Biophys. Acta.* **1051,** 199–202.

20

Molecular Cloning
of Protein Phosphatase Type 2C Isoforms
from Retinal cDNA

**Dagmar Selke, Susanne Klumpp, Benjamin Kaupp,
and Arnd Baumann**

1. Introduction

Protein phosphatases are important for the maintenance of functional activity of many proteins. Phosphorylation of proteins by protein kinases is a common posttranslational modification that modulates the signaling capability of the target polypeptides. Neurotransmitter receptors, which belong to the superfamily of GTP (G) protein-coupled receptors, are uncoupled from their signaling pathway(s) after phosphorylation by receptor-specific protein kinases.

Phosphorylation of the photopigment rhodopsin terminates the biochemical amplification cascade mediating visual transduction in vertebrate photoreceptor cells *(1)*. Regeneration of functionally active rhodopsin requires, besides exchange of the chromophore, the presence of protein phosphatases (PP). Two protein phosphatases were already identified in retinal tissue: PP1 and PP2A, but only PP2A was found to dephosphorylate rhodopsin *(2)*. As outlined in a previous contribution to this volume *(3)*, a novel biochemical purification protocol now led to the identification of PP2C isoforms in bovine retinal tissue. The substrate(s) of the PP2C isoforms in the retina are not known, yet.

In order to further study the function of PP2C in signaling cascades in vertebrate retinal tissue, the authors initiated a PCR-based molecular cloning approach to isolate cDNA sequences for both isozymes. Amplification of DNA-fragments by PCR, however, requires sequence information of homologous genes either at the amino acid or nucleic acid level. cDNA encoding PP2C isoforms have already been cloned from various tissues of mammalian species

From: *Methods in Molecular Biology, Vol. 93: Protein Phosphatase Protocols*
Edited by: J. W. Ludlow © Humana Press Inc., Totowa, NJ

(4–6), although not from bovine. Thus, degenerate primers were synthesized based on available sequence information.

Here the authors describe experimental details of the cloning strategy that include cDNA synthesis from retinal RNA, amplification of specific fragments by PCR, and isolation of full-length clones from cDNA libraries. The protocols given in this chapter are generally applicable and will enable the reader to isolate the desired sequences from any suitable type of tissue. For additional information in molecular biological techniques, the reader is referred to more comprehensive textbooks *(7,8)*.

2. Materials

1. Plasmid pBluescript SK⁻ (available from Stratagene, Germany).
2. LB-medium (1 L): 10 g of tryptone, 5 g of NaCl, 5 g of yeast extract; adjust to pH 7.4 and autoclave.
3. LB-agar (1 L): Add 15 g of agar to LB-medium and autoclave.
4. Ampicillin stock solution: 100 mg/mL in water.
5. cDNA synthesis: kits containing all buffers, enzymes, and sterile water required to synthesize 1st and 2nd strand cDNA are available from several manufacturers. The authors routinely obtain good results with cDNA synthesis kit from Life Technologies (Gibco/BRL), Germany.
6. Restriction enzymes, Klenow enzyme, T4 Polynucleotide Kinase, and T4 DNA ligase are supplied with appropriate buffers (10X conc.) by many manufacturers. Follow their instructions for storage and handling.
7. Agarose.
8. 3 M lithium chloride (LiCl), autoclave.
9. 3 M sodium acetate (NaAc), autoclave.
10. 100% ethanol (p.a. grade).
11. 70% ethanol.
12. Calf intestine alkaline phosphatase: supplied by several manufacturers, the authors use Boehringer enzyme, which is supplied with appropriate buffer (10X conc.).
13. TE buffer: 10 mM Tris-HCl, pH 7.5, 1 mM ethylendiaminetetra-acetic acid disodium salt (Na$_2$EDTA), autoclave.
14. Frozen competent XL1 Blue cells are available commercially from Stratagene.
15. Phenol: 1 kg phenol equilibrated with 80 mL of 3 M Tris-base and several changes of TE buffer. To minimize oxidation add 1.2 g 8-Hydroxyquinoline. Equilibrated phenol is stored at –20°C. Phenol is toxic and causes severe burns and should be handled with care.
16. Chloroform (p.a. grade).
17. Phenol/chloroform: 1:1 mixture of solution 15 and 16.
18. 10% sodium dodecylsulfate (w/v; SDS).
19. 5 M potassium acetate (KAc), autoclave.
20. Solution I: 50 mM glucose, 10 mM Na$_2$EDTA, 25 mM Tris-HCl, pH 8.0, autoclave.

21. Solution II: 0.2 M NaOH, 1% (v/v) SDS.
22. Solution III: mix 60 mL KAc (5 M) and 11.5 mL of glacial acetic acid, adjust pH to 4.8 with HCl conc., and add water to a final volume of 100 mL, autoclave.
23. RNase: the authors prefer RNase cocktail of RNase A and RNase T1 available from Ambion, Germany.
24. Diethyl Pyrocarbonate (DEPC): water and all buffers, except those containing salts with primary amines (i.e., Tris), used for RNA purification and cDNA synthesis should be treated with 0.01% DEPC (v/v) for 6–12 h. DEPC is highly toxic and carcinogen, therefore work under a fume hood while handling.
25. NZY-medium (1 L): 10 g of NZ-amine, 5 g of NaCl, 5 g of yeast extract, 2 g of $MgSO_4 \times 7 H_2O$; adjust to pH 7.4, and autoclave.
26. NZY-agar (1 L): Add 15 g of agar to NZY-medium and autoclave.
28. NZY-top agar (1 L): Add 6 g of agarose to NZY-medium and autoclave.
29. SSC 20 X (1 L): 3 M NaCl, 0.3 M sodium citrate, adjust to pH 7.5, and autoclave.
30. Denhardts 100X (1 L): 20 g of bovine serum albumin (fraction V), 20 g of Ficoll 400, 20 g of polyvinylpyrrolidone; store frozen in aliquots at –20°C.
31. Herring testes DNA (10 mg/mL): 1 g of lyophilized herring testes DNA, purchased by Sigma (St. Louis, MO) cat. no. D-6898, is dissolved in 100 mL of TE, autoclaved, and kept frozen in 1 mL aliquots at -20°C.
32. 0.1 M magnesium chloride ($MgCl_2$), autoclave.
33. 1 M magnesium sulfate ($MgSO_4$), autoclave.
34. 10 mM magnesium sulfate ($MgSO_4$); make up from 1 M stock solution and autoclave.
35. 20% maltose (w/v), dissolve in water, sterilize by filtration and store at 4°C.
36. XL1 Blue Mg^{2+}-cells: inoculate a single colony of bacterial strain XL 1 Blue (available from Stratagene, Germany) in 50 mL LB-medium supplemented with 10 mM $MgSO_4$ and 0.2% maltose for 6–8 h at 37°C in a shaking incubator. Determine OD_{600} of culture photometrically. Collect cells in a sterile Falcon tube by centrifugation 5000g, 10 min., 4°C. Resuspend cells to 10 OD/mL in 10 mM $MgSO_4$. For plating λ-phages, the authors use 50 μL and 300 μL of this suspension per 9 cm and 15 cm plate, respectively.
37. Nylon membranes: available from several manufacturers; the authors routinely use Qiabrane membranes from Qiagen, Germany.
38. Always use double distilled, autoclaved water.

3. Methods

3.1. cDNA Synthesis

1. Purify poly (A)$^+$ RNA from bovine retina (*see* **Note 1**). Precipitate 5 μg of poly (A)$^+$ RNA by adding 1/10 vol of RNAgrade 3 M NaAc and 2.5 vol of 100% ethanol. Incubate at –20°C for at least 2 h. Spin down RNA precipitate at 12,000g, 15 min., 4°C in a microfuge. Discard supernatant and wash pellet with 100 μL of 70% ethanol. Spin down at 12,000g, 5 min, 4°C in a microfuge. Air dry pellet for 10 min.
2. Resuspend poly (A)$^+$ RNA in 8.0 μL of DEPC-treated water. Add 2.0 μL of 0.5 μg/μL oligo-dT primers (*see* **Note 2**) and denature for 10 min at 60°C in a water bath. Quickly chill the probe on ice. Spin down briefly in a microfuge for 5 s.

3. Add 20.0 µL of DEPC-treated water, 5.0 µL of 100 m*M* DTT, 10.0 µL of 5X concentration 1st strand buffer, and 2.5 µL of dNTP-mixture (10 m*M* each). Incubate at 37°C for 2 min.

4. Add 2.5 µL of 200 U/µL MMLV reverse transcriptase. Incubate at 37°C for 60 min (*see* **Note 3**). Then place tube on ice and keep 5.0 µL of reaction mixture for controls and store at −20°C (*see* **Note 4**).

5. For synthesis of 2nd strand cDNA add 294.5 µL of sterile water. Then add 7.5 µL of dNTP-mixture (10 m*M* each), 40.0 µL of 10X concentration 2nd strand buffer, 10.0 µL of 10 U/µL DNA polymerase I, 1.75 µL of 2 U/µL RNase H, and 1.25 µL of 10 U/µL *E. coli* DNA ligase. Incubate at 16°C for 2 h.

6. Add 2.0 µL of 5 U/µL T4 DNA polymerase and incubate at 16°C for 5 min. Then add 25.0 µL of 0.25 *M* EDTA to stop enzymatic reaction.

7. Add 400 µL of phenol/chloroform and mix by vortexing. Separate phases by centrifugation: 12,000*g*, 2 min., room temperature in a microfuge. Transfer aqueous phase to a fresh tube and repeat phenol-extraction.

8. Add 400 µL of chloroform to aqueous phase, mix by vortexing, and perform centrifugation as in **step 7**. Transfer aqueous phase to a fresh tube and repeat chloroform-extraction.

9. Use ~0.5% of 2nd strand cDNA to check amount and length of synthesized cDNA molecules on 0.75% agarose gel (*see* **Note 5**). cDNA is ready for PCR or for construction of cDNA library.

3.2. Amplification of PP2C Fragments by PCR

1. Choose appropriate oligonucleotide primers based on vertebrate PP2C cDNA sequences (*see* **Note 6**). Oligonucleotide primers are commercially available from several manufacturers. Usually they are purified by HPLC-chromatography and can be used without further purification.

2. Determine concentration of oligonucleotide: 1 A_{260} = 30 µg/mL. For PCR adjust concentration to 10 ng/nucleotide X µL (*see* **Note 7**).

3. Work on ice during set up of PCR. Use 20 ng of 2nd strand cDNA, add 71.0 µL of water, 10.0 µL of PCR reaction-buffer (purchased by supplier of enzyme), 4.0 µL of 0.1 *M* $MgCl_2$, 1.0 µL of dNTP-mixture (20 m*M* each, PCR grade), 2.5 µL of each oligonucleotide (*see* **Note 6**), 1.0 µL of 2.5 U/µL Taq polymerase, and overlay with a drop of mineral oil. Run a control PCR without addition of template DNA.

4. Run 1st PCR: 5 cycles at 94°C, 50 s, 45°C, 50 s, 72°C, 50 s, 35 cycles at 94°C, 50 s, 50°C, 50 s, 72°C, 50 s.

5. Purify amplified products by one phenol/chloroform and one chloroform-extraction. Precipitate DNA after addition of 1/10 vol of 3 *M* LiCl and 3 vol of 100% ethanol for 1 h at −20°C. Spin down: 12,000*g*, 15 min., 4°C in a microfuge, dry, and resuspend in 10.0 µL of TE buffer.

6. Set up nested PCR with internal primers (*see* **Note 6**) using 2.0 µL from 1st PCR as a template. Use buffers as described in **Subheading 3.2.3.** Run 2nd PCR: 40 cycles at 94°C, 50 s, 54°C, 50 s, 72°C, 50 s.

7. Transfer reaction mixture to a fresh tube. Use 10 μL of mixture to analyze for synthesis of fragments on 1.5% agarose gel.
8. Purify amplified DNA by one phenol/chloroform and one chloroform-extraction and ethanol-precipitation.
9. When synthesis was successful, proceed with subcloning the PCR fragment.

3.3. Subcloning of PCR Fragments into pBluescript-Vector

1. Spin down PCR-fragments (*see* **Subheading 3.2.8.**): 12,000*g*, 10 min., 4°C in a microfuge, dry, and resuspend pellet in 11.0 μL of water.
2. Add 1.5 μL of 10X concentrated incubation buffer, 1.5 μL of dNTP-mixture (1 m*M* each), and 1.0 μL of 2 U/μL Klenow enzyme to fill in recessive or protruding ends. Incubate at 37°C for 30 min.
3. Purify by one phenol/chloroform and chloroform-extraction. Add 1.5 μL of 3 *M* LiCl and 45.0 μL of 100% ethanol to aqueous phase to precipitate DNA. Spin down precipitate: 12,000*g*, 10 min., 4°C, in a microfuge, and dry.
4. Resuspend pellet in 11.0 μL of water. Add 1.5 μL of kinase reaction buffer (provided by the supplier of the enzyme), 1.5 μL of 10 m*M* ATP, and 1.0 μL of 10 U/μL T4 Polynucleotide Kinase. Incubate at 37°C for 30 min. Inactivate enzyme at 65°C for 10 min.
5. Separate fragments on 1.5% agarose gel. Cut out bands from the gel with a scalpel and transfer to Eppendorf cup. To prepare fragments from agarose gels, the authors use a centrifugation method *(9)* which is fast, convenient, and gives reliable yields of ~50% for fragments up to 2 kb in length.
6. Digest 5 μg of vector DNA made up to 30.0 μL in the appropriate buffer with 15–20 U of Eco RV for 2 h at 37°C. Check that the DNA is completely digested by running a 1.5 μL sample on 0.75% agarose gel.
7. Extract with phenol/chloroform followed by chloroform, add 3 μL of 3 *M* LiCl and 90 μL of 100% ethanol. Precipitate the DNA in a microfuge: 12,000*g*, 15 min, 4°C. Dry the pellet and resuspend in 10 μL of TE buffer. Add 16 μL of water, 3.0 μL of 10X phosphatase buffer, and 1 U of calf intestine alkaline phosphatase. Incubate at 37°C for 60 min. Make up to 25 m*M* EGTA and incubate at 65°C for 20 min. Extract with phenol/chloroform followed by chloroform, and precipitate the DNA with LiCl/ethanol.
8. Spin down the vector DNA in a microfuge, dry, and resuspend in TE buffer. Run a sample of vector DNA and PCR fragment alongside standard DNA samples and estimate the concentrations. Set up 15.0 μL ligation reaction, in the buffer provided by the manufacturer, containing equimolar concentrations of phosphatase-treated vector and of fragment to be inserted. In case of pBluescript vector (2954 bp) take 50 ng DNA and 5 ng of 300-bp fragment. Add 1 U of T4 DNA ligase and incubate at 16°C for 2–4 h. Run control ligation containing phosphatase-treated vector DNA only.
9. Transform ligated DNAs into competent XL1 Blue cells. Plate the transformation mixes onto LB-agar plates containing 100 μg/mL ampicillin.
10. Ampicillin resistant colonies are grown overnight in 5 mL LB-medium containing 100 μg/mL ampicillin. To prepare miniprep DNA, spin down 1.5 mL of cul-

ture in a microfuge tube. Discard supernatant (autoclave!) and resuspend the pellet in 100 μL of solution I by vortexing. Add 200 μL of solution II and mix by vortexing. Immediately add 150 μL of solution III, vortex, and spin down debris at 12,000*g*, 3 min, room temperature in a microfuge. Transfer supernatant to a fresh tube containing 200 μL of phenol/chloroform. Mix and spin for 2 min. Aqueous phase is transferred to a fresh tube containing 200 μL of chloroform. Mix and spin for 1 min. Transfer aqueous phase to fresh tube containing 1 mL of 100% ethanol. Mix well and spin: 12,000*g*, 3 min, room temperature. Wash the pellet in 70% ethanol, dry, and resuspend in 25.0 μL of TE buffer.

11. Use 2.0 μL aliquots of this DNA for restriction analysis to determine the presence of insert. Restriction analysis should be performed in the presence of RNase. Store DNA of positive clones in the refrigerator or ethanol precipitated at −20°C.

12. Approximately 2 μg of DNA are denatured by adding 4.0 μL of 2 *M* NaOH in a total volume of 10 μL at 37°C for 30 min. Then add 3.0 μL of of 3 *M* NaAc, 7.0 μL of water, and 60 μL of 100% ethanol, and place at −20°C for 1 h. Spin down precipitate: 12,000*g*, 4°C, 30 min., wash thoroughly with 70% ethanol, dry, and proceed with sequencing reactions as described by the supplier of sequencing kit.

3.4. Screening for Full-Length cDNA Clones

1. A bovine retina cDNA library cloned into λZapII-vector (Stratagene) was used to isolate full-length cDNA-clones. 5X 10^4 recombinant phages are incubated with 300 μL XL1-Blue Mg^{2+}-cells (10 OD$_{600}$/mL) for 15 min at 37°C. After addition of 12 mL NZY-top agarose, phages are plated onto 15-cm NZY-agar plates and incubated at 37°C overnight. Plaque lifts onto nylon membranes are performed essentially as described in Sambrook et al. *(7)*.

2. The cloned PCR fragment that contains the correct sequence of PP2C, is cut out of the vector using *Eco* RI and *Hin*dIII restriction enzymes. Fragments are separated on a 1.5% agarose gel, the desired fragment is cut out of the gel and purified by centrifugation through glass wool (*see* **Subheading 3.3.6.**). For radioactive labeling, use a commercially available labeling kit. Purify labeled probe either by ethanol precipitation or by gel filtration on Sephadex G75 column. Hybridization is done overnight in 50% formamide (v/v), 5X SSC, 5X Denhardts, 0.1% SDS, 100 μg/mL denatured herring testes DNA, and 1.5 10^6 cpm/mL labeled probe at 42°C. Washing is performed in 1X SSC, 0.1% SDS once at room temperature for 5 min, and twice at 65°C for 30 min each. Exposure to X-ray films is for 6 h to overnight at −80°C. Signals are correlated with plaque regions on the agar plates and phages are further purified by successive fine screens. Finally, cDNA clones are obtained by in vivo excision following the protocol of Stratagene. cDNA clones are characterized by restriction analysis. The longest clones are chosen for sequencing.

4. Notes

1. During preparation and handling of RNA, special care must be taken not to contaminate probes with RNases that are always around. Whenever possible, use

sterile plasticware. Glass should be baked for 12 h at 180°C. Water and buffer solutions, except those containing primary amines, should be treated with 0.01% DEPC (v/v) for 6–12 h. DEPC is destroyed by autoclaving for 1 h at 121°C. When preparing RNA for the first time it is recommended to use a kit that contains all buffers to isolate total- and poly(A)$^+$ RNA. Kits are available from several manufacturers; the authors obtained good results with Fast Track (Invitrogen) and Oligotex Direct mRNA kits (Quiagen, Germany).

2. The length of synthesized 1st strand cDNA is dependent on the molar ratio of poly(A)$^+$ and primers used to initiate synthesis. Using a reduced primer concentration leads to generation of longer 1st strand cDNA products. This is especially important when transcripts of >5 kb are to be cloned. For cDNA synthesis it is also recommended to use a commercially available kit. The authors obtained good results with cDNA synthesis system (Life Technologies GmbH, Germany). The kit contains all buffers, enzymes, and primers sufficient for several experiments.

3. If necessary, one can add another 200 U of reverse transcriptase after 30 min at 37°C.

4. After synthesis of 1st strand cDNA, one can immediately set up a PCR to amplify the gene of interest. Therefore, it is recommended to keep a portion of this fraction at –20°C. If one decides to prepare a cDNA library, 2nd strand cDNA synthesis should be performed directly after synthesis of 1st strand cDNA.

5. Separation of synthesized fragments on agarose gels containing ethidium bromide can be analyzed by UV-illumination. A smear will be visible that should reach lengths of ~5–7 kb as monitored by control DNA of known fragment size run in parallel.

6. The successful amplification of a sequence from cDNA preparations strongly depends on the oligonucleotide primers chosen for the experiment. In case that sequences of homologous genes are already known, search for strings of amino acid residues that are highly conserved between species. A stretch of 6–8 amino acids with ~70% identity will be fine. Translate the amino acid sequence into nucleotide sequence and take into account the redundancy of the genetic code. Thus, for the respective 3' end of the oligonucleotide, such amino acid residues should be preferred that are encoded by only one or a few nucleotide triplets. Choose primers, about 500 bp apart from each other. This easily allows detection of amplification products by agarose gel electrophoresis. In case the transcript is rare or that highly degenerated oligonucleotides have to be used in the 1st PCR, prepare a second set of primers which are located within the amplified sequence. A 2nd PCR should always result in amplification of desired fragment. The authors used the following pairs of oligonucleotide primers for 1st and 2nd PCR, respectively: no. 1 (sense) 5'-GAAATGGA(GA)GA(TC)GCACATAC and no. 2 (antisense) 5'-TC(GA)TT(GT)CCCAT(GAC)AC(GA)TCCCA; no. 3 (sense) 5'-GT(GATC)TA (TC)GA(TC)GG(GATC)CA(TC)GC(GATC)G and no. 4 (antisense) 5'-CC(TC)TT(GATC)CC (GA)TG(GATC)AC(GA)CA(AT)TT. no. 1 and no. 3 correspond to amino acids EMEDAHT and VYDGHAG, no. 2 and no. 4 are complementary to amino acid sequences WDVMGNE and

KCVHGKG, respectively. These sequences are highly conserved between PP2C from rabbit liver and skeletal muscle as well as from rat kidney and liver *(4–6)*. To determine the annealing temperature of oligonucleotide primers, use equation: T_m (°C) = 4 X (G + C) + 2 X (A + T), where (G + C) is the number of guanosine and cytosine nucleotides and (A + T) is the content of adenosine and thymidine nucleotides of the respective oligonucleotide. When using degenerate oligonucleotides for PCR, calculate the lowest T_m value, i.e., oligonucleotide with highest (A + T) content. Choose annealing temperature which is 2°C below the calculated T_m value.

7. Dilute oligonucleotides to a final concentration of 10 ng per nucleotide and μL. For conventional PCR the authors take 1 μL of this solution per 100 μL reaction mixture. While using degenerate oligonucleotides this value should be increased 5–10 fold.

References

1. Stryer, L. (1991) Visual excitation and recovery. *J. Biol. Chem.* **266,** 10,711–10,714.
2. Fowles, C., Akhtar, M., and Cohen, P. (1989). Interplay of phosphorylation and dephosphorylation in vision: Protein phosphatases of bovine rod outer segments. *Biochemistry* **28,** 9385–9391.
3. Klumpp, S. and Selke, D. Separation of protein phosphatase type 2C isozymes by chromatography on Blue Sepharose, in *Methods in Molecular Biology* vol. 93 (Walker, J. M., ed.), Humana, Totowa, NJ.
4. Tamura, S., Lynch, K. R., Larner, J., Fox, J., Yasui, A., Kikuchi, K., Suzuki, Y., and Tsuiki, S. (1989) Molecular cloning of rat type 2C protein phosphatase mRNA. *Proc. Natl. Acad. Sci. USA* **86,** 1796–1800.
5. Wenk, J., Trompeter, H.-I., Pettrich, K.-G., Cohen, P. T. W., Campbell, D. G., and Mieskes, G. (1992) Molecular cloning and primary structure of a protein serine/threonine phosphatase 2C isoform. *FEBS Lett.* **297,** 135–138.
6. Mann, D. J., Campbell, D. G., McGowan, C. H., and Cohen, P. T. W. (1992) Mammalian protein serine/threonine phosphatase 2C: cDNA cloning and comparative analysis of amino acid sequences. *Biochim. Biophys. Acta* **1130,** 100–104.
7. Sambrook, J., Fritsch, E. F., and Maniatis,T. (1989) *Molecular Cloning: A Laboratory Manual*, 2nd ed., Cold Spring Harbor, Cold Spring Harbor Laboratory.
8. Ausubel, F. M., Brent, R., Kingston, R. E., Moore, D. D., Seidman, J. G., Smith, J. A., and Struhl, K., eds. (1987) in *Current Protocols in Molecular Biology*, 3 vols., John Wiley & Sons, Inc.
9. Heery, D. M., Gannon, F., and Powell,R. (1990) A simple method for subcloning DNA fragments from gel slices. *Trends Genetics* **6,** 173.

21

Analysis of Protein Interactions Between Protein Phosphatase 1 and Noncatalytic Subunits Using the Yeast Two-Hybrid Assay

Nadja T. Ramaswamy, Brian K. Dalley, and John F. Cannon

1. Introduction

A large variety of methods exist to identify protein-protein interactions. The two hybrid method is unique in that interactions are assayed in vivo. The method can be used "analytically" or "preparatively." In the analytical application, one surveys interactions between two proteins known to interact by comparing interaction of mutant proteins. The preparative two hybrid method allows one to identify putative proteins that interact with a given protein (called the "bait") by screening libraries. This chapter will focus only on the analytical application of the two hybrid assay to analyze interactions of mutant protein phosphatase 1 (PP1) molecules with some PP1 noncatalytic subunits. For examples of the latter screening method, readers are referred to original papers that employed the method to isolate PP1 interacting proteins (1–3).

The modular nature of eukaryotic transcription factors is exploited in the two hybrid assay, which was first described by Fields and Song (4). Transcription activation domains (AD) and DNA binding domains (DB) of transcription factors are separated and treated independently. These domains are individually fused to each partner protein of an interacting pair to make hybrid fusion proteins. When these two hybrid proteins associate by their intrinsic affinity, the AD is juxtaposed near a cis DNA sequence that the DB recognizes and transcription is activated. The "readout" of this assay, therefore, is from transcription of a reporter gene. Variations in this assay are made in the choices of AD, BD, and reporter gene. This assay has been performed traditionally in yeast, but is not restricted to this organism (5). To simplify description of this

From: *Methods in Molecular Biology, Vol. 93: Protein Phosphatase Protocols*
Edited by: J. W. Ludlow © Humana Press Inc., Totowa, NJ

method, these variations will not be discussed. Instead, the system that the authors have used extensively will be described.

GAL4 is a Saccharomyces cerevisiae transcriptional activator that regulates genes in response to extracellular galactose *(6)*. GAL4 contains one AD (of two) within amino acids 768–881. Because both hybrid proteins must enter the nucleus to interact and activate transcription, a nuclear localization signal (NLS) from SV40 T antigen is fused to GAL4(768–881) in the plasmid vectors used. GAL4 DB is encoded in amino acids 1–147, which also includes a NLS. GAL4 binds 5' to several galactose-regulated genes in yeast to an DNA sequence called UAS_{GAL}. The authors use the GAL1 UAS_{GAL}-*LacZ* reporter, which produces *LacZ* encoded β-galactosidase activity proportional to the GAL4 activity reconstituted from interactions of the two GAL4 hybrid proteins.

GLC7 encodes the sole PP1 enzyme in *S. cerevisiae (7)*. It is essential for viability. The diverse roles played by GLC7 were identified by characterizing *glc7* mutants. These studies indicate that indicated GLC7 functions in controlling mitosis *(8)*, meiosis *(7,9)*, glycogen accumulation *(7,10)*, translation initiation *(11)*, cell morphology *(9,12)*, and glucose repression *(13,14)*.

A number of GLC7 interacting proteins were initially identified either by analysis of mutants (GLC8, SDS22, GAC1, REG1) or by application of two hybrid screening *(2,3)*. Although homologs of some of these proteins were known from biochemistry of mammalian PP1, many more of them have been uniquely or initially identified using yeast genetics. Three yeast proteins, which interact with GLC7, will be used to illustrate the methodology. They are GAC1 *(15)* homologous to muscle glycogen binding subunit RGL *(16)*, GLC8 *(7)* homologous to inhibitor-2 *(17)*, and SDS22 *(18,19)*, a protein initially discovered in yeast, which has a human homolog *(20)*.

Two hallmarks of PP1 enzymes are extensive homologies (e.g., greater than 80% amino acid identity between *S. cerevisiae* and human PP1) and numerous interacting proteins. These are potentially related because conservation of many interacting proteins could be the selective force that maintained homology in PP1 sequences. Recognizing that each PP1 residue contains "high information content" the authors systematically characterized the contribution of GLC7 residues in interprotein interaction and GLC7 function. Random *glc7* mutants were isolated based on their phenotypes in yeast *(9)*. Mutant GLC7 proteins have been recently characterized for their interactions with some PP1 binding proteins *(21)* using the two hybrid assay described here. The authors previously used this method with mutant RAS proteins to show their interactions with RAS effectors and regulatory proteins *(22)*. A "reverse genetic" approach is now feasible using the three-dimensional structure of PP1 or related protein phosphatases *(23,24)*. Residues anticipated to modulate interprotein binding or PP1 function could be mutated by oligonucleotide mutagenesis and then scored for binding or function traits in yeast. The combination of procedures described in this chapter to analyze binding of noncatalytic subunits to mutant PP1 proteins is diagrammed in **Fig. 1**.

2. Materials

2.1. Plasmid Vectors

1. Plasmid vector pAS1 (GenBank U46855) expresses GAL4-DB and uses the yeast *TRP1* selectable marker and pACT2 (GenBank U29899) expresses GAL4-AD and uses the *LEU2* selectable marker *(1)* (*see* **Note 1**). *GLC7* or mutant *GLC7* genes are cloned into pAS1 and *GAC1*, *GLC8*, or *SDS22* into pACT2 such that GAL4 fusions are expressed (*see* **Note 2**).
2. Control plasmids: pSE1111 (pACT2-*SNF4*), pSE1112 (pAS1-*SNF1*). *S. cerevisiae* SNF1 and SNF4 positively interact in the two hybrid assay *(25)*. Transformants with both plasmids are always included as a positive control and interaction with either SNF1 or SNF4 is used as a negative control (*see* **Note 3**).

2.2. Strains

1. Strains JC981 and JC993 were used in this study (*see* **Note 4**). They should be stored at −70°C in sterile 15% glycerol for long-term, but can be used from YEPD plates for 1 mo stored at 4°C. Yeast vectors with *TRP1*, *LEU2*, or *HIS3* can be selected in these strains. The *URA3* marker is unavailable because it was used to select integration of *GAL1-LacZ*.
2. JC981: *MATα gal4 gal80 his3 trp1-901 ura3-52 leu2-3,112 URA3::GAL1-LacZ*
3. JC993: *MATα gal4 gal80 his3 trp1-901 ura3-52 leu2-3,112 URA3::GAL1-LacZ*

2.3. Culture Media

1. Rich medium (YEPD): 1% yeast extract (Difco, Detroit, MI), 2% glucose, 2% Bacto-Peptone (Difco). This medium is used for nonselective growth. All plates have 2% agar included before autoclaving.
2. 100X complete amino acid mix: 2 g of arginine, 1 g of histidine, 6 g of isoleucine, 6 g of leucine, 4 g of lysine, 1 g of methionine, 6 g of phenylalanine, 5 g of threonine, and 4 g of tryptophan per L of water, filter sterilized (0.45 μm pore). Leucine- or tryptophan-deficient mixes were made by omitting them individually (-Leu or -Trp) or together (for -Leu-Trp media). Make only 100–200 mL of these stocks and store at room temperature (*see* **Note 5**).
3. Omission media plates: Add 6.7 g of yeast nitrogen base without amino acids (Difco 0919-15-3), 20 g of agar and 20 g of glucose to one liter water, and autoclave, then add, 10 mL of 100X amino acid mix (-Leu, -Trp, or -Leu-Trp), 1 mL of 1% adenine, 4 mL of 1% uracil, and 5 mL of 1% tyrosine (each filter sterilized) before pouring.
4. Liquid minimal media: Mix 6.7 g of yeast nitrogen base without amino acids, 10 g of succinic acid, 4.5 g NaOH, and 20 g of glucose completely in 1 L water and autoclave, after which add, 10 mL of 100X amino acid mix (-Leu, -Trp, or -Leu-

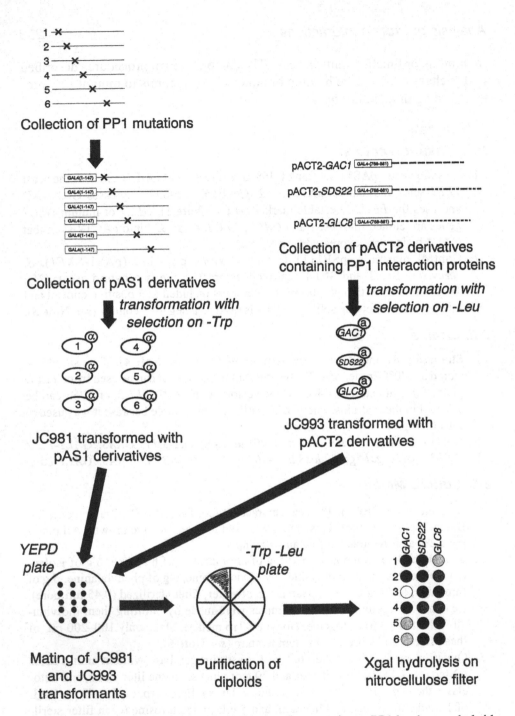

Fig. 1. Scheme to analyze binding of interacting proteins to PP1 by the two-hybrid assay. In this example six mutant forms of PP1 were generated by mutagenesis. Mutant PP1 DNA sequences were cloned into pAS1 such that they made fusions to the GAL4-BD. Coding sequences of interacting proteins were fused to GAL4-AD in ACT2.

Trp), 1 mL of 1% adenine, 4 mL of 1% uracil, and 5 mL of 1% tyrosine. Sodium succinate is not metabolized and it buffers well at pH 5.8 promoting better yeast growth.

2.4. Materials

1. CTDNA: Dissolve 500 mg calf thymus DNA (Sigma, St. Louis, MO, D-1501) in 100 mL water (takes about 2 h). Sonicate until viscosity is greatly reduced (about 10–15 min at 50% setting). Distribute 10 mL portions into 25-mL centrifuge tubes, add potassium acetate to 0.3 M and 15 mL ethanol to each. Incubate at –70°C for 15 min, centrifuge at 10,000g for 10 min, decant ethanol, and air dry. Store at –20°C. When needed, add 10 mL of autoclaved water to the contents of one tube, mix, and store at 4°C.
2. LiAc/TE/PEG: 100 mM lithium acetate, 10 mM Tris-HCl, pH 8.0, 1 mM ethylene diamine tetraacetic acid (EDTA), 40% polyethylene glycol (MW 3350), filter sterilize.
3. Nitrocellulose: The authors have used Micron Separations Inc. or Schleicher and Schuell supported or unsupported with equal success.
4. 3MM filter paper (Whatmann).
5. Z buffer: 60 mM $Na_2HPO_4 \cdot 7H_2O$, 40 mM $NaH_2PO_4 \cdot H_2O$, 10 mM KCl, 1 mM $MgSO_4 \cdot 7H_2O$, 0.27% β-mercaptoethanol.
6. Xgal: 100 mg/mL 5-bromo-4-chloro-3-indolyl-β-D-galactopyranoside (Diagnostic Chemicals Limited) dissolved in dimethylformamide, store at –20°C.
7. 95% ethanol.
8. Filtration manifold: BRL cat. no. 31050-016 or equivalent.
9. 30° and 37°C incubators with roller or shaker.
10. Petri dishes: 100 × 15 mm round and 100 × 100 × 15 mm square.
11. 0.1% SDS.
12. 1 M sodium carbonate.
13. Chloroform.
14. Liquid nitrogen.
15. Spectrophotometer.
16. ONPG: 4 mg/mL. o-nitrophenyl-β–D-galactopyranoside in water. Store at –20°C.

3. Methods

3.1. Transformation of Yeast Strains

Transformation of yeast was performed as described previously *(26)*.

1. Grow strains JC981 and JC993 overnight in liquid YEPD at 30°C with agitation. Transfer approx 10^8 cells (0.5 mL) to a microcentrifuge tube, spin 10 s, and decant the supernatant.

Fig. 1. (*continued*) These two collections of hybrid fusions were separately transformed into yeast. Yeast transformants were mated in all pair-wise combinations and diploids purified on a medium that selects for both hybrid plasmids. Relative affinities of the two hybrids in diploids was assayed by β-galactosidase hydrolysis of Xgal.

2. Add 20 mL of CTDNA and 1 μg of the respective plasmid DNA (either pAS1 or pACT2 fusions) to cells and vortex vigorously (*see* **Note 6**).
3. Add 0.5 mL of LiAc/TE/PEG to the cells and vortex.
4. Leave cells overnight at room temperature without agitation.
5. Plate 50 μL of the settled cells onto appropriate omission media (-Trp media for pAS vectors and —Leu media for pACT2 vectors) and incubate for 2–3 d at 30°C to allow the growth of transformants (*see* **Note 7**).

3.2. Mating Transformants to Generate Diploids

It is necessary to get the two plasmids containing the GAL4-DB and GAL4-AD fusions into the same cells to score the interaction. Plasmids can be transformed serially (first one, then the other), simultaneously (selection on -Leu-Trp, with low efficiency), or individually, followed by mating. The authors use the latter method and describe it here. It has the advantage of producing many combinations of hybrids with little effort. It is useful to simultaneously assay the interaction of mutant GLC7 proteins with several interacting proteins. JC981 and JC993 are haploids of opposite mating types. Transform JC981 with pAS1 derivatives (pAS1-*GLC7wt* or pAS1-*GLC7mut*) and JC993 with pACT2 derivatives (pACT2-*GAC1*, -*GLC8*, or -*SDS22*). Haploids of opposite mating types can conjugate to form diploids. Diploids formed by mating JC981 and JC993 transformed as described can be selected on -Leu-Trp media.

1. After JC981 transformant colonies are about 2–3 mm diameter on the selective minimal medium plate, transfer two or more with a sterile inoculating loop or toothpick to a YEPD plate.
2. Transfer a similar number of JC993 transformants to the same spots on the plate as the JC981's and mix on the surface gently. The resulting mixture should occupy less than 9 mm².
3. Incubate at 30°C for 6–24 h.
4. Purify diploids by streaking mating mixture in sectors (4–8 sectors per 100 × 15 mm plate) on -Leu-Trp plates.
5. Incubate at 30°C for 2–3 d.

3.3. Qualitative Two Hybrid Assay

The β-galactosidase hydrolysis of Xgal produces an insoluble blue product that remains associated with the cells. Relative affinities of interacting proteins can, therefore, be compared via intensities of blue. The following method is a refinement of a published method *(27)*. A filtration manifold is used to control the area that each strain assayed occupies on the nitrocellulose filter. Many strains can be assayed simultaneously under essentially identical conditions (*see* **Note 8**).

1. Suspend approx 8 μL of diploid (prepared above) or doubly transformed haploid cells (matchhead quantity) in 0.5 mL of distilled water.

2. Cut 3 MM and nitrocellulose to size to fit filtration manifold. An 8 × 10 array takes a 9 cm² (*see* **Note 9**).
3. Moisten 3 MM and nitrocellulose with distilled water.
4. Place nitrocellulose on top of 3 MM and sandwich in filtration manifold.
5. Apply 200 µL of resuspended cells to nitrocellulose in the presence of vacuum. It is important to make sure that the cells are packed tightly onto the nitrocellulose and that all liquid has sucked through before the manifold is disassembled.
6. Lyse cells in liquid nitrogen for approx 1 min.
7. Place thawed nitrocellulose in a petri dish containing 3 MM moistened with Z buffer with 10 µL of X-gal stock added per mL.
8. Incubate at 37°C until blue coloration observed (because of the hydrolysis of Xgal). This can take up to 24 h (*see* **Note 10**).
9. Store air-dried filters at room temperature covered with stacks of dry 3 MM to prevent curling.
10. Photograph using daylight film, a blue lens filter, and tungsten lights. A blue background around the filters can prevent commercial developers from producing green images.

3.4. Quantitative β-Galactoside Liquid Assays

These assays are performed by a modification of a previously described method *(28)* and allow numerical comparison of interactions.

1. Grow diploids in -Leu-Trp omission media with agitation at 30°C. After 24 h of growth of the diploids in omission media, dilute fivefold (1 mL into 4 mL fresh medium) and grow until absorbance at 600 nm (using a medium blank) is about 1.0 (record the actual absorbance, A_{600}).
2. Centrifuge 1.5 mL at 10,000g for 10 s to pellet, of the culture, discard medium, and freeze at −20°C.
3. Thaw cells at room temperature and suspend in 1 mL of Z buffer.
4. Permeabilize cells by the addition of 50 µL of 0.1% SDS and 3 drops of chloroform followed by vigorous vortexing for 15 s.
5. Start the reaction by the addition of 200 µL ONPG and incubate at 37°C until the appearance of a yellow coloration.
6. Stop the reaction by the addition of 0.5 mL of 1 *M* sodium carbonate.
7. Pellet cell debris by centrifuging 10,000g for 1 min.
8. Measure absorbance of the supernatant at 420 nm (A_{420}).
9. Miller units of β-galactosidase activity are calculated as follows:

$$\text{Miller unit} = 1000 \cdot A_{420}/(A_{600} \cdot \text{time [min]} \cdot \text{volume [mL]})$$

4. Notes

1. These are *E. coli*/yeast shuttle vectors. The β-lactamase gene confers resistance to 50 µg/mL in *E. coli*. All plasmid preparations are from *E. coli* using standard methods *(29)*. Biosynthetic genes are used as yeast selectable markers. There-

pAS1:

GAL4(1-147) - HA- CAT ATG GCC ATG GAG GCC CCG GGG ATC CGT CGA CCT GCA

pACT2:

GAL4(768-881) - HA- CAT ATG GCC ATG GAG GCC CCG GGG ATC CGA ATT CGA GCT CGA GAG ATC T

Fig. 2. Multiple cloning site region of pAS1 and pACT2 *(1)*. Both vectors include GAL4 residues at the amino termini followed by the hemagglutinin epitope (HA). The locations of unique restriction enzyme cleavages are noted within the multiple cloning site region. The reading frame of GAL4 and HA are noted by the triplet groupings of nucleotides.

fore, the yeast strains used must contain appropriate auxotrophic mutations (*trp1* or *leu2*, for example). These plasmids are high-copy yeast vectors (5–50 copies/cell). The GAL4 fusions in each are expressed constitutively from the yeast *ADH1* promoter. Other potential vectors are available *(30)*, some commercially (Clonetech, Palo Alto, CA).

2. *GLC7* genes were amplified by the polymerase chain reaction (PCR) using primers that included *Bam*HI sites *(31)* and ligated into pAS1 at the unique *Bam*HI site. Plasmid pAS1 has unique sites for *Nde*I, *Nco*I, *Sfi*I, *Sma*I, *Bam*HI, and *Sal*I that can be used to fuse fragments to GAL4-DB. The *GAC1*, *GLC8*, and *SDS22* genes were either cloned as PCR generated restriction fragments or suitable fragments were removed from other plasmids. Plasmid pACT2 has unique *Nco*I, *Sfi*I, *Sma*I, *Bam*HI, *Eco*RI, and *Xho*I sites that generate GAL4-AD fusions. It is not necessary to include all residues of either interacting protein although it is essential to maintain the GAL4 reading frame (**Fig. 2**). The authors included all residues of the proteins described.

3. For the collection of pAS1-*GLC7* mutant fusions, pAS1-*GLC7* (wild-type) and pACT2-*SNF4* is used as a negative control. The positive control, pAS1-*SNF1* and pACT2-*SNF4*, results in a similar β-galactosidase activity as the pAS1-*GLC7* (wild-type)/pACT2-*SDS22* or pAS1-*GLC7* (wild-type)/pACT2-*GAC1* interactions, but pAS1-*GLC7* (wild-type)/pACT2-*GLC8* is substantially lower and requires longer incubation times for suitable Xgal hydrolysis.

4. Y187 and Y190 contain *ade2* mutations that make colonies turn red *(1)*. Blue Xgal hydrolysis from weak two hybrid interactions are difficult to visualize in *ade2* red cells. JC981 and JC993 were derived from Y190 and Y187 *(22)*. They are *ADE2*+ and grow better than their ancestors. Other strains can be used *(30)*. The yeast strain and vector utilized has a profound effect on the signal from almost every two hybrid.

5. It is natural for the tryptophan and histidine in these mixes to turn brown or yellow, respectively. They can be used despite the coloration. All amino acids are natural "L" isomers.

6. The plasmid DNA does not have to be extensively purified. Small scale preparations are suitable. The presence of RNA actually enhances transformation efficiency. Any DNA preparation that can be digested with restriction enzymes will likely work.

7. Initial experiments should include empty vectors and no DNA control transformations. The former should yield the greatest transformation efficiency and growth rate, whereas the later should have no transformants. Toxic fusion proteins are common because of high expression and nuclear localization of some proteins can cripple normal cell functions. Toxicity is evident from reduced transformation efficiency and reduced growth rate of transformants (smaller colony diameter) compared with cells transformed with empty vectors. If one hybrid protein is toxic, cotransformation with both DNAs (1 μg each, selection on -Leu-Trp plates) might alleviate growth inhibition.

8. It is important to proceed from transformation to mating and assay as quickly as possible for reproducible results. Yeast transformed with pAS1 or pACT2 derivatives are stable if grown on selective media, but frequently fresh transformants yield greater apparent affinities than transformants that have grown several generations since transformation.

9. At least two duplicate filters are prepared for each collection of interactions assayed. They are allowed to incubate on Z buffer for different times. Short incubation times will emphasize the strong interactions and longer incubations the weaker interactions. Color development of the negative control constrain the upper limit of incubation time. Little additional Xgal hydrolysis occurs after 24 h.

10. The method described here compares affinities of proteins known to interact. Despite this knowledge, some hybrid proteins fail to yield β-galactosidase activity greater than the negative control. This is particularly an issue for proteins with low affinities or transient association. A possible solution for this problem is to switch fusions. For example, pACT2-*GLC7wt* and pAS1-*GLC8* failed to yield significant β-galactosidase activity. Switching to the pAS1-*GLC7* and pACT2-*GLC8* produced sufficient (albeit low) activity to recognize variations in mutant GLC7 affinities. There are many examples where switching fusions showed that one pair of hybrids yielded more β-galactosidase activity than the other. Another solution is to use a truncation of one of the proteins. This is entirely empirical. Testing a variety of overlapping RAF or NF1 domains for RAS interaction showed that apparent interactions went from strong to undetectable *(22)*.

References

1. Durfee, T., Becherer, K., Chen, P. L., Yeh, S. H., Yang, Y., Kilburn, A. E., Lee, W. H., and Elledge, S. J. (1993) The retinoblastoma protein associates with the protein phosphatase type 1 catalytic subunit. *Genes Dev.* **7,** 555–569.
2. Tu, J., Song, W., and Carlson, M. (1996) Protein phosphatase type 1 interacts with proteins required for meiosis and other cellular processes in *Saccharomyces cerevisiae. Mol. Cell. Biol.* **16,** 4199–4206.

3. Frederick, D. L. and Tatchell, K. (1996) The *REG2* gene of *Saccharomyces cerevisiae* encodes a type 1 protein phosphatase-binding protein that functions with Reg1p and the Snf1 protein kinase to regulate growth. *Mol. Cell. Biol.* **16,** 2922–2931.

4. Fields, S. and Song, O. (1989) A novel genetic system to detect protein-protein interactions. *Nature* **340,** 245–246.

5. Dang, C. V., Barrett, J., Villa-Garcia, M., Resar, L. M., Kato, G. J., and Fearon, E. R. (1991) Intracellular leucine zipper interactions suggest c-Myc hetero-oligomerization. *Mol. Cell. Biol.* **11,** 954–962.

6. Johnston, M. (1987) A model fungal gene regulatory mechanism: the *GAL* genes of *Saccharomyces cerevisiae*. *Microbiol. Rev.* **51,** 458–476.

7. Cannon, J. F., Pringle, J. R., Fiechter, A., and Khalil, M. (1994) Characterization of glycogen-deficient *glc* mutants of *Saccharomyces cerevisiae*. *Genetics* **136,** 485–503.

8. Hisamoto, N., Sugimoto, K., and Matsumoto, K. (1994) The *GLC7* type 1 protein phosphatase of *Saccharomyces cerevisiae* is required for cell cycle progression in G2/M. *Mol. Cell. Biol.* **14,** 3158–3165.

9. Cannon, J. F., Clemens, K. E., Morcos, P. A., Nair, B. M., Pearson, J. L., and Khalil, M. (1995) Type 1 protein phosphatase systems of yeast. *Adv. Prot. Phosphatases* **9,** 215–236.

10. Feng, Z., Wilson, S. E., Peng, Z.-Y., Schelnder, K. K., Reimann, E. M., and Trumbly, R. J. (1991) The yeast *GLC7* gene required for glycogen accumulation encodes a type 1 protein phosphatase. *J. Biol. Chem.* **226,** 23,796–23,801.

11. Wek, R. C., Cannon, J. F., Dever, T. E., and Hinnebusch, A. G. (1992) Truncated protein phosphatase GLC7 restores translational activation of *GCN4* expression in yeast mutants defective for the eIF-2 alpha kinase GCN2. *Mol. Cell. Biol.* **12,** 5700–5710.

12. Black, S., Andrews, P. D., Sneddon, A. A., and Stark, M. J. (1995) A regulated *MET3-GLC7* gene fusion provides evidence of a mitotic role for *Saccharomyces cerevisiae* protein phosphatase 1. *Yeast* **11,** 747–759.

13. Tu, J. and Carlson, M. (1994) The GLC7 type 1 protein phosphatase is required for glucose repression in *Saccharomyces cerevisiae*. *Mol. Cell. Biol.* **14,** 6789–6796.

14. Tu, J. and Carlson, M. (1995) REG1 binds to protein phosphatase type 1 and regulates glucose repression in *Saccharomyces cerevisiae*. *EMBO J.* **14,** 5939–5946.

15. François, J. M., Thompson-Jaeger, S., Skroch, J., Zellenka, V., Spevak, W., and Tatchell, K. (1992) *GAC1* may encode a regulatory subunit for protein phosphatase type 1 in *Saccharomyces cerevisiae*. *EMBO J.* **11,** 87–96.

16. Tang, P. M., Bondor, J. A., Swiderek, K. M., and DePaoli-Roach, A. A. (1991) Molecular cloning and expression of the regulatory (RG1) subunit of the glycogen-associated protein phosphatase. *J. Biol. Chem.* **266,** 15782–15789.

17. Foulkes, J. G. and Cohen, P. (1980) The regulation of glycogen metabolism. Purification and properties of protein phosphatase inhibitor-2 from rabbit skeletal muscle. *Eur. J. Biochem.* **105,** 195–203.

18. MacKelvie, S. H., Andrews, P. D., and Stark, M. J. (1995) The *Saccharomyces cerevisiae* gene *SDS22* encodes a potential regulator of the mitotic function of yeast type 1 protein phosphatase. *Mol. Cell. Biol.* **15,** 3777–3785.
19. Hisamoto, N., Frederick, D. L., Sugimoto, K., Tatchell, K., and Matsumoto, K. (1995) The *EGP1* gene may be a positive regulator of protein phosphatase type 1 in the growth control of *Saccharomyces cerevisiae. Mol. Cell. Biol.* **15,** 3767–3776.
20. Renouf, S., Beullens, M., Wera, S., Van, E. A., Sikela, J., Stalmans, W., and Bollen, M. (1995) Molecular cloning of a human polypeptide related to yeast sds22, a regulator of protein phosphatase-1. *FEBS Lett.* **375,** 75–78.
21. Ramaswamy, N. T., Li, L., Khalil, M., and Cannon, J. F. (1997) Regulation of yeast glycogen metabolism and separation by *Glc7p* protein phosphatase. *Genetics* (submitted).
22. Dalley, B. K. and Cannon, J. F. (1996) Novel, activated RAS mutations alter protein-protein interactions. *Oncogene* **13,** 1209–1220.
23. Goldberg, J., Huang, H. B., Kwon, Y. G., Greengard, P., Nairn, A. C., and Kuriyan, J. (1995) Three-dimensional structure of the catalytic subunit of protein serine/threonine phosphatase-1. *Nature* **376,** 745–753.
24. Egloff, M.-P., Cohen, P. T. W., Reinemer, P., and Barford, D. (1995) Crystal structure of the catalytic subunit of human protein phosphatase 1 and its complex with tungstate. *J. Mol. Biol.* **254,** 942–959.
25. Yang, X., Hubbard, E. J. A., and Carlson, M. (1992) A protein kinase substrate identified by the two-hybrid system. *Science* **257,** 680–682.
26. Elble, R. (1992) A simple and efficient procedure for transformation of yeasts. *BioTechniques* **13,** 18–20.
27. Breeden, L. and Nasmyth, K. (1985) Regulation of the yeast *HO* gene. *Cold Spring Harb. Symp. Quant. Biol.* **50,** 643–650.
28. Yocum, R. R., Hanley, S., West, R., and Ptashne, M. (1984) Use of *lacZ* fusions to delimit regulatory elements of the inducible divergent *GAL1-GAL10* promoter in *Saccharomyces cerevisiae. Mol. Cell. Biol.* **4,** 1985–1998.
29. Maniatis, T., Fritsch, E. F., and Sambrook, J. (1982) *Molecular cloning: A Laboratory Manual.* Cold Spring Harbor Laboratory, Cold Spring Harbor, New York.
30. Bartel, P. L. and Fields, S. (1995) Analyzing protein-protein interactions using two-hybrid system. *Meth. Enzymol.* **254,** 241–263.
31. Stuart, J. S., Frederick, D. L., Varner, C. M., and Tatchell, K. (1994) The mutant type 1 protein phosphatase encoded by *glc7-1* from *Saccharomyces cerevisiae* fails to interact productively with the GAC1-encoded regulatory subunit. *Mol. Cell. Biol.* **14,** 896–905.

22

Identifying Protein Phosphatase 2A Interacting Proteins Using the Yeast Two-Hybrid Method

Brent McCright and David M. Virshup

1. Introduction

Protein phosphatase 2A (PP2A) is one of the major intracellular serine/threonine protein phosphatases. PP2A in vivo is an ABC heterotrimer whose substrate specificity is regulated in large part by the variable B subunit. Several additional polypeptides have been identified that modulate the activity of the heterotrimer. It seems likely that more regulatory subunits and interacting polypeptides exist that may be of low abundance or expressed in a tissue-specific manner. One attractive approach to finding these novel interacting polypeptides is to utilize the two-hybrid interaction method.

Since its introduction in 1989 *(1)* the two hybrid method (also known as the interaction trap) has become a widely utilized tool to identify protein-protein interactions using the power of yeast genetics. There are now available multiple vectors and a large number of commercially available libraries to facilitate two-hybrid assays. The method, which is well-described in a number of reviews *(2,3)*, is based on the finding that a number of transcription factors have modular domains; their sequence-specific DNA-binding domains can function separated from their transcriptional activation domains. Thus, a subunit of PP2A may be expressed as a fusion protein with the specific DNA binding domain of a transcription factor, e.g., the LexA protein. This is used as the "bait." The "prey" is expressed from a separate plasmid as a fusion with a yeast transcriptional activation domain. If the two fusion proteins (the bait and the prey) physically interact in the yeast nucleus, the transcriptional activation domain is apposed to the DNA-binding domain, and a functional transcription factor is formed. Transcriptional activation is then selected or screened for by

From: *Methods in Molecular Biology, Vol. 93: Protein Phosphatase Protocols*
Edited by: J. W. Ludlow © Humana Press Inc., Totowa, NJ

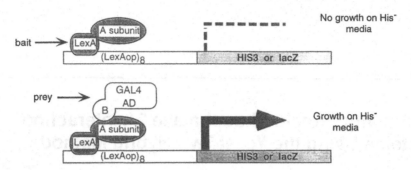

Fig. 1. Cartoon depiction of the essentials of the two-hybrid method, as illustrated by the interaction of PP2A B subunits with a LexA-A subunit fusion protein. B = PP2A B subunit. GALY AD = transcriptional activation domain of GAL4. See text for details.

assaying for the transcription of reporter genes, often HIS3 and lacZ, cloned downstream of the LexA operators (**Fig. 1**).

One important caveat in the application of the two-hybrid method is that there must be an independent assay to prove that the interacting cDNAs encode polypeptides that interact in vivo. PP2A is ideal in this regard, as good probes to identify the A and C subunits are available, and enzyme assays can be done to assess the effect of the expressed polypeptides on PP2A activity. Using the two-hybrid method as outlined in this chapter, the authors have isolated five members of the PP2A B'/RTS1/B56 gene family and demonstrated their interaction with the PP2A holoenzyme *(4,5)*.

This chapter describes one detailed method for searching for PP2A interacting polypeptides. Variations in this methodology that may prove fruitful include searching cDNA libraries from diverse sources, using modified PP2A baits such as the Aβ or the mutant A subunits as described in **ref. 6**, and coexpressing modifying proteins such as the Cα and Cβ subunits.

2. Materials
2.1. Yeast Strains and Plasmids

1. *Saccharomyces cerevisiae* strains suitable for the two-hybrid screen. These strains carry, integrated into their genome, two genes, usually lacZ and HIS3, driven by multimerized LexA binding sites (**Fig. 1**). Alternative strains use Gal4 binding sites, and may use LEU2 as the reporter gene *(7)*. These strains are also auxotrophic for at least 3 markers, e.g., tryptophan, leucine, and histidine. These selectable markers are used for maintaining the bait and prey plasmids, and for selection of interacting clones. In addition, it is desirable to have two similar strains that are of opposite mating types. The authors used the yeast strain L40 to screen the cDNA library *(8)*. L40 has the partial genotype MATa his3Δ200, trp1-1, leu2-3,112, ade2, LYS2::(lexAop)$_4$-HIS3, URA3::lexAop)$_8$-lacZ *GAL4*. As

lower case denotes mutant genes, this strain requires histidine, tryptophan, leucine, and adenine for survival. To perform mating assays the yeast strain AMR70 (constructed by Rolf Sternglanz) was used, which has the partial genotype MATα his3 lys2 trp1 leu2 URA3::(lex$_{op}$)$_8$-lacZ GAL4. These strains are widely available. *S. cerevisiae* strains that utilize Gal4 operator sequences instead of LexA are commercially available.

2. A bait plasmid. The bait plasmid expresses the DNA binding domain of a transcription factor, followed by cloning sites to insert cDNA encoding your bait. It must also contain a selectable marker (e.g., TRP1) compatible with the yeast strains being used. The authors used the plasmid pBTM116 to express their bait, the Aα subunit of PP2A. This plasmid, constructed by Paul Bartel and Stan Fields, uses the bacterial protein LexA as the DNA binding domain *(8)*. Commercially available variations use the DNA-binding domain of the Gal4 protein. The bait protein is able to bind to its cognate DNA sequence engineered upstream of reporter genes carried in the yeast. However, without a transcriptional activation domain, there is no production of His3 or β-galactosidase, and the yeast will neither grow on media lacking histidine, nor metabolize chromogenic substrates of β-galactosidase. When a polypeptide carrying a transcriptional activation domain binds to the bait construct, it brings a transcriptional activation domain to the DNA and transcription of HIS3 and LacZ begins (**Fig. 1**).

3. A cDNA library contained in a plasmid designed for the two-hybrid screen. This plasmid should express the activation domain of a transcription factor (e.g., GAL4 or VP16) fused to the library cDNAs. To express such interacting proteins, the authors have used the plasmid pGAD-GH, that expresses polypeptides from a HeLa cDNA library fused to the Gal4 transcriptional activation domain (GAL4-AD). This prey plasmid must also contain a selectable marker, e.g., LEU2, and it may be desirable for it to contain a nuclear localization signal. Several companies now sell two hybrid systems (Clontech, Palo Alto, CA; Origene Technologies, Rockville, MD; Invitrogen, Carlsbad, CA); alternatively, these components can be obtained from other labs, or constructed by individual labs to meet specific requirements.

2.2. Plates and Media for Two-Hybrid Screen

Detailed descriptions of yeast media are available from a variety of sources, including the Cold Spring Harbor Laboratory manual *(9)* and Methods in Enzymology, vol. 194 *(10)*.

2.3. Drop-Out Media

Synthetic "drop-out" media is defined media with one or several essential nutrients "dropped out." This media is required for selection and maintenance of plasmids in transformed yeast. A detailed description of its preparation is available in standard texts *(9,10)*. Pre-mixed drop-out media is available from Bio101 (La Jolla, CA).

2.3.1. Synthetic Liquid Media

1. -trp -ura (-WU); to expand L40 cells with bait plasmid pBTM116.
2. -trp -ura -leu (-WUL); allows growth of cells with both pBTM116 and the library plasmid pGAD encoding Leu2.
3. -trp -his -ura -leu -lys (-WHULK); allows growth of cells carrying both the bait and prey vectors only if they also interact and drive expression of the HIS3 gene.

2.3.2. Solid Media

1. -trp -ura (-WU); to select transformants of L40 containing the bait vector pBTM116.
2. -trp -leu -ura (-WUL); to quantitate transformants and to select for diploid cells in the mating assay.
3. -trp -his -ura (-WHU); to test for bait activation of HIS3 transcription.
4. -trp -his -ura -leu -lys (-WHULK); to select for interacting clones in the first step of the library search, allows growth of cells carrying both the bait and prey vectors only if they also interact and drive expression of the HIS3 gene.

For Red-White screening of yeast carrying ADE2, the plates are supplemented with tryptophan, the adenine concentration is reduced 10-fold to 10 mg/L, and glucose is reduced to 1%.

2.4. Equipment Needed

2.4.1. General Equipment Needed for the Two-Hybrid Method

1. 30°C incubator.
2. Shaking incubator, adjustable to 30°C.
3. Rotating wheel at room temperature or 30°C.
4. For replica plating, 4" velvet squares (autoclaved) and a cylinder 9 cm in diameter, at least 5 cm tall, fabricated from aluminum or Plexiglas. A hose clamp is used to hold the velvet in place on the cylinder.
5. Appropriate media and plates (**Subheadings 2.2.** and **3.3.**).
6. Bait and prey plasmids of interest.
7. 1 M Lithium acetate (sterile).
8. DMSO (high quality, not yellow in color).
9. 10 mg/mL sheared salmon sperm DNA.
10. 50% PEG-3350 (sterile filtered).
11. TE (10 mM Tris-HCl, 1 mM EDTA, pH 8.0).

2.5. β-Galactosidase Activity Assay-Solution Method (9)

1. Z buffer: 60 mM Na$_2$HPO$_4$, 40 mM NaH$_2$PO$_4$, 10 mM KCl, 1 mM MgSO$_4$, pH 7.0.
2. Yeast lysis buffer: 0.1 M NaCl, 50 mM Tris-HCl, pH 8.0, 1 mM EDTA, 1 mM PMSF, 2 μg/mL pepstatin, and 2 μg/mL leupeptin.
3. 4 mg/mL ONPG (o-nitrophenyl β-D galactopyranoside) in H$_2$O.
4. Glass beads, acid washed (Sigma, St. Louis, MO; others).

5. 1 M Na_2CO_3.
6. Bradford reagent (Biorad, Hercules, CA).

2.6. β-Galactosidase Activity Assay-Filter Method (11,12)

1. Nitrocellulose filters.
2. Aluminum foil folded in the shape of a round saucer.
3. Liquid nitrogen.
4. Z buffer (*see* **Subheading 2.5.**).
5. X-gal (5-bromo-4-chloro-3-indolyl-β-D-galactoside): 50 mg/mL in dimethyl formamide.

2.7. Isolation of cDNA Plasmid from Positive Yeast Clones (9)

1. Lysis buffer: 2.5 M LiCl, 50 mM Tris-HCl, pH 8.0, 4% Triton X-100, 62.5 mM EDTA.
2. Acid washed glass beads.
3. Phenol/chloroform/isoamyl alcohol 25:24:1.
4. Ice-cold ethanol, 100% and 70%.

2.8. DNA Analysis and Grouping of Two-Hybrid cDNAs

1. Dot blot analysis-96 well-dot blotting manifold, nylon membrane (.45 μm), Prime-It kit (Stratagene), [$\alpha^{32}P$] dCTP, Sepharose G-50.
2. PCR and restriction digest analysis -Taq polymerase, dNTPs, primers, restriction enzymes that have four base recognition sites (HaeIII, HinfI, and so on).

2.9. Assay for Interaction of Cloned Subunits with PP2A Holoenzyme

Mammalian and bacterial expression systems to express and purify the putative interacting protein.

3. Methods

The bait and cDNA library plasmids are chosen on the basis of the anticipated binding characteristics of the desired interacting polypeptide. If one is searching for PP2A regulatory B subunits, the bait used could be either the A subunit or the catalytic subunit. In addition, partial products (putative domains) could be used to either utilize specific regions of the subunits as bait, or to inactivate functional domains that may interfere with the two-hybrid screen. For the authors' two-hybrid screen, they used the full-length PP2A Aα subunit fused to LexA as the bait. In addition, they used lamin fused to LexA (LexA-lamin) as a negative control and truncated and mutated forms of the Aα subunit to further analyze the putative B subunits. To facilitate mating assays (described below) the ADE2 gene was also cloned into the bait vector (**Fig. 2**).

Two general types of cDNA libraries are available, those with inserts of ~500-bp average size, and those with longer, potentially full-length inserts.

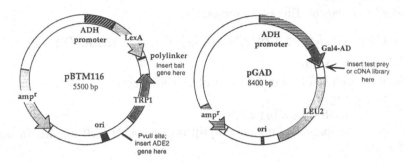

Fig. 2. pBTM116 is a representative bait vector, whereas pGAD is a widely used prey vector.

For list of suppliers, *see* **Subheading 2.1.3.** The shorter inserts will contain short polypeptides and 'domains' lacking potentially inhibitory or toxic sequences. The disadvantages of libraries with small inserts is that more clones must be screened, large domains will be missed, and polypeptide fragments may have interaction specificities irrelevant to the parental protein, leading to an increased frequency of false positives. The authors chose in their screen to utilize a library from HeLa cDNA with near full-length inserts *(13)*.

Libraries may be amplified by electroporation into high-efficiency competent cells, e.g., DH10B cells from Life Technologies, and grown in bulk culture before large scale plasmid purification following standard protocols. Be sure your transformation efficiency is high enough to represent the complexity of the library several times over, i.e., if the library has 2×10^6 unique clones, make sure you have at last 8–10×10^6 primary transformants. This may require measuring transformation efficiency and transforming the library several times.

3.1. Cloning of the Bait and Controls into the Two-Hybrid Constructs

For test preys and positive controls the authors used the PP2A Cα subunit cloned in frame with the GAL4 activation domain (Cα-GAL4), PR72-GAL4, and Bα-VP16. All these bait and control proteins were cloned in frame using PCR. A low number of PCR cycles (usually 10–12), and a high fidelity polymerase such as Pfu DNA polymerase (Stratagene), should be used to minimize the chance for PCR error.

Before a large-scale library screen is performed, it is imperative that the bait be characterized as fully as possible. It only takes a week or two to perform the following controls, and add confidence to future results. Alternatively, one can waste months analyzing the clones obtained in a large-scale transformation done using improper conditions. The following preliminary experiments should be performed.

1. Show the bait fusion protein is expressed, either by immunoblot, immunoprecipitation, or by interaction with a known positive control bait, e.g., A subunit bait and B subunit prey (*see* **Note 1**).
2. Does the bait transactivate on its own or interact with a negative control prey, e.g., GAL4AD-lamin? Does it interact with a known interacting protein? If the bait alone transactivates HIS3 expression, then a high background will be obtained. In this case, it may be necessary to add low levels of the His3 enzyme inhibitor 3-aminotriazole into the plates.

3.2. Small Scale Yeast Transformations for Testing Controls

The following method was adapted from the method of Schiestl and Gietz *(14,15)*.

1. Prepare an overnight culture of L40 in 10 mL YPAD (or -WU synthetic media if transforming yeast already containing the bait plasmid) by overnight growth in a shaker at 30°C.
2. In the morning, dilute to OD_{600} of 0.4 in 50 mL YPAD and let grow an additional 2–4 h.
3. Wash the yeast by centrifugation at 1500g at room temperature. Resuspend the pellet in 40 mL TE and recentrifuge.
4. Resuspend the washed cells in 2 mL 100 mM lithium acetate, 0.5X TE, and incubate for 10 min at room temperature.
5. Mix well 100 μL of the cell suspension, 1 μg plasmid, and 100 μg sheared, denatured salmon sperm DNA prepared by the method of **ref. *14***.
6. Add 700 μL 100 mM lithium acetate, 40% PEG-3350 in TE, mix well, and incubate at 30°C for 30 min.
7. Add 88 μL DMSO, mix well, and heat shock 7 min at 42°C.
8. Wash the cells twice by spinning briefly in a microcentrifuge, removing the supernatant, and resuspending in 1 mL TE. The second time, resuspend in 100 μL and plate on selective medium. Incubate the plates at 30°C. Colonies will be visible in 2 d.

Two plasmids can be transformed together, therefore, the test baits and preys can be simultaneously transformed. By testing the doubly transformed yeast for growth on -leu, -trp and -WHULK plates, two-hybrid interactions can be assayed.

It is helpful to try a small-scale library transformation of yeast containing the bait to estimate what the efficiency of transformation is, that is, how many +leu, +trp colonies per μg of library DNA are obtained. The expected efficiency is 10,000–100,000 transformants per μg of library DNA. In addition, this small-scale library transformation can be used to estimate the number of positives by selecting for +his colonies.

Interacting polypeptides can be selected on -his plates, and the interaction confirmed and quantitated by assessing lacZ expression. Qualitative lacZ assays are performed by the β-galactosidase filter method on large numbers of putative positives, whereas quantitative assays are performed with the ONPG solution assay. The amount of background lacZ expression for the bait plasmid alone should first be quantified by ONPG assay (*see* **Subheading 3.3.**). A positive control may be constructed, depending on the bait being used. The authors have demonstrated strong interaction between the A subunit bait and several B subunit preys.

3.3. Quantitative Analysis of Two-Hybrid Interaction by ONPG Assay

The stronger the two-hybrid interaction, the more transcription of the reporter genes HIS3 and lacZ. Thus, a quantitative measure of the amount of β-galactosidase activity present in a yeast extract is a rough estimate of the strength of the two-hybrid interaction *(16)*. This can be performed using the colorimetric β-galactosidase substrate ONPG.

1. A 10-mL overnight culture is grown in appropriate media, usually -trp -leu (-WL).
2. Harvest the cells by centrifugation at 1500*g* for 5 min, wash once and resuspend in 700 µL lysis buffer in 1.5 mL microfuge tube.
3. Add 300-µL glass beads and vortex vigorously. To keep the extracts from getting hot, it is best to vortex for 15–30 s, then place on ice for 30 s. Repeat at least three times.
4. Pellet cellular debris and beads by centrifuging at 14,000*g* for 5 min in a microcentrifuge.
5. Collect supernatant and determine protein concentration.
6. Add 30 µg of protein to Z buffer to make 1 mL total.
7. Add 200 µL of 4 mg/mL ONPG.
8. Incubate at 30°C until a yellow color is observed (5 min to several hours).
9. Stop reaction by adding 0.5 mL of 1 M Na$_2$CO$_3$.
10. Read absorbance at 420 nm using a spectrophotometer.
11. Calculate β-galactosidase activity in Miller units using the following formula: Activity (in U/min/µg) = OD$_{420}$/time (min)/µg protein *(9,17)*.

A more rapid large scale method may be the permeabilized cell assay *(18)*. Yeast are grown in -WUL to an OD$_{600}$ of 1.0, diluted between 1:5 to 1:100 into Z buffer (final volume of 1 mL), and permeabilized by the addition of 50 µL 0.1% SDS and 3 drops of chloroform followed by vortexing and centrifugation. Extracts are assayed as above and normalized to OD$_{600}$ or protein concentration.

3.4. Large-Scale Yeast Transformation and Selection for Interacting Clones on -His Plates

1. For large scale library transformations, take a saturated 2 mL culture of L40 cells with the bait plasmid in -WU and inoculate into 100 mL of -WU. Let this grow

overnight shaking in a 30°C incubator. Inoculate into 1 L of prewarmed YPAD. Grow an additional 3 h, giving the cells a chance to double. Few cells will lose the bait plasmid in only 3 h.

2. Pellet the cells by centrifugation at 2500 rpm in a fixed angle rotor (e.g., a Sorvall GSA) for 5 min at room temperature. Resuspend the pellet in 500 mL TE and repellet.

3. Resuspend the pellet in 20 mL 100 m*M* LiAc/0.5X TE.

4. Mix together 1 mL of 10 mg/mL denatured salmon sperm DNA and 400 μg library plasmid DNA then add to the yeast mixture.

5. Add 140 mL 100 m*M* LiAc/40% PEG-3350 in TE, mix well, and incubate 30 min at 30°C.

6. Place cell/DNA mix in a sterile 2-L beaker, and cover with foil. Add 17.6 mL DMSO and swirl to mix. To heat shock, place the beaker in a 42°C water bath for 6 min with occasional mixing. At the end of the heat shock, rapidly cool by adding 400 mL YPA (without glucose) and shift to a room temperature water bath.

7. After 10 min, pellet the cells as above, wash the pellet once with 500 mL YPAD, and resuspend in 1 L prewarmed YPAD.

8. Allow the cells to recover 1 h at 30° C in YPAD with gentle shaking.

9. Take 1 mL of the culture, gently pellet, resuspend in 1 mL -WUL, and plate 1 and 10 μL on -WUL plates. The number of colonies multiplied by the dilution factor will give the number of primary transformants, hopefully in the $10–100 \times 10^6$ range. To achieve full coverage of a cDNA library, aim for a four- to fivefold over-representation of the library.

10. Pellet the liter of cells in YPAD, wash once in -WUL, and resuspend in 1 L prewarmed -WUL. Note that when using the lexA system, it appears necessary to allow the cells to recover by shaking at 30°C in -WUL for from 4–16 h in -WUL before plating on -WHULK plates. Otherwise, the plating efficiency is very low. This recovery step may not be necessary if using a GAL4 bait.

11. Wash the cells twice in -WHULK and resuspend the final pellet in 10 mL of -WHULK.

12. Plate 100-200 μL/plate on 25–100 15 cm -WHULK plates. Make 10^{-6} and 10^{-7} dilutions and plate on -WUL plates to calculate the plating efficiency and the number of doublings since **step 9**. Too many nonviable cells on a plate will hinder the growth of His+ cells.

13. Approximately 48 h later, pick the His+ colonies (i.e., those with putative interacting pairs) to a grid on -WHULK plates for further analysis.

3.5. Testing for Specific Interaction by X-Gal Assay

Interacting pairs will express both His3 and β-galactosidase. Production of β-galactosidase is qualitatively assessed by a filter X-Gal assay. For the filter assay (adapted from **refs. *11*** and ***12***):

1. Prepare a -WUL or -WHULK plate with the arrayed colonies to be tested (**step 13**, above).

2. Lay a dry nitrocellulose filter (0.45 µm supported nitrocellulose circles work well) onto the colonies for 5 min.
3. To lyse the yeast now adsorbed to the filter, place colony-side up on an aluminum foil boat floating in liquid nitrogen. After 30 s, submerge the filter for an additional 5 s. Remove and allow to thaw at room temperature, colony side up.
4. To detect the presence of β-galactosidase, to 1.5 mL of Z buffer, add 15 µL of a 50 mg/mL X-gal stock. Place a #1 Whatman filter circle in a Petri dish, soak it with the 1.5 mL of X-gal:Z buffer mix, and place the nitrocellulose filter colony side up on top of the filter paper. Cover the dish and incubate at 30°C. Strong interactions will lead to the hydrolysis of visible quantities of X-gal within 30 min, although the reaction can be left for longer times.

Pick all the HIS3$^+$ colonies expressing lacZ to a new grid. In the authors' library screen, 90% of the HIS3$^+$ colonies were also lacZ$^+$.

3.6. Testing for Specific Interactions by Mating Assay

After colonies have gone through the β-galactosidase assay, it is necessary to screen for nonspecific interactions by testing the putative positive prey plasmids for nonspecific interactions against other bait plasmids. Interacting preys should interact only with the specific bait and not with alternative, unrelated baits. This assay will weed out library positives that for example, interact with the DNA binding domain of the bait, or with the reporter genes directly. To test the ability of the interacting preys to interact with alternative baits, it is necessary to isolate the prey plasmids of interest separate from the bait plasmid. This requires the ability to either select or screen for yeast that have lost the bait plasmid, but still carry the prey plasmid of interest. If the bait plasmid carries a marker that can be selected against, URA3 or CAN, yeast colonies of interest can simply be plated on the appropriate medium (e.g., -leu, + 5FOA) and only cells with the prey plasmid will grow. This approach is not available for pBTM116. The authors have utilized an alternate, albeit more laborious strategy, instead cloning the ADE2 gene into their bait plasmid (*see* map). In this case, the candidate colonies are grown for 2 d in the absence of selection for the bait plasmid (in 2 mL of YPAD or -leu +trp liquid medium) and then plated on -leu, low adenine plates. Colonies with the bait plasmid with the ADE2 gene are white, whereas colonies that have lost the ADE2 gene (and hence the bait plasmid) accumulate a metabolite in the adenine biosynthetic pathway and are red. This method requires a large number of selective plates, i.e., one per colony. In the authors' hands, only about 1–5% of the colonies per plate will lose the bait after this procedure.

A modification of this procedure suggested by Stan Hollenberg (personal communication) is to spot a very small number of cells directly onto -leu, +trp plates, then restreak from red parts of the colony onto a second -leu, +trp plate.

Note that the number of cell divisions determines the number of cells that lose the bait plasmid, so if necessary increase the number of cell divisions by either plating fewer cells the first time, or allowing the colonies to grow longer.

The L40 yeast now containing only the prey plasmid (i.e., are leu- and red on low adenine media) are then mated with the tester strain (e.g, AMR70) carrying one of several bait plasmids. For a negative control, the authors have tested their library positives against LexA-lamin; additionally, they have used LexA-PP2A A subunits with various mutations to allow classification of interactions. For instance, A baits with a mutation in loop 5 *(6,19)* do not interact with known B subunits *(4)*. whereas those with carboxy-terminal truncations do not interact with the PP2A catalytic subunit.

3.6.1. The Mating Assay

1. Prepare a lawn of yeast (AMR70, Matα) containing one of the various test baits, including at a minimum the original bait and a nonspecific bait such as LexA-lamin, on -WU media.
2. Prepare an array of the L40 strain (Mata) carrying the various prey plasmids, but cured of the original bait, on -leu plates.
3. Replica plate the AMR70 and L40 cells together on YPAD plates and incubate overnight at 30°C to allow mating to take place.
4. Replica plate from YPAD to -WUL to select for diploid yeast and -WHULK to select for His3 production. Bait-prey interaction and, thus, production of lacZ can also be assayed by filter assay from the -WUL plate with the β-galactosidase filter assay. False positive prey will show interaction with nonspecific baits such as LexA-lamin, whereas true positives will continue to interact with the original bait. In the author's assay, 80% of the HIS3+/lacZ+ colonies retested positive with the original bait in the mating assay.

3.7. Isolation of Interacting cDNA Library Plasmids (the Yeast Miniprep)

1. Yeast colonies containing the prey plasmid of interest are grown overnight in 5 mL-leu liquid media.
2. Pellet cells at 2500 rpm for 5 min.
3. Resuspend the pellet in 450 µL lysis buffer and transfer to a 1.5 mL tube.
4. Add 150 µL glass beads and 300 µL phenol:chloroform:isoamyl alcohol (25:24:1). Vortex at high speed for at least 1 min.
5. Centrifuge 1 min top speed in a microcentrifuge.
6. Transfer the aqueous phase to a fresh tube and precipitate with 2 vol ice-cold EtOH. Wash pellet once with cold 70% EtOH, then resuspend in 25 µL 10 m*M* Tris-HCl, 1 m*M* EDTA, pH 8.0.
   This DNA can still contain nucleases and may not be stable even at −70°C. Further purification by repeat phenol extraction and precipitation is necessary if not used for bacterial transformation PCR, or dot blotting immediately.

3.8. Analysis and Classification of Library Plasmids

A two-hybrid screen may yield only a few positives that pass through all the screening, or it may produce hundreds of positives. In the former case, it may be relatively easy to sort through the recovered plasmids to determine how many times each library clone was recovered. However, if hundreds of positives are found, it may be necessary to sort the plasmids into specific families before further analysis. Restriction digest analysis, DNA sequencing, and dot-blot hybridization can be used to further analyze the library clones.

3.8.1. Dot-Blot Hybridization

The authors found that yeast plasmid preps immobilized on nylon membranes could be probed multiple times with ^{32}P-labeled inserts from selected positives. If a positive was represented multiple times in the library, only one representative clone need be further analyzed. This is also a very useful technique to determine whether you have recovered known PP2A interacting polypeptides.

1. Make 3 duplicate sets of dot blots by spotting yeast minipreps on nylon membranes in a 96-well format, and UV crosslinking. Always include as a positive control the library plasmids that will be used to probe the membrane.
2. After transformation of recovered library plasmids into *E. coli*, make minipreps of several randomly chosen clones. It may be fruitful to probe with known interacting proteins if you expect your screen will turn these up.
3. Isolate the library insert by restriction digest, agarose gel electrophoresis, and Geneclean.
4. Radiolabel the insert by nick-translation or random hexanucleotide priming.
5. Hybridize the probes to the membranes, wash and expose membranes to film. Clones that hybridize to multiple colonies are more likely to be interesting than those that are unique and were isolated only once.

In the authors' two-hybrid screen, 170/212 positive clones contained an insert from the same gene. Thus, the dot-blot approach reduced the complexity of subsequent analysis substantially (*see* **Note 2**).

3.8.2. Analysis of Library Plasmids by Restriction Digest and PCR

A related method to identify library clones derived from the same gene relies on PCR and restriction analysis. This approach will work well with a more limited number of positives.

1. Make DNA minipreps from yeast containing the library isolates.
2. Use primers flanking the cDNA insertion site to amplify the insert in a standard 30-cycle PCR reaction.
3. Digest the PCR products with restriction enzymes that recognize four base sites and separate on 8% polyacrylamide-TBE gel.

4. Stain with ethidium bromide and photograph. Clones obtained from the same gene will share many common restriction fragments.

3.8.3. Sequencing of Plasmids Isolated by the Two-Hybrid Screen

The same primers used for PCR can be used for sequencing both ends of two-hybrid clones. Inserts may be sequenced after recovery from the yeast, or by PCR amplification and direct sequencing of the PCR product. Yeast miniprep DNA will not be of sequencing quality; passage through *E. coli* is required.

3.9. Biochemical Assays for In Vivo Interaction

A two-hybrid interaction *per se* does not prove that the polypeptides of interest directly interact in vivo. Additional biochemical assays are required to demonstrate that a polypeptide encoded by a library isolate in fact has a physiologic effect on PP2A (*see* **Note 3**). The most direct demonstration that a new subunit of PP2A has been cloned is to show that the new subunit is associated with other subunits of PP2A in vivo. Copurification and coimmunoprecipitation are strong demonstrations of interaction. Alternatively, the new gene may encode a regulator of the PP2A holoenzyme (*see* **Note 4**). Demonstration of an effect of the recombinant protein on enzyme activity may then be required (*see* **Note 5**). Note that in all cases, these assays are best performed using full-length protein, rather than a nonfull-length library isolate.

3.9.1. To Demonstrate In Vivo Interaction

1. The two-hybrid clone of interest is subcloned into a mammalian expression vector that contains a constitutive promoter (e.g., CMV) and an epitope such as the hemagglutinin tag.
2. The constructs are expressed by transient transfection into the cell line of choice. It may be prudent to avoid COS cells, as they express SV40 small t antigen that is known to interact with the A subunit of PP2A.
3. 36–48 h following transfection, the cells are lysed and subjected to immunoprecipitation with the anti-epitope antibody. The epitope-tagged candidate PP2A subunit should immunoprecipitate with other PP2A subunits, as analyzed by immunoblotting.
4. PP2A phosphatase activity in the immunoprecipitate can be determined by measuring the release of trichloroacetic acid-soluble ^{32}P from [^{32}P]-labeled substrates in the presence or absence of 1 nmol okadaic acid.

4. Notes

1. Failure to detect an interaction in the two-hybrid method may be a result of a number of causes, including a poorly expressed or unstable fusion protein, steric hindrance by the fusion portion of the polypeptide, and failure of the interacting

polypeptides to translocate to the nucleus. In the authors' two-hybrid screen, using the Aα subunit of PP2A as bait, Cα and Bα subunits interacted only weakly with the bait. In their library screen, these genes were not isolated at all. Thus, not all interesting interacting proteins will be identified by the two-hybrid method.

2. An alternative to dot-blot hybridization using yeast miniprep DNA is to do yeast colony hybridization. This method uses zymolase to break down the cell wall of the yeast and allow the plasmid DNA to bind to a nylon membrane *(9,17)*. The colony lifts can then be probed for identity to each other or a known PP2A subunit.

3. If there are only one or two promising candidates, you may want to isolate full-length cDNA clones from the appropriate libraries before attempting to demonstrate biochemical interactions. This will save you time if they interact positively in the secondary screen.

4. If the newly identified subunit only weakly interacts with PP2A, it may be a regulator rather than a stable subunit. To test this, it may be necessary to over-express the protein, e.g., in *E. coli*, and then add it back to PP2A holoenzyme and measure changes in enzyme activity on selected substrates.

5. How to decide which clones to focus attention on is perhaps the most challenging part of the two-hybrid screen. Some interacting clones may make immediate biological sense, whereas others may have a more obscure relevance. The authors chose to focus on clones that appeared to constitute a gene family. However, solitary isolates may be caused by important interactions with low abundance proteins. It does appear that some genes are isolated frequently with multiple disparate baits; a listing is available on the World Wide Web, at www.fccc.edu/research/labs/golemis/TwoHybridReferences.html

6. The following World Wide Web sites also have helpful information.
Two-hybrid methods and media, using a slightly different system: xanadu.mgh.harvard.edu/brentlabweb/finleyandbrent1996.html
Transformation protocols: www.umanitoba.ca/faculties/medicine/ human_genetics/gietz/Trafo.html#TRAFO
Yeast methods: www.fhcrc.org/~gottschling/homepage.html and www.cgl.ucsf.edu/sacs_home/HerskowitzLab/protocols/protocol.html
Suppliers: Clontech (ww.clontech.com), Invitrogen, Origene Technologies (www.origene.com), and Bio 101 supplies specialty yeast media (www.bio101.com) sell two-hybrid interaction plasmids, libraries, and yeast strains. Presumably they also provide an instruction manual.

Acknowledgments

The authors thank Stan Hollenberg for a critical reading of and helpful comments on the manuscript. The work described herein was supported by NIH grant R01 CA71074 and the Jason Overman Cancer Research Fund.

References

1. Fields, S. and Song, O. (1989) A novel genetic system to detect protein-protein interactions. *Nature* **340**, 245–246.

2. Bartel, P., Chien, C. T., Sternglanz, R., and Fields, S. (1993) Elimination of false positives that arise in using the two- hybrid system. *Biotechniques* **14,** 920–924.
3. Fields, S. and Sternglanz, R. (1994) The two-hybrid system: an assay for protein-protein interactions. *Trends Genet.* **10,** 286–292.
4. McCright, B. and Virshup, D. M. (1995) Identification of a new family of protein phosphatase 2A regulatory subunits. *J. Biol. Chem.* **270,** 26,123–26,128.
5. McCright, B., Rivers, A. M., Audlin, S., and Virshup, D. M. (1996) The B56 family of protein phosphatase 2A regulatory subunits encodes differentiation-induced phosphoproteins that target PP2A to both nucleus and cytoplasm. *J. Biol. Chem.* **271,** 22,081–22,089.
6. Ruediger, R., Roeckel, D., Fait, J., Bergqvist, A., Magnusson, G., and Walter, G. (1992) Identification of binding sites on the regulatory A subunit of protein phosphatase 2A for the catalytic C subunit and for tumor antigens of simian virus 40 and polyomavirus. *Molec. Cell. Biol.* **12,** 4872–4882.
7. Mendelsohn, A. R. and Brent, R. (1994) Applications of interaction traps/two-hybrid systems to biotechnology research. *Curr. Opin. Biotechnol.* **5,** 482–486.
8. Vojtek, A., Hollenberg, S., and Cooper, J. (1993) Mammalian Ras interacts directly with the serine/threonine kinase Raf. *Cell* **74,** 205–214.
9. Kaiser, C., Michaelis, S., and Mitchell, A. (1994) *Methods in Yeast Genetics,* Cold Spring Harbor Laboratory Press, Cold Spring Harbor, New York.
10. Guthrie, C. and Fink, G. R. (eds) (1991) *Guide to Yeast Genetics and Molecular Biology* vol. 194. in *Methods in Enzymology.* (Abelson, J. N. and Simon, M. I., eds.) Academic, San Diego, CA.
11. Chevray, P. M. and Nathans, D. (1992) Protein interaction cloning in yeast: Identification of mammalian proteins that react with the leucine zipper of Jun. *Proc. Natl. Acad. Sci. USA* **89,** 5789–5793.
12. Breeden, L. and Nasmyth, K. (1985) Regulation of the yeast HO gene. *Cold Spring Harb. Symp. Quant. Biol.* **50,** 643–650.
13. Hannon, G. J., Demetrick, D., and Beach, D. (1993) Isolation of the Rb-related p130 through its interaction with CDK2 and cyclins. *Genes Devel.* **7,** 2378–2391.
14. Schiestl, R. H. and Gietz, R. D. (1989) High efficiency transformation of intact yeast cells using single stranded nucleic acids as a carrier. *Curr. Genet.* **16,** 339–346.
15. Gietz, D., St. Jean, A., Woods, R. A., and Schiestl, R. H. (1992) Improved method for high efficiency transformation of intact yeast cells. *Nucleic Acids Res.* **20,** 1425.
16. Estojak, J., Brent, R., and Golemis, E. A. (1995) Correlation of Two-Hybrid Affinity Data with In Vitro Measurements. *Molec. Cell. Biol.* **15,** 5820–5829.
17. Sherman, F., Fink, G. R., and Hicks, J. B. (1986) *Methods in Yeast Genetics,* Cold Spring Harbor Laboratory, Cold Spring Harbor, New York.
18. Yocum, R. R., Hanley, S., Robert West, J., and Ptashne, M. (1984) Use of *lacZ* fusions to delimit regulatory elements of the inducible divergent *GAL1-GAL10* promoter in *Saccharomyces cerevisiae. Molec. Cell. Biol.* **4,** 1985–1998.
19. Ruediger, R., Hentz, M., Fait, J., Mumby, M., and Walter, G. (1994) Molecular model of the A subunit of protein phosphatase 2A: Interaction with other subunits and tumor antigens. *J. Virol.* **68,** 123–129.

23

Protein Phosphatase 2A Regulatory Subunits

cDNA Cloning and Analysis of mRNA Expression

Julie A. Zaucha, Ryan S. Westphal, and Brian E. Wadzinski

1. Introduction

The protein serine/threonine phosphatase 2A (PP2A) family of holoenzymes has been implicated in the regulation of many cellular signaling molecules including metabolic enzymes, cell-surface receptors, cytosolic protein kinases, and transcription factors (reviewed in **refs. 1–5**). PP2A is a multimeric enzyme, composed of a catalytic subunit associated with two regulatory subunits (reviewed in **refs. 1–5**). The core structure of PP2A is a heterodimer consisting of the 36 kDa catalytic subunit (C) tightly complexed with the 65 kDa A regulatory subunit (A). This heterodimer complexes with one of multiple B subunits (ranging from 54–130 kDa). Currently, three B subunit families have been identified (B or B55, B' or B56, and B'' or B72). Whereas no amino acid sequence homology exists between the different families, isoforms within the same family are highly homologous. The physiological significance of this heterogeneity is not known; however, recent data suggest that the B regulatory subunit has a role in determining the substrate selectivity of the catalytic subunit. It also is postulated that the B regulatory subunit participates in localization of the catalytic subunit to distinct cellular microenvironments (reviewed in **refs. 1–5**).

Most of the phosphatase B subunits have been identified by classical biochemical techniques, library screening strategies, and more recently, the yeast two-hybrid system **(6)**. Whereas the B subunits identified to date likely represent the majority of B subunits, recent data from the authors' laboratory suggest there are additional B subunits not yet cloned. Homology-based reverse

From: *Methods in Molecular Biology, Vol. 93: Protein Phosphatase Protocols*
Edited by: J. W. Ludlow © Humana Press Inc., Totowa, NJ

transcriptase-polymerase chain reaction (RT-PCR) cloning provides an excellent method to clone known phosphatase subunit cDNAs, as well as novel and low abundance species *(7,8)*. In this method, a degenerate oligonucleotide primer corresponding to a region of amino acid sequence homology in the known B subunits (**Fig. 1A**) is used in the reverse transcription reaction to generate a mixture of B subunit cDNAs from a sample of total RNA. Sets of degenerate oligonucleotide primers (**Fig. 1B**) are then used for PCR amplification of the cDNAs, and the partial cDNA clones are ligated into a bacterial plasmid. The DNA sequences are determined by dideoxynucleotide sequencing and analyzed for homology to known B subunit cDNAs. The authors have successfully utilized this strategy to clone rat homologs of three known B subunits (Bα, Bβ, and Bγ) and one novel B subunit from rat brain.

To study the tissue distribution and expression levels of these B subunit mRNAs, the cloned cDNAs were used to generate probes for ribonuclease (RNase) protection analyses. This method is very sensitive and specific for detecting and quantitating mRNA levels of closely related B subunit isoforms. In these analyses, radiolabeled antisense riboprobes corresponding to B subunit RT-PCR products are hybridized to total RNA from selected tissues. Following ribonuclease A digestion, the protected fragments are analyzed by gel electrophoresis and autoradiography. Messenger RNA transcripts that differ in one or more nucleotides from the B subunit riboprobe are digested by ribonuclease A *(9)*, indicating the high specificity of this method. This chapter outlines the use of homology-based RT-PCR cloning coupled with RNase protection analysis to isolate B subunit cDNAs and determine mRNA levels of known and novel B subunit family members. Moreover, the described methodology can be utilized to study the expression of other phosphatase regulatory subunits, as well as the expression of homologous phosphatase catalytic subunits.

2. Materials

2.1. RT/PCR

1. Total RNA (prepared from selected tissues using TriReagent [Molecular Research Center, Cincinnati, OH] following the manufacturer's recommended protocol) (*see* **Note 1**).
2. Nuclease-free water (use for all solutions).
3. Degenerate sense and antisense oligonucleotides (synthesized on an Expedite Model 8909 DNA synthesizer, Perceptive Biosystems, Framingham, MA): 50 µ*M*.
4. MgCl$_2$: 25 m*M*.
5. 10X reverse transcription buffer: 100 m*M* Tris-HCl, pH 8.8, 500 m*M* KCl, 0.1% Triton X-100.

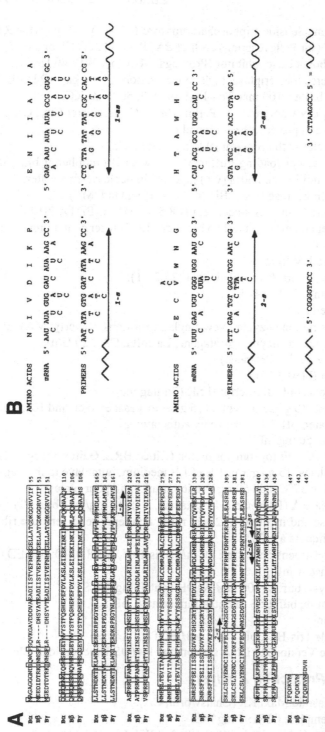

Fig. 1. Designing Degenerate Oligonucleotide Primers for RT-PCR. **(A)** Alignment of the amino acid sequences for Bα, Bβ, and Bγ. Boxes denote regions of amino acid sequence identity. The arrows identify amino acid residues to which degenerate synthetic oligonucleotide primers were prepared for RT-PCR cloning. **(B)** Degenerate oligonucleotide primers corresponding to the indicated amino acid sequences are depicted. The primers are either in the sense (s) or antisense (as) orientation. The 5′-ends of the sense and antisense primers contained Kpn I and EcoR I restriction sites, respectively.

281

6. Deoxyribonucleoside triphosphate mixture (dNTPs): 10 mM (for RT reaction) and 2 mM (for PCR reaction) each of dATP, dCTP, dGTP, and dTTP.

7. RNasin ribonuclease inhibitor (Promega, Madison, WI): 40 U/μL.

8. AMV reverse transcriptase, high concentration (Promega): 10 U/μL.

9. 10X PCR buffer: 100 mM Tris-HCl, pH 8.3, 500 mM KCl.

10. AmpliTaq DNA polymerase (Perkin-Elmer Cetus, Foster City, CA): 5 U/μL.

11. Nuclease- and protease-free mineral oil.

12. Thermocycler with programmable heating block.

13. 10X agarose gel loading buffer: 0.4% (w/v) Bromophenol blue, 0.4% (w/v) xylene cyanol FF, and 60% (v/v) glycerol in nuclease-free water.

14. SeaKem LE agarose (FMC BioProducts, Rockland, ME).

15. TAE buffer: 40 mM Tris-acetate, pH 8.5, 2 mM Na$_2$EDTA·2H$_2$O.

16. TAE buffer containing ethidium bromide: TAE buffer containing 0.5 μg ethidium bromide/mL.

17. Short-wave UV light.

18. Phenol/chloroform/isoamyl alcohol (25:24:1).

19. 3 M sodium acetate, pH 5.2.

20. Absolute ethanol.

21. EcoR I restriction enzyme (New England BioLabs, Beverly, MA): 20 U/μL.

22. Kpn I restriction enzyme (Stratagene, La Jolla, CA): 20 U/μL.

23. Multi-Core buffer (Promega).

24. Geneclean II (BIO 101).

25. Bacterial plasmid: Bluescript II SK (Stratagene).

26. LB medium: 10 g bacto-tryptone, 5 g bacto-yeast extract, and 10 g NaCl in 1 L of Milli Q water, pH 7.0; sterilize by autoclaving.

27. Ampicillin: 50 mg/mL.

28. Competent *E. coli* for transformation (Gibco BRL, Gaithersburg, MD).

29. LB agar plates: 15 g agar in 1 L of LB medium; autoclave to melt agar and sterilize; cool to 37°C, add 50 μg ampicilin/mL, and pour plates.

30. Ribonuclease A (Pharmacia, Piscataway, NJ): 10 mg/mL; boil 30 min to inactivate DNase and incubate 1–2 h at room temperature to renature ribonuclease; store in aliquots at −20°C.

31. Bacteria resuspension buffer: 50 mM Tris-HCl, pH 8.0, 10 mM EDTA, 100 μg ribonuclease A/mL.

32. Bacteria lysis buffer: 200 mM NaOH containing 1% SDS.

33. Neutralization buffer: 3 M potassium acetate, pH 5.5.

34. Isopropanol.

35. TE: 10 mM Tris-HCl, pH 8.0, 1 mM EDTA.

36. Sequenase Version 2.0 DNA sequencing kit (US Biochemical, Cleveland, OH).

2.2. RNase Protection Analysis

1. Qiagen Plasmid Maxi Kit (Chatsworth, CA).

2. Plasmid template DNA: 1 mg/mL.

3. Acc65 I restriction enzyme (Promega): 10U/μL.

4. Phenol/chloroform/isoamyl alcohol (25:24:1).
5. Absolute ethanol.
6. Ammonium acetate: 10 M.
7. Diethylpyrocarbonate (DEPC)-treated water (use for all solutions).
8. 5X transcription buffer: 200 mM Tris-HCl, pH 7.5, 30 mM MgCl$_2$, 10 mM spermidine, 50 mM NaCl.
9. DTT: 100 mM.
10. Ribonucleotides: ATP (10 mM), GTP (10 mM), CTP (10 mM), UTP (200 μM).
11. RNasin ribonuclease inhibitor (Promega): 40 U/μL.
12. T7 RNA Polymerase (New England Biolabs): 50 U/μL.
13. T3 RNA Polymerase (Pharmacia): 60 U/μL.
14. [α-^{32}P]UTP: 10 mCi/mL (800 Ci/mmol).
15. RNase-free DNase (Promega): 1 U/μL.
16. Glycogen: 30 mg/mL.
17. RNA hybridization solution: 40 mM PIPES, pH 6.4, 400 mM NaCl, 1 mM EDTA, 80% formamide.
18. Total RNA (prepared as described in **Subheading 2.1.**).
19. Yeast tRNA: 10 mg/mL.
20. Ribonuclease A: *see* **Subheading 2.1.**
21. Ribonuclease digestion buffer: 10 mM Tris-HCl, pH 7.5, 300 mM NaCl, 5 mM EDTA, 25 μg ribonuclease A/mL.
22. SDS: 20%.
23. Proteinase K (Gibco BRL): 10 mg/mL.
24. Gel loading buffer (to prepare 1 mL): 920 μL 100% formamide, 20 μL 12.5% (w/v) Bromophenol blue (in 50% ethanol), 10 μL 25% (w/v) xylene cyanol (in water), 50 μL 10 X TBE buffer (890 mM Tris-base, 890 mM boric acid, 20 mM EDTA).
25. Acrylamide/urea gel: 4 or 6% acrylamide/7 M urea.
26. Kodak X-OMAT AR film.
27. 1 Kb DNA ladder (Gibco/BRL): 1 mg/mL.
28. 10X T4 polynucleotide kinase buffer: 700 mM Tris-HCl, pH 7.6, 100 mM MgCl$_2$, 50 mM DTT.
29. T4 polynucleotide kinase (New England BioLabs): 10 U/μL.
30. [γ-^{32}P]ATP: 3000 Ci/mmol (10 mCi/mL).

3. Methods (*see* Notes 2 and 3)

3.1. RT/PCR

3.1.1. Designing PCR Primers (see *Notes 4–6*)

1. Based on conserved regions of amino acid sequence within the phosphatase B regulatory subunits (**Fig. 1A**), sense and antisense degenerate oligonucleotide primers are synthesized (**Fig. 1B**). The sets of primers are designed to yield PCR products of approx 100–700 bp. To enable unidirectional cloning of PCR products into *E. coli* plasmids (*see* **Subheading 3.1.5.**), the 5'-ends of the primers contain either EcoR I or Kpn I restriction sites.
2. Prepare stock solutions (50 μM) of each primer. Store at –20°C.

3.1.2. RT Reaction

The following protocol is a modification of the protocol provided by Promega for synthesis of single-stranded cDNA from RNA.

1. In a nuclease-free 0.5 mL microcentrifuge tube, add 2 μg of substrate RNA, 0.5 μL of antisense primer (e.g., 1-*as* or 2-*as*), and nuclease-free water to bring the volume to 10 μL.
2. Incubate for 10 min at 65°C to disrupt RNA secondary structure, then cool on ice to allow annealing of primer and RNA.
3. Add the following reagents in order: 4 μL 25 mM MgCl$_2$, 2 μL 10X reverse transcription buffer, 2 μL 10 mM dNTPs, 0.5 μL RNasin, and 1.5 μL AMV reverse transciptase or 1.5 μL nuclease-free water (control). Mix gently.
4. Incubate for 1 h at 42°C then terminate reaction by heating for 10 min at 99°C. Samples can be stored at –20°C if not used immediately.

3.1.3. PCR (see **Notes 7** and **8**)

1. Combine the following components in a nuclease-free microcentrifuge tube on ice: 10 μL 10X PCR buffer, 63.5 μL nuclease-free water, 10 μL 2 mM dNTPs, 10 μL 25 mM MgCl$_2$, 2 μL each of the appropriate sense primer and antisense primer, and 2 μL of the appropriate reverse transcriptase reaction. Mix gently.
2. Heat tube to 72°C then add 0.5 μL of 5 U/μL AmpliTaq DNA polymerase.
3. Overlay solution with 2 drops of mineral oil.
4. Perform PCR in a thermocycler as follows: 35 cycles of 2 min at 94°C (to denature template DNA), 2 min at 50°C (to anneal primers to template DNA), and 2 min at 72°C (for elongation); followed by 72°C for 7 min.
5. Remove 13.5 μL of the lower aqueous phase and mix with 1.5 μL of 10X agarose gel loading buffer.
6. Prepare a 1.2% agarose gel in TAE buffer and load sample in a well. Following electrophoresis in TAE buffer containing ethidium bromide, visualize PCR products in the gel under short-wavelength UV light.

3.1.4. Isolation of PCR Products

1. Remove 80 μL of PCR product and combine with 120 μL of nuclease-free water and 200 of μL phenol/chloroform/isoamyl alcohol. Vortex well and centrifuge for 5 min.
2. Transfer 190 μL of the top aqueous layer to another microcentrifuge tube and add 19 μL of 3 M sodium acetate and 525 μL of absolute ethanol. Mix and store at –80°C for at least 1 h followed by centrifugation for 10 min.
3. Discard supernatant and add 1 mL of 70% ethanol, vortex gently, centrifuge for 5 min, and discard supernatant. Repeat 70% wash. Resuspend air-dried pellet in 50 μL of nuclease-free water.
4. Combine 50 μL of the extracted PCR product with 8 μL of Multi-Core buffer, 3 μL of Kpn I, 3 μL of EcoR I, and 16 μL of nuclease-free water. Incubate for

Table 1
Expression of Phosphatase B Regulatory Subunits
in Adult Rat Brain[a]

Phosphatase subunit	1-*as*/2-*s* (264 bp)	1-*as*/2-*s* (663 bp)	2-*as*/1-*s* (240 bp)
B_α	11	10	8
B_β	21	10	3
B_γ	9	9	9
B_{novel}	1		3

[a]The number of sequenced clones corresponding to the different phosphatase B regulatory subunits is listed for each set of degenerate oligonucleotide primers. The primers are either in the sense *(s)* or antisense *(as)* orientation. Sequences were verified by DNA sequencing of the PCR products ligated into Bluescript II SK.

at least 2 h and purify digested DNA fragment by agarose gel electrophoresis and Geneclean II (according to manufacturer's recommended directions).

3.1.5. Cloning and Sequencing of PCR Products

1. Ligate the DNA fragment into Bluescript II SK linearized with EcoR I and Kpn I.
2. Transform high-efficiency competent *E. coli* (e.g., HB-101, DH5α, and so on) and plate onto LB agar plates containing ampicillin. Incubate overnight at 37°C.
3. Inoculate 5 mL of LB medium containing ampicillin (50 µg/mL) with single colonies of *E. coli* and incubate at 37°C overnight with vigorous shaking.
4. Purify plasmid DNA using a modification of the alkaline lysis mini-prep protocol *(10)*. Aliquot 1.5 mL of the overnight culture into a 2-mL microcentrifuge tube and centrifuge for 2 min to pellet the bacteria. Resuspend pellet in 300 µL of resuspension buffer and add 300 µL of lysis buffer. Mix by inverting tube several times and incubate at room temperature for 5 min. Add 300 µL of neutralization buffer, mix, and centrifuge for 15 min to pellet the cell debris and chromosomal DNA. Transfer the supernatant to another microcentrifuge tube and repeat centrifugation. Transfer the supernatant again. Add 0.8 vol of isopropanol, mix, and centrifuge for 15 min to pellet plasmid DNA. Discard supernatant and add 1 mL of 70% ethanol to the pellet. Mix and centrifuge for 5 min. Discard supernatant and resuspend air-dried pellet in 20 µL of TE buffer.
5. Check an aliquot (2 µL) of plasmid DNA for the presence of PCR product insert by restriction enzyme digestion (*Eco*RI and *Kpn*I) and analysis on agarose gel electrophoresis as described above (*see* **Subheading 3.1.3.**).
6. Determine the DNA sequence of ligated PCR product using Sequenase Version 2.0 DNA sequencing kit and the Bluescript SK II primers flanking the cloning site.
7. **Table 1** summarizes the results obtained from RT-PCR cloning of rat brain total RNA using the indicated sets of primers.

3.2. RNase Protection Analysis

3.2.1. Riboprobe Design and Synthesis (see **Notes 9–11**)

1. Plasmid DNA is purified from 150 mL *E. coli* culture using Qiagen Plasmid Maxi Kit following manufacturer's recommended protocol.

2. Plasmids containing template DNA cloned in the reverse orientation downstream of the T3 bacteriophage promoter are linearized using Acc65 I to produce a 5' overhang (*see* **Note 12**). Incubate 200 μL of digestion reaction containing 20 μg of plasmid DNA and 20 μL of Acc65 I in the manufacturer's recommended buffer (150 mM NaCl, 6 mM Tris-HCl, pH 7.9, 6 mM MgCl$_2$, 1 mM DTT) for 5 h at 37°C. Following this incubation, an aliquot (5 μL) of the digestion reaction should be analyzed by agarose gel electrophoresis (*see* **Section 3.1.3.**). If any undigested plasmid remains, add additional enzyme, buffer, and water. Continue incubation at 37°C to ensure complete digestion.

3. Following plasmid digestion, add an equal vol of phenol/chloroform/isoamyl alcohol to the mixture, vortex and centrifuge for 5 min.

4. Transfer the upper aqueous phase to a new microcentrifuge tube. Add 2.5 vol of absolute ethanol and 1/5 vol of 10 M ammonium acetate. Mix solution and store on dry ice for 10–30 min followed by centrifugation for 15 min. Discard supernatant and dry pellet (air-dry or Speed Vac). Resuspend linearized plasmid to a final concentration of 1 mg/mL in DEPC-treated water.

5. Into a nuclease-free microcentrifuge tube, add in the following order: 2 μL 5X transcription buffer, 0.5 μL 100 mM DTT, 0.5 μL 10 mM ATP, 0.5 μL 10 mM GTP, 0.5 μL 10 mM CTP, 0.5 μL 200 μM UTP, 0.5 μL RNasin, 0.5 μL T3 RNA polymerase, 4 μL [α-^{32}P]UTP (40 μCi), and 0.5 μL 1 mg/mL linearized plasmid containing template DNA.

6. Incubate for 60–90 min at 37°C.

7. Add 5 μL of RNase-free DNase to digest template DNA and incubate for an additional 20 min at 37°C.

8. To the reaction mixture, add 85 μL of DEPC-treated water and 100 μL of phenol/chloroform/isoamyl alcohol. Mix by vortexing and separate phases by centrifugation for 5 min.

9. Transfer the upper aqueous phase (100 μL) to another microcentrifuge tube and add 20 μL of 10 M ammonium acetate, 2 μL of 30 mg glycogen/mL, and 300 μL of absolute ethanol. Vortex and store on dry ice for 10–30 min followed by centrifugation for 15 min. Discard supernatant, resuspend pellet in 100 μL of DEPC-treated water and repeat ethanol precipitations twice, omitting the addition of glycogen. Resuspend final air-dried pellet in 100 μL of RNA hybridization solution.

10. Transfer 1 μL of radiolabeled riboprobe to a scintillation vial containing 3 mL of scintillation fluid, mix well, and count in a scintillation counter. Approximately 100,000–1,000,000 cpm per μL should be expected.

3.2.2. Hybridization

1. Into a nuclease-free microcentrifuge tube, combine 3–10 μg of desired RNA and the appropriate amount of carrier yeast tRNA for a total of 20 μg RNA.

2. Add approx 500,000–700,000 cpm of radiolabeled riboprobe and the appropriate amount of hybridization solution to bring the volume to 30 μL.
3. Heat the samples at 85°C for 10 min and slowly cool to 45°C over 90 min. Continue incubation at 45°C for at least 14 h.

3.2.3. RNase Digestion

1. Add 350 μL of RNase digestion buffer to each hybridization reaction and incubate for 45 min at 30°C.
2. Inactivate the RNase A by adding 10 μL of 20% SDS and 10 μL of proteinase K to each sample. Incubate for 30–45 min at 37°C.
3. Add 400 μL of phenol/chloroform/isoamyl alcohol to each sample. Mix by vortexing and separate phases by centrifugation for 5 min.
4. To the upper aqueous phase (400 μL), add 2 μL of 30 mg/mL glycogen and 1 mL of absolute ethanol. Vortex and store on dry-ice for 10–30 min followed by centrifugation for 15 min. Discard supernatant and wash pellet with 150 μL of 70% ethanol. Resuspend air-dried pellet in 5 μL of gel loading buffer and 5 μL of DEPC-treated water. Vortex for 30 s.
5. Heat samples for 10 min at 85°C prior to loading on an acrylamide/urea gel.

3.2.4. Electrophoresis and Analysis of Protected Fragment (see *Notes 13–17*)

1. Electrophorese the samples at 25 mA on a 4% or 6% acrylamide gel (0.4 mm) containing 7 *M* urea for approx 45 min (or until the Bromophenol blue dye runs off the gel bottom).
2. Following electrophoresis, dry the gel and expose to film for 3–14 h to visualize the radiolabeled protected fragment.
3. Quantitation of the relative amounts of mRNA can be obtained by phosphoimager analysis, using cyclophilin as an internal control.
4. **Figure 2** shows an autoradiogram of a representative RNase protection assay.

4. Notes

1. Although total RNA prepared by TriReagent (Molecular Research Center) has been used in this protocol, RNA prepared by other methods (e.g., cesium chloride, Qiagen total RNA kit, etc.) also have been used successfully in the described experiments. RT/PCR cloning can be performed on total RNA (0.5–2.0 μg) prepared from specific tissues, tissue regions, or cells.
2. The Notes the authors found to be most relevant to the described methodology are included in this section. More general notes on RT-PCR and RNase protection analyses can be found in **refs.** *11* and *12*.
3. Perform all centrifugations in a table-top microcentrifuge at maximum speed.
4. Ideally, the degenerate primers should have the following characteristics: approx 18–20 nucleotides; a Tm higher than 50°C; and a degeneracy of less than 600. For example, the 1-*as* oligonucleotide primer, GC(GATC)AC(GATC)GC(GAT)AT (GAT)AT(GA)TT(CT)TC, has a degeneracy of 4 × 4 × 3 × 3 × 2 × 2 or 576.

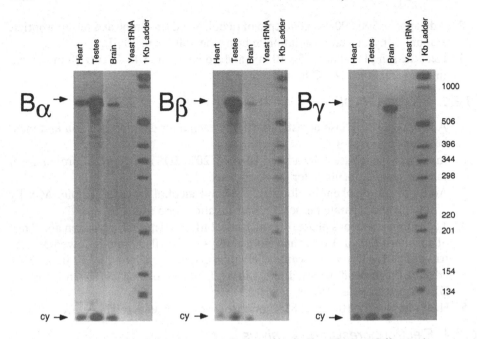

Fig. 2. Analysis of B Subunit mRNA Expression in Rat Tissues. Ribonuclease protection assays were performed using the indicated radiolabeled riboprobes and 7 μg
total RNA from adult rat heart, testes, and brain. Bα is expressed in all three tissues,
Bβ is expressed in both rat testes and brain, and Bγ appears to be only expressed in the
brain. Protected fragments corresponded to the predicted size of the PCR product (650
nucleotides) and no fragments were observed with the yeast tRNA control. A radiolabeled cyclophilin riboprobe was used as an internal control. The sizes of the radiolabeled DNA standards are indicated.

5. The oligonucleotide primers used in this method contain restriction sites at
   their 5'-ends; however, primers lacking restriction sites can be used if PCR
   products are ligated into *E. coli* plasmids by T-A overhang cloning or blunt-
   end cloning *(13)*.
6. Different restriction sites can be added to the primers as long as these sites are
   not present in the cDNA of interest and a suitable bacterial plasmid containing
   the desired restriction sites within the multiple cloning site (MCS) is available. If
   the cloned DNA is to be used to prepare riboprobes (*see* **Subheading 3.2.1.**) for
   RNase protection analyses, the MCS in the expression plasmid should be flanked
   by either the T3 or T7 promoter.
7. PCR amplification of an aliquot of control RT reaction (no reverse transcriptase)
   should not yield a PCR product. If a PCR product(s) is observed, contaminating
   template DNA and/or genomic DNA are most likely present in the reaction
   mixture. In this situation, prepare fresh reagents and use a different preparation of RNA.

8. To check for potential PCR errors, it is important to sequence both strands of DNA from multiple RT/PCR clones. The sequence of the PCR product must be identical to the cDNA of interest for use in RNase protection assays.

9. Other *E. coli* plasmids containing the cloned DNA of interest also can be utilized for this protocol; however, the appropriate restriction enzyme (i.e., to linearize the plasmid) and bacteriophage RNA polymerase must be chosen so that transcription yields an antisense riboprobe. In the present protocol, Kpn I and EcoR I restriction sites were incorporated into the 5'-end and 3'-end, respectively, of the B subunit PCR product and the product was ligated into Bluescript II SK. The resulting plasmid was linearized with Acc65 I, followed by transcription using T3 RNA polymerase to yield an antisense riboprobe complementary to the B subunit mRNA. Alternatively, the B subunit PCR product could be cloned into Bluescript II KS, linearized with Acc65 I, and transcribed using T7 RNA polymerase to produce an antisense riboprobe.

10. The sequence of the riboprobe must be exactly complementary to the sequence of the mRNA of interest because ribonuclease A can partially cleave RNA-RNA complexes with only one nucleotide mismatch; complete cleavage of these complexes occurs with two or more adjacent nucleotide mismatches. Utilization of a riboprobe with one or more mismatches can lead to the detection of multiple protected fragments and, consequently, uninterpretable results. The detection of multiple protected fragments when using the correct riboprobe suggests the possible presence of alternatively spliced mRNAs.

11. Riboprobes should be used the day they are prepared.

12. Restriction sites that generate 3' overhangs should not be used for linearizing the template DNA because these overhangs are initiation sites for the polymerase and can lead to synthesis of RNA complementary to the probe.

13. A hybridization reaction containing yeast tRNA (20 μg) and no substrate RNA is a negative control.

14. To quantitate the relative levels of specific mRNA in various samples, an internal control must be included to normalize for variations in RNA amounts loaded onto the gel. Ideally, the control riboprobe should be smaller than the riboprobe corresponding to the mRNA of interest. A portion of the cyclophilin gene is commonly used; it is transcribed as described in **Subheading 3.2.2.** except that 3 μL of DEPC-treated water and 1 μL of [α-^{32}P]UTP (10 μCi) are added to the transcription reaction. The radiolabeled cyclophilin riboprobe should be added to all hybridization reactions at one-fourth the radioactivity of the B subunit riboprobe.

15. The percentage of acrylamide in the gels used for electrophoresis depends on the size of the riboprobes; larger riboprobes (400 nucleotides or longer) are analyzed on a 4% gel while smaller riboprobes are analyzed on a 6% gel.

16. The size of the protected fragment is determined by comparison with the migration of radiolabeled DNA standards on the same gels. To prepare standards, combine 1 μg of the 1 Kb ladder (GIBCO BRL), 1 μL of 10X T4 kinase buffer, 1 μL of T4 kinase, 3 μL of [γ-^{32}P]ATP, and 3 μL of nuclease-free water. Incubate for 30–60 min at 37°C, add 90 μL of nuclease-free water, extract with phenol/chlo-

roform/isoamyl alcohol, and ethanol precipitate as described in **Subheading 3.2.1.** Resuspend the final pellet in 10 μL of nuclease-free water and use approx 20,000 cpm per lane.

17. Early termination of the RNA polymerase reaction and high background in the hybridization are common problems observed with RNase protections. These problems often can be circumvented by using nuclease-free solutions and tubes and by selecting a smaller riboprobe (i.e., 100–300 nucleotides). An additional source of background labeling could be caused by incomplete digestion of template DNA which copurify and hybridize with the riboprobe and result in smeared bands appearing on the autoradiogram. If this occurs, try a fresh tube of DNase, a new preparation of template DNA, or gel purify the radiolabeled riboprobe *(13)*.

Acknowledgments

The authors would like to thank Dr. Ron Emeson and his laboratory colleagues for assisting them with the RNase protection assays. They also wish to thank Yuk-Ying Wong for her assistance in preparing this chapter. This research was supported by a grant from the National Institute of Health (GM51366). J. A. Z. is supported by a National Institutes of Health Training Grant (T32 HL07323).

References

1. Cohen, P. (1989) The structure and regulation of protein phosphatases. *Annu. Rev. Biochem.* **58,** 453–508.
2. Mumby, M. C. and Walter, G. (1993) Protein serine/threonine phosphatases: structure, regulation, and function in cell growth. *Physiol. Rev.* 73,673–73,699.
3. DePaoli-Roach, A. A., Park, I. K., Cerovsky, V., Csortos, C., Durbin, S. D., Kuntz, M. J., Sitikov, A., Tang, P. M., Verin, A., and Zolnierowicz, S. (1994) Serine/threonine protein phosphatases in the control of cell function. *Adv. Enzyme Regul.* **34,** 199–224.
4. Shenolikar, S. (1994) Protein serine/threonine phosphatases—new avenues for cell regulation. *Annu. Rev. Cell Biol.* **10,** 55–86.
5. Wera, S. and Hemmings, B. A. (1995) Serine/threonine protein phosphatases. *Biochem. J.* **311,** 17–29.
6. McCright, B. and Virshup, D.M. (1995) Identification of a new family of protein phosphatase 2A regulatory subunits. *J. Biol. Chem.* **270,** 26,123–26,128.
7. Wadzinski, B. E., Heasley, L. E., and Johnson, G. L. (1990) Multiplicity of protein serine-threonine phosphatases in PC12 pheochromocytoma and FTO-2B hepatoma cells. *J. Biol. Chem.* **265,** 21,504–21,508.
8. Chen, M. X. and Cohen, P. T. (1993) Identification of novel protein (serine/threonine) phosphatase genes in *Saccharomyces cerevisiae, Drosophila melanogaster,* and man by polymerase chain reaction, in *Protein Phosphorylation: A Practical Approach* (Hardie, D. G., ed.), IRL, Oxford, pp. 253–264.

9. Sambrook, J., Fritsch, E. F., and Maniatis, T. (1989) Enzymes used in molecular cloning, in *Molecular Cloning: A Laboratory Manual* (Nolan, C., et al., eds.) Cold Spring Harbor Laboratory, Cold Spring Harbor, p. 5.81.
10. Engebrecht, J. and Brent, R. (1995) Minipreps of plasmid DNA, in *Current Protocols in Molecular Biology* (Ausubel, F. M. et al., eds.) Wiley, New York, pp. 1.6.1.–1.6.2.
11. Beverley, S. M. (1995) Enzymatic amplification of RNA by PCR in *Current Protocols in Molecular Biology* (Ausubel, F. M. et al., eds.) Wiley, New York, pp. 15.4.1–15.4.6.
12. Gilman, M. (1995) Ribonuclease protection assay, in *Current Protocols in Molecular Biology* (Ausubel, F. M. et al., eds.) Wiley, New York, pp. 4.7.1.–4.7.8.
13. Finney, M. (1995) Molecular cloning of PCR products, in *Current Protocols in Molecular Biology* (Ausubel, F. M. et al., eds.) Wiley, New York, pp. 15.7.1.–15.7.3.

24

Synthetic Lethal Screening in Protein Phosphatase Pathways

Jianhong Zheng and John F. Cannon

1. Introduction

Synthetic lethal screening is a useful genetic strategy that has a variety of applications in identifying new genes in yeast or other organisms. The method starts with a gene, *YFG* (your favorite gene), and isolates yeast mutants that need that gene for viability. These mutants are subsequently used to identify the mutant yeast gene or isolate genes from other species that complement the yeast mutation. This method has proven successful in identifying yeast homologs based on the function of foreign genes *(1)*. In synthetic lethal screening, only functional homology is required, rather than DNA or amino acid homology. Therefore, this approach is also called "Cloning by Function." Additionally, synthetic lethal screening has also been productive in identifying genes, which function within the same or parallel pathways as *YFG* in complicated regulatory systems *(2–4)*.

The general scheme for this approach involves first expressing *YFG* on a yeast plasmid, mutagenizing a yeast strain transformed with the plasmid, and then screening for a plasmid dependent yeast mutant. **Figure 1** shows an outline of procedures for a typical experiment. The parent strain (*ade2 ade3 leu2 ura3*) is transformed with a "screening" plasmid (2 micron *ADE3 LEU2 YFG*). Plasmid dependence can be scored by a colony sectoring assay *(5,6)*. This assay is based on changes in cellular pigmentation that occur because of mutations in adenine biosynthetic genes. Mutant *ade2* yeast cells accumulate a red pigment, but *ade2 ade3* cells do not. Thus, a strain that is *ade2 ade3*, but carries *ADE3* on an autonomously replicating plasmid, forms red colonies. Cells that lose the plasmid are white. Yeast transformed with unstable plasmids form red colonies with white sectors. Desired mutants that are inviable upon plasmid loss

From: *Methods in Molecular Biology, Vol. 93: Protein Phosphatase Protocols*
Edited by: J. W. Ludlow © Humana Press Inc., Totowa, NJ

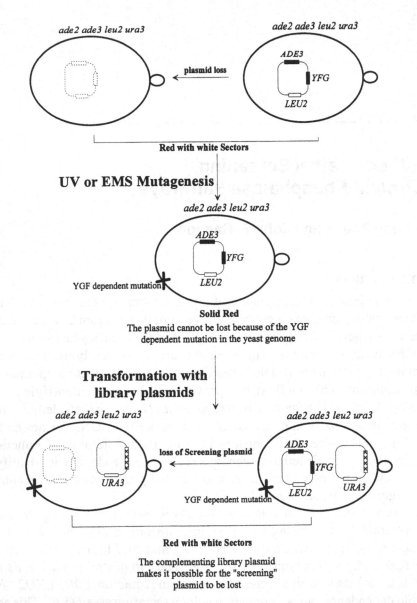

Fig. 1. Synthetic lethal screening strategy. *YFG* is a gene of interest that encodes a component of a protein phosphatase pathway. It is a gene that is not normally found in yeast for the example in this figure. In the upper right, a *ade2 ade3* yeast strain is transformed with a *YFG LEU2 ADE3* unstable plasmid (**Subheading 3.1.**). Because *YFG* is nonessential for yeast viability, this plasmid is lost easily and yields red colonies with white sectors. After mutagenesis (**Subheading 3.2.**) mutant colonies that require *YFG* are recognized by their nonsectoring trait (**Subheading 3.3.**). *YFG*-dependent mutations are characterized genetically to determine dominance (**Subheading 3.4.**). Recessive mutations are further characterized. Genes from yeast or other organisms that complement the *YFG*-dependent mutations are isolated from libraries (**Subheading 3.5.**). Plasmids that harbor genes that complement *YFG*-depen

will produce red, nonsectoring colonies. Genes that complement these mutations can then be isolated by transforming a library into the mutant strains and
screening for the ability to sector again.

Besides desired *YFG*-dependent mutations, three other mechanisms yield
nonsectoring red colonies: plasmid integration; *ade3* to *ADE3* gene conversion; and dependence on elements of the screening plasmid other than *YFG*
(2). Fortunately, all three background sources can be distinguished from desired
mutants by a single test. The critical experiment is to introduce a "testing"
plasmid that expresses *YFG*, but does not carry *ADE3* into the candidate strains.
If any mutant is dependent on *YFG*, it should regain the sectoring trait when
the testing plasmid with *YFG* is introduced, but remain solid red when the
testing plasmid without *YFG* is introduced.

The synthetic lethal screening method has been used in the authors' laboratory to study GLC7, the yeast type 1 phosphatase (PP1) regulatory pathway.
One example is the screening of *GLC8* dependent mutants. *GLC8* encodes a
yeast homolog of mammalian protein phosphatase inhibitor-2 *(7)*. The genetic
and biochemical analysis of *GLC8* function shows that it does not inhibit either
rabbit or yeast PP1 and instead plays a positive role *(7,8)*. Screening for *GLC8*
dependent mutants resulted in the identification of mutations in *GLC7*. These
results show that mutant forms of this PP1 enzyme require GLC8 activity to function. There are other examples of successful application of this method in the study
of phosphatase/kinase signal transduction pathway *(3,4)*. For example, *PTC1* (a
protein serine/threonine phosphatase gene) was identified in a synthetic lethal
screen because *PTC1* mutants require the *PTP1* or *PTP2* gene (protein tyrosine
phosphatase gene) for growth. This result indicates that tyrosine phosphatases and
PTC1 play an essential role in the same signal transduction pathway.

The following protocol presents an outline for the reasoning and steps
involved in a typical synthetic lethal screening. Readers can use it as a guide in
designing experimental strategies and should find synthetic lethal screening a
very productive approach in identifying interesting genes in many systems.

2. Materials

2.1. Yeast Strains

The parental strains for screening are derived from CH1305 and CH1462
(1) (*see* **Note 1**).

1. CH1305: *MATa ade2 ade3 leu2 ura3 lys2 can1.*
2. CH1462: *MATa ade2 ade3 leu2 ura3 his3 can1.*

Fig. 1. (*continued*) dent mutations yield colonies that regain the sectoring trait. These plasmids are isolated from yeast and further characterized (**Subheading 3.6.**).

2.2. Media

1. YEPD: 1% yeast extract (Difco, Detroit, MI), 2% glucose, 2% Bacto-Peptone (Difco). Add 2% agar for plates.
2. 100x complete amino acid mix: 2 g of arginine, 1 g of histidine, 6 g of isoleucine, 6 g of leucine, 4 g of lysine, 1 g of methionine, 6 g of phenylalanine, 5 g of threonine and 4 g of tryptophan per liter of water, filter sterilized.
3. Liquid omission media: 20 g of glucose, 6.7 g of yeast nitrogen base without amino acids, dissolved in 1 L of distilled water and autoclave. Add 10 mL of 100X amino acid mix, 1 mL of 1% adenine, 4 mL of 1% uracil, and 5 mL of 1% tyrosine (each made in 100 mM NaOH and filter sterilized). Media deficient in uracil, leucine, or lysine (-URA, -LEU, -LYS) are made by making and using -Leu or -Lys 100X amino acid mix or by omitting uracil after autoclaving. Add 2% agar for plates.
4. -URA low adenine plate: essentially the same as -URA plate, but use 0.6 mL of 1% adenine per L.
5. ETOH plates: 1% yeast extract, 2% Bacto-Peptone, 2% agar, autoclave, and add 2% (v/v) absolute ethanol before pouring plates. Store in sealed bags for no more than one month at room temperature.

2.3. Plasmids

1. Screening plasmid: An unstable yeast plasmid that carries *ADE3*, *LEU2*, and *YFG* (*see* **Note 2**).
2. Testing plasmid: derived from pRS316 or pRS426 plasmids *(9,10)*, with *YFG* and *URA3*, but no *ADE3*.
3. Plasmid library (*see* **Note 3**).

2.4. Reagents for Transformation

1. LiAc/TE: 100 mM lithium acetate, 10 mM Tris-HCl, pH 8.0, 1 mM ethylene diamine tetraacetic acid (EDTA), filter sterilized.
2. CTDNA: Dissolve 500 mg of calf thymus DNA (Sigma, St. Louis, MO, D-1501) in 100 mL of water. Sonicate until viscosity is greatly reduced (about 10–15 min at 50% setting). Distribute 10-mL aliquots into 25-mL centrifuge tubes, add potassium acetate to 0.3 M and 15 mL ethanol to each. Incubate at −70°C for 15 min, centrifuge at 10,000g for 10 min, decant ethanol and air dry. Store at −20°C. When needed, add 10 mL of autoclaved water to the contents of one tube. This makes 5 mg/mL of DNA stock. Mix and store at 4°C.
3. LiAc/TE/PEG: 100 mM lithium acetate, 10 mM Tris-HCl, pH 8.0, 1 mM EDTA, 40% polyethylene glycol (MW 3350), filter sterilized.
4. TE: 10 mM Tris-HCl, pH 8.0, 1 mM EDTA, filter sterilized.

2.5. Reagents for EMS Mutagenesis

1. 50 mM potassium phosphate buffer pH 7.0, autoclaved.
2. Methanesulfonic acid ethyl ester (EMS), (Sigma cat. no. M0880).

3. Stop solution: 10% (w/v) sodium thiosulfate, made fresh before use and filter sterilized.

2.6. Reagents for DNA Preparation

1. Sorbitol/EDTA: 1 M sorbitol, 100 mM EDTA.
2. Zymolyase: zymolyase-20T (from Seikagaku America, Ijamsville, MD).
3. Tris/EDTA/SDS: 400 mM Tris base, 250 mM EDTA, 2.5% SDS.
4. KAc: 5 M potassium acetate, unbuffered and autoclaved.

2.7. Equipment

1. Low power dissection microscope.
2. 254 nm UV source (lamp, tissue culture hood, or transilluminator).
3. Haemacytometer.

3. Methods

This section describes experimental details for the major parts of the synthetic lethal screening (**Fig. 1**).

3.1. Transformation of Yeast Using the Plate Method

Because of simplicity and high efficiency, the plate method for yeast transformation *(11)* is used in this step and elsewhere. An alternative method is the "high efficiency yeast transformation" method (*see* **Subheading 3.5.**).

1. Take a loopful of cells from a fresh plate of CH1305 or CH1462 (less than 2 wk in refrigerator).
2. Resuspend in 0.4 mL LiAc/TE/PEG solution.
3. Add 10 µL of CTDNA and 1 µg of the respective "screening" plasmid DNA to cells and vortex vigorously (*see* **Note 4**).
4. Leave cells overnight at room temperature without agitation.
5. Plate 50 µL of the settled cells onto appropriate omission media (-Leu media for pTSV30A derived plasmids) and incubate for at least 3 d at 30°C to allow the growth of transformants.

3.2. Mutagenesis

1. Inoculate 5 mL of liquid -LEU omission media with transformants from **Subheading 3.1.** Incubate overnight at 30°C with vigorous shaking. It usually takes about 20 h for CH1305 and CH1462 cultures to reach a density of 3–5 × 10⁷ cells/mL.
2. Count cells in haemacytometer to determine cell density.
3. Mutagenize with UV light or EMS.

3.2.1. UV Mutagenesis

1. Make dilutions of cells in sterile water.
2. Spread 100 µL of appropriate dilution of the cells to give 3000–30,000 cells on each YEPD plate.

3. Expose the cells to UV light by opening the lid. A pilot experiment is always necessary to determine the survival rate after UV treatment. In the authors' hands, exposure to a UV light with an intensity of 240 mW for 40 s gives 90% of killing (*see* **Note 5**).

4. Incubate the plates at 30°C for 5–7 d. Plates should be incubated in the dark to avoid photoreactivation.

3.2.2. EMS Mutagenesis

1. Wash the cells twice in potassium phosphate buffer by centrifuging gently to pellet and resuspend in 5 mL of the same buffer.

2. Add 300 µL of EMS to 5 mL of cells in a screw cap tube, tighten the cap well and vortex vigorously.

3. Incubate for 10 min (or appropriate time) at 30°C. A pilot experiment is always needed to determine the survival rate. In the authors' hands, 10 min at 30°C gives a killing efficiency of about 90%.

4. Stop mutagenesis by adding 5 mL of sodium thiosulfate solution and mix well.

5. Collect the cells by gentle centrifugation and wash twice with sterile water.

6. Carry out all the aforementioned operations in a hood and avoid inhaling volatile vapors.

7. Make appropriate dilutions and plate 300 cells on each YEPD plate. Incubate at 30°C for 5–7 d.

3.3. Isolation of the Plasmid Dependent Mutants

1. After 5–7 d of incubation at 30°C, the colonies should show red color. Most of the colonies will be white with red sectors. Identify colonies with uniform red color. If the colonies are small and the plate is crowded, a dissecting microscope may be needed to recognize nonsectoring colonies.

2. Streak solid red colonies on YEPD plates to purify single colonies to confirm the nonsectoring trait. This is an important step because cells from some apparently nonsectoring colonies sector when restreaked out on a whole plate. It is normal to have a few completely white colonies from real nonsectoring strains (*see* **Note 6**).

3. As mentioned in the introduction, there are at least three sources of background interference: plasmid integration, gene conversion, and mutants that are dependent on the screening plasmid backbone instead of *YFG* itself. Transform the solid red cells with the "testing" plasmid pRS316-*YFG* and pRS316 vector alone (*see* **Note 7**). Select for the transformants on -URA low adenine plates (*see* **Note 8**). Make sure leucine is supplemented in the media to allow the loss of the screening plasmid. Incubate at 30°C for 5–7 d and score the sectoring trait.

4. Recognize plasmid integrants and gene convertants because they do not regain the ability to sector. Mutant transformants that can sector with pRS316-*YFG*, but not with pRS316 alone are true *YFG*-dependent mutants, which are saved. Sectoring varies from strain to strain. So pay attention not only to the ones that regain sectoring trait completely but also to the ones that only partially sector (*see* **Note 9**).

3.4. Dominance Testing and Complementation Analysis

Only genes for recessive synthetic lethal mutations can be readily cloned by complementation *(12)*. Therefore, determining the dominance of nonsectoring mutants is necessary. Additionally, this method can be extended to classify nonsectoring mutants into complementation groups if mutants were isolated in CH1305 and CH1462 *(2,13)*.

1. Using an inoculating loop or toothpicks, transfer nonsectoring mutants isolated in CH1305 to very small patches (about 9 mm^2) on YEPD plate.
2. Transfer and mix a similar quantity of CH1462 to each patch, keeping the mating mixture area as small as possible.
3. Incubate overnight at 30°C.
4. Select diploids by streaking mating mixtures on -Lys -Leu plates and incubate 3–4 d at 30°C (*see* **Note 10**).
5. Streak diploids on YEPD plates to score sectoring trait. Diploids that sector contain recessive synthetic lethal mutations, nonsectoring diploids contain dominant mutations.

3.5. Cloning by Complementation

Transform the *YFG*-dependent mutant strains with library DNA (*see* **Note 3**) using the "High efficiency Yeast Transformation" method (*see* **Note 11**).

1. Inoculate cells into liquid YEPD medium and grow overnight to 1–2 × 10^7 cells/mL.
2. Dilute to 2 ×10^6 cells/mL in fresh YEPD and grow to 1 × 10^7 cells/mL.
3. Harvest cells, wash in 1 mL of TE/LiAc and pellet the cells by centrifuging 1 min at 10,000g in a microcentrifuge.
4. Resuspend in 50 µL of TE/LiAc. Mix with 50 µg of CT DNA and 1 µg of library plasmid DNA.
5. Add 300 µL of LiAc/TE/PEG solution and incubate at 30°C for 30 min.
6. Heat shock in a 42°C water bath for 15 min.
7. Spin down in microcentrifuge at 10,000g for 10 s.
8. Resuspend in 1 mL of TE and dilute appropriately to give about 300 transformants/plate.
9. Plate about 300 transformants on each -URA low adenine minimal plate (*see* **Note 12**).
10. Incubate at 30°C for 5–7 d. Look for transformants that are sectoring or white. Both kinds of colonies have a good chance of carrying a complementing plasmid.
11. Pick all sectoring or white transformants and streak each out on a -URA low adenine plate. The purpose of this step is to confirm the sectoring trait and to allow the complete loss of the "screening" plasmid to get single white colonies that carry only the library plasmids.
12. Pick one white colony from each plate and grid on -URA plate. Let grow for one day at 30°C.

13. Replica plate to -LEU, -URA and ETOH plates. The ones that can grow on ETOH but not on -LEU should be saved. Colonies that cannot grow on ETOH plates are petites (*see* **Note 6**) and are discarded.
14. Prepare plasmid DNA from yeast transformants that grow on ETOH, but not on -LEU (*see* **Subheading 3.7.**). Repeat transformation of the nonsectoring mutant and score their nonsectoring trait. If the purified plasmid contains DNA that complements the synthetic lethal mutation then these secondary transformants should all be sectoring.
15. If the complementing plasmid is from a yeast library, candidate genes on the plasmid can be quickly recognized by determining DNA sequences of the yeast sequences at the vector/insert junction. These sequences are then compared with the yeast genomic database (available on the world wide web at http://genome-www.stanford.edu/Saccharomyces/).

3.6. Purification of Plasmid DNA from Yeast

The plasmids used are *E. coli*/yeast shuttle vectors. Therefore, all plasmid preparations are actually from *E. coli* using common methods *(14)*. Getting plasmids from yeast into *E. coli* is a matter of preparing a crude yeast DNA preparation followed by transforming *E. coli* to ampicillin resistance. Steps of yeast DNA preparation are described.

1. Grow yeast in 10 mL of YEPD broth for about 24 h at 30°C.
2. Harvest cells by gentle centrifugation, resuspend in 1 mL of Sorbitol/EDTA and transfer to 1.5 mL microcentrifuge tubes.
3. Centrifuge for 10 s at 10,000g, resuspend in 300 µL Sorbitol/EDTA containing 14 mM 2-mercaptoethanol and 300 µg/mL zymolyase.
4. Incubate at 37°C for 1 h, Centrifuge for 10 s at 10,000g.
5. Resuspend pellet in 400 µL of Tris/EDTA/SDS and heat at 65°C for 30 min.
6. Add 100 µL of KAc and incubate on ice for 1 h.
7. Centrifuge at 10,000g for 15 min, transfer supernatant to a new tube.
8. Fill the tube with room temperature ethanol, mix by inversion ten times, and centrifuge 1 min.
9. Dry pellets, dissolve nucleic acids in 100 µL TE (might take overnight at 4°C).
10. Transform *E. coli* with 5 µL of this crude DNA.

4. Notes

1. If *YFG* is a gene not normally found in yeast, then CH1305 and CH1462 can be used without alteration. Adenovirus E1A is an example of a *YFG* gene not found in yeast *(15)*. If such genes are used, providing a yeast promoter to ensure expression is necessary. Yeast genes for calcineurin, BEM1, SPA2, tyrosine phosphatases, and GLC8 were deleted before screening began *(3,4,8,16,17)*. Standard yeast genetic procedures are used to construct appropriate strains *(13,18)*.
   Strains CH1305 and CH1462 are a pair of strains that have opposite mating types. Because of their nonisogenic backgrounds, they yield different collections of

mutations. The authors have been more successful with CH1305 and derivative strains. For initial screens, they suggest using both strains to maximize non-sectoring mutation variety.

2. The authors use pTVS30A (a gift from John Pringle, University of North Carolina, Chapel Hill) which is unstable and high-copy because of a two-micron origin of replication. Other plasmids have been used *(2,5,15)*. These are *E. coli*/yeast shuttle vectors. The β-lactamase gene confers resistance to 50 μg/mL in *E. coli*. All plasmid preparations are from *E. coli* using standard methods *(14)*. Biosynthetic genes are used as yeast selectable markers. Therefore, the yeast strains used must contain appropriate auxotrophic mutations (*ura3* or *leu2*, for example).

3. Libraries of plasmids that express genes in yeast. If the experimental goal is to identify the yeast genes, which have the synthetic lethal mutations, then use a yeast genomic library cloned into a low copy vector *(12)*. If the goal is to isolate genes from other organism, which complement the yeast mutations, then suitable expression libraries are needed. The authors have previously used human cDNA libraries for this purpose *(19)*.

4. The plasmid DNA does not have to be extensively purified. Small scale preparations are suitable. The presence of RNA actually enhances transformation efficiency. Any DNA preparation that can be digested with a restriction enzyme will likely work.

5. If mutants are frequent, use a mutagenesis that kills 90% of the cells. For rare mutants, 1% survival will be more efficient. Plan on 20 or more plates with approx 300 surviving colonies for each mutagenesis. Ideally, there will be one to four nonsectoring colonies per plate.

6. Nonsectoring mutant clones normally produce about 2% completely white colonies. These white colonies are "petite" mutants that have lost the ability to respire, which is necessary to produce the red pigment. These colonies should be ETOH⁻ Leu⁺ (fail to grow on ethanol plates and grow on leucine-deficient media). The sectoring assay used in this experiment is very sensitive. Nonsectoring mutants include those that absolutely need the plasmid to survive, but also ones that have a significant growth advantage (ones that grow slowly without *YFG*).

7. The "Screening" plasmid that the authors use (*see* **Note 2**) is a high-copy unstable plasmid, whereas the testing plasmid could be a low copy, *CEN* plasmid (e.g., pRS316-based). It is possible that some mutants may have been plasmid-dependent because they required the high-copy *YFG*. This situation would lead to partial sectoring when transformed with low-copy *YFG*.

8. When using a minimal media to score the sectoring trait, limiting the adenine concentration is necessary. The authors have found that 6 μg/mL of adenine is optimal for both the red color development and healthy growth of yeast cells. Some cells take a long time to turn red on regular minimal plates that contain 10 μg/mL of adenine.

9. One disadvantage of this screening is it has a significant level of undesirable nonsectoring mutations. These include gene conversions, plasmid integration, uninteresting plasmid backbone-dependent mutants. The ratio between the *YFG*

dependent mutants and these undesirables varies from each screen. It depends on *YFG* and the relative target size of *YFG* dependent mutations. The authors have seen yields of *YFG* dependent mutants as high as 20% of the nonsectoring colonies and as low as 0.5%.

10. Because CH1305 and CH1462 are opposite mating types they can mate to generate diploid yeast. Nonsectoring mutants isolated in CH1305 will maintain the *LEU2* screening plasmid and, therefore, will be Leu⁺ Lys⁻. CH1462 is Leu⁻ Lys⁺. Therefore, after mating (which takes about 5 h), diploids can be purified from their parents because diploids will be Leu⁺ Lys⁺ and will grow on media deficient in both lysine and leucine.

11. The plate method for transformation is used in most experiments because of its simplicity. When a high efficiency is required (e.g., when screening a library), the high efficiency transformation method should be used. However, it was found that some strains produce many white revertants when grown in liquid medium. This is a serious problem and makes screening for complementing plasmids impossible. One solution to this problem is to use fresh solid red colonies collected from a YEPD plate and use them directly for transformation using the plate method.

12. If good yeast libraries are used *(12)*, an average of 1 in 2000 transformants will carry a plasmid that complements the mutant. Screening approx 5000 transformants from a yeast library is necessary to find most complementing plasmids. Obviously, more complex libraries require screening more transformants *(19)*.

References

1. Kranz, J. E. and Holm, C. (1990) Cloning by function: an alternative approach for identifying yeast homologs of genes from other organisms. *Proc. Natl. Acad. Sci. USA* **87**, 6629–6633.

2. Bender, A. and Pringle, J. R. (1991) Use of a screen for synthetic lethal and multicopy suppressee mutants to identify two new genes involved in morphogenesis in *Saccharomyces cerevisiae*. *Mol. Cell. Biol.* **11**, 1295–1305.

3. Maeda, T., Tsai, A. Y. M., and Saito, H. (1993) Mutations in a protein tyrosine phosphatase gene (PTP2) and a protein serine/threonine phosphatase gene (PTC1) cause a synthetic growth defect. *Mol. Cell. Biol.* **13**, 5408–5417.

4. Costigan, C., Gehrung, S., and Snyder, M. (1992) A synthetic lethal screen identifies SLK1, a novel protein kinase homolog implicated in yeast cell morphogenesis and cell growth. *Mol. Cell. Biol.* **12**, 1162–1178.

5. Koshland, D., Kent, J. C., and Hartwell, L. H. (1985) Genetic analysis of the mitotic transmission of minichromosomes. *Cell* **40**, 393–403.

6. Hieter, P., Mann, C., Snyder, M., and Davis, R. W. (1985) Mitotic stability of yeast chromosomes: a colony color assay that measures nondisjunction and chromosome loss. *Cell* **40**, 381–392.

7. Cannon, J. F., Pringle, J. R., Fiechter, A., and Khalil, M. (1994) Characterization of glycogen-deficient glc mutants of *Saccharomyces cerevisiae*. *Genetics* **136**, 485–503.

8. Cannon, J. F., Clemens, K. E., Morcos, P. A., Nair, B. M., Pearson, J. L., and Khalil, M. (1995) Type 1 protein phosphatase systems of yeast. *Adv. Prot. Phosphatases* **9**, 215–236.

9. Sikorski, R. S. and Hieter, P. (1989) A system of shuttle vectors and yeast host strains designed for efficient manipulation of DNA in *S. cerevisiae. Genetics* **122**, 19–27.

10. Christianson, T. W., Sikorski, R. S., Dante, M., Shero, J. H., and Hieter, P. (1992) Multifunctional yeast high-copy-number shuttle vectors. *Gene* **110**, 119–122.

11. Elble, R. (1992) A simple and efficient procedure for transformation of yeasts. *BioTechniques* **13**, 18–20.

12. Rose, M. D. (1987) Isolation of genes by complementation in yeast. *Methods. Enzymol.* **152**, 481–504.

13. Sherman, F., Fink, G. R., and Hicks, J. B. (1986) *Laboratory Course Manual for Methods in Yeast Genetics.* Cold Spring Harbor Laboratory, Cold Spring Harbor, NY.

14. Maniatis, T., Fritsch, E. F., and Sambrook, J. (1982) *Molecular cloning: A Laboratory Manual.* Cold Spring Harbor Laboratory, Cold Spring Harbor, NY.

15. Zieler, H. A., Walberg, M., and Berg, P. (1995) Suppression of mutations in two *Saccharomyces cerevisiae* genes by the adenovirus E1A protein. *Mol. Cell. Biol.* **15**, 3227–3237.

16. Garrett-Engele, P., Moilanen, B., and Cyert, M. S. (1995) Calcineurin, the Ca2+/calmodulin-dependent protein phosphatase, is essential in yeast mutants with cell integrity defects and in mutants that lack a functional vacuolar H(+)-ATPase. *Mol. Cell. Biol.* **15**, 4103–4114.

17. Peterson, J., Zheng, Y., Bender, L., Myers, A., Cerione, R., and Bender, A. (1994) Interactions between the bud emergence proteins Bem1p and Bem2p and Rho-type GTPases in yeast. *J. Cell Biol.* **127**, 1395–1406.

18. Guthrie, C. and Fink, G. R. (1991) *Guide to Yeast Genetics and Molecular Biology.* Academic Press, New York.

19. Thon, V. J., Khalil, M., and Cannon, J. F. (1993) Isolation of human glycogen branching enzyme cDNAs by screening complementation in yeast. *J. Biol. Chem.* **268**, 7509–7513.

25

The Search for the Biological Function
of Novel Yeast Ser/Thr Phosphatases

Joaquin Ariño, Francesc Posas, and Josep Clotet

1. Introduction
1.1. Isolation of Novel Ser/Thr Phosphatases in Yeast

The genome of the yeast S. *cerevisiae* contains a fairly large number of genes encoding Ser/Thr protein phosphatases. Among these genes can be found the homologs of the classical phosphatases described in mammalian tissues in the early 1980s. For instance, the catalytic subunit of PP1 is encoded by the gene *GLC7*, whereas two forms of PP2A are encoded by genes *PPH21* and *PPH22*. As it happens in other organisms, the yeast catalytic subunits of PP1, PP2A, and PP2B (calcineurin) are quite related in sequence. In addition to these phosphatases, yeast cells contain other phosphatase genes that, while related to PP1 or PP2A in their primary structure, are functionally different. The main features of these genes (*1–13*) are described in **Table 1**.

In some cases, these novel forms of PP have been found by classical complementation assays, as it was the case of *SIT4 (PPH1)*. In other cases, the approach that led to their identification and cloning was the screening of libraries with heterologous probes or the amplification of genomic fragments by PCR using oligonucleotides encoding conserved regions in the catalytic subunits of PP1, PP2A, and PP2B. This later approach was used in the author's laboratory to identify and clone two of these novels forms, namely PPG1 and PPZ1 (*5,6*).

1.2. The Search for the Biological Function
of Novel Ser/Thr Phosphatases: An Example of Reverse Genetics

Phosphorylation of proteins at Ser/Thr residues is a major regulatory mechanism in a large variety of cellular functions. Therefore, the identification of a

From: *Methods in Molecular Biology, Vol. 93: Protein Phosphatase Protocols*
Edited by: J. W. Ludlow © Humana Press Inc., Totowa, NJ

Table 1
Structure and Known Function
of Novel Yeast Ser/Thr Protein Phosphatases

Gene name	Size, aa	Structural features	Known functions	Ref.
SIT4 (PPH1)	311	Contains only catalytic core More related to PP2A than to PP-1	Required for G1 to S-phase transition Homolog of Drosophila PPV	1,2
PPH3	308	Contains only catalytic core More related to PP2A than to PP-1	Some functional PP-2A-like activity	3,4
PPG1	368	50 residue C-terminal extension More related to PP2A than to PP-1	Involved in glycogen accumulation	5
PPZ1/PPZ2	692/710	C-terminal halves are related to PP-1 and 92% identical to each other N-terminal extension rich in Ser and basic residues	Null mutant has increased tolerance to Na+ and Li+ cations Null mutant has increased sensitivity to caffeine leading to cell lysis Overproduction suppresses Mpk1 null mutation	6–10
PPQ1 (SAL6)	549	Ser/Asn rich amino terminal domain More relate to PP-1 than to PP-2A	Affects translational accuracy Enhance efficiency of omnipotent suppressors	11,12
PPT1	513	Tetratricopeptide repeat motifs	Nuclear localization	13

gene product as a putative Ser/Thr phosphatase gives little clue about its function. However, a growing number of yeast pathways are known or suspected to be regulated by phospho-dephosphorylation reactions, particularly those related to signal transduction events (14).This allows for the design of several relatively simple tests using yeast strains either overexpressing the gene or carrying a nonfunctional copy, to learn whether or not a phosphatase could be involved in a given pathway. These tests include the ability to grow in different carbon sources, the sensitivity to some conditions (heat-shock, presence of mating factors or caffeine, salt stress, and so forth) as well as the accumulation of certain metabolites as glycogen. By using these tests, the authors have been able to associate the novel phosphatases PPG and PPZ1/PPZ2 to certain cellular functions, thus opening the way for further studies. The authors have selected three protocols that allows for the determination of parameters corresponding to cellular pathways under the regulation of the aforementioned phosphatases. It is important to note that this involvement is not necessarily exclusive. For instance, PP1 and PP2A are also involved in the regulation of glycogen metabolism *(15–17)* and PP2B (calcineurin) is connected to the sodium stress signaling pathway *(18,19)*. Therefore, the following protocols might be useful for a rather wide spectrum of phosphatase studies.

2. Materials

2.1. Determination of Glycogen Content

1. Yeast culture at the desired point of growth (*see* **Note 1**).
2. Filtering system (i.e., Gelman Sciences, Ann Arbor, MI) connected to a water vacuum pump and nitrocellulose membranes (0.45 μm, Gelman Sciences).
3. Capped 10-mL plastic tubes.
4. Reagents and materials for preparation of acid extracts: 10% perchloric acid solution, 0.5 mm diameter acid washed glass beads (Sigma, St. Louis, MO).
5. Reagents and materials for glycogen precipitation: Whatman 31 ET filter paper, microwave oven (optional), 66% ethanol (prechilled at −20°C), acetone (optional).
6. Reagents and materials for glycogen hydrolysis: 1.5-mL Eppendorf tubes, shaker at 37°C and microfuge. Amyloglucosidase solution: we use the enzyme obtained from Aspergillus niger supplied by Boehringer Mannhein as an ammonium sulfate suspension to prepare a 14 U/mL solution in 0.4 *M* acetate buffer, pH 4.8. This enzyme solution should be prepared freshly by direct dilution of the commercial preparation (usually 10x).
7. Measurement of the glucose released: Spectrophotometer, cuvets and hexokinase, or glucose oxidase-based kits for determination of glucose (Boehringer Mannheim, Germany.

2.2. Determination of Intracellular Sodium and Lithium

1. Stock solution of LiCl or NaCl (usually 5 *M*, sterile).
2. Filtering system and 0.45-μm nitrocellulose filters.
3. Cell washing solution: 1.5 *M* sorbitol, 20 m*M* MgCl$_2$.
4. Thermal block or bath set at 95°C and microfuge.
5. Flame photometer.

2.3. Monitoring Cell Lysis by Using Alkaline Phosphatase as Marker

1. Yeast cultures.
2. Caffeine stock solution (0.1 *M*). Sterilize by filtration.
3. Microfuge and Eppendorf tubes.
4. Spectrometer.
5. Alkaline phosphatase activity assay kit (Boehringer Manheim).

3. Methods

3.1. Determination of Glycogen Content

A semiquantitative estimation of the accumulation of glycogen in yeast cells can be made by exposing colonies to iodine vapors for 2–3 min. Cells containing glycogen stain brownish and the intensity of the color provides with an idea about the glycogen content of a given strain. It should be noted that certain strains, as those carrying *ade* mutations, also appear as a red-brownish colonies when grown for several days. This might interfere with the evaluation of the iodine-staining. In any case, a quantitative determination of glycogen content is needed in most cases.

The protocol described here is based in the precipitation of glycogen onto a square of filter paper in the presence of ethanol and digestion of the precipitated polysaccharide with amyloglucosidase. The released glucose is determined spectrophotometrically. The method is an adaptation of a previously described one for measuring glycogen content in cultured cells *(20)*. Significant differences have not been observed when measuring glycogen with this method or the classical one *(21,22)*, whereas it avoids the time-consuming neutralization step. The released glucose can be measured by using either the hexokinase-glucose 6-P dehydrogenase *(21,23)* or the glucose oxidase methods *(24)*. A routine commercially available kit from Boehringer Mannheim is used and the final samples processed, without further treatment, in a standard clinical autoanalyzer (Cobas Bio, Roche). Of course, measurement of glucose can be performed manually in an standard spectrophotometer using the same kits. Content of glycogen is usually expressed as mg of glucose per gram of cells.

1. Collect yeast cells (about 200 mg wet weight) by filtration of the appropriate volume of culture (*see* **Note 2**). Proceed as quickly as possible to avoid glycogen breakdown during manipulation.
2. Remove quickly the cell cake with a spatula (*see* **Note 3**) and deposit it in the bottom of 10-mL precooled tubes containing 1 mL of cold 10% perchloric acid and 0.5 mL of glass beads. Mix vigorously for 2 s in a vortex and immediately place the tube in liquid nitrogen. Samples can be processed immediately or stored at −70°C for future determination.
3. Samples are thawed on ice and cells disrupted mechanically by vigorous vortex (pulses of 1 min followed by 1 min on ice, repeated five times). Spin samples at low speed (1000g for 3 min) to remove glass beads and cellular debris and save 500 µL of supernatant.
4. Spot 100 µL of the acid supernatant onto a 2 × 1 cm rectangle of Whatman 31 ET filter paper. These papers must be unequivocally labeled using a pencil (not a pen, as the ink will fade away). Drop the filter paper in a beaker containing cold 66% ethanol (store the ethanol at −20°C before use). Maintain stirring for 10 min. A protective mesh must be used to avoid disintegration of the filter papers during

stirring. There are a number of easily available kitchenware devices that can fit in the beaker and serve as a protective mesh. In this step, glycogen precipitates onto the paper and low molecular weight saccharides are washed away.

5. Remove the ethanol and replace it with fresh 66% ethanol (not necessarily cold). Maintain stirring for 30 min. Repeat this step once more.

6. Remove the paper squares with the help of forceps (do not discard the ethanol: it can be used for the first wash in the next experiment!) and place them in a tray. Let them air dry either at room temperature, in an oven, or by placing the tray in an standard microwave oven at a low power setting for 5 min. The last method is very fast but the use of metallic trays must be avoided. Air dry at room temperature can be expedited by washing the paper in a small volume of acetone (caution, highly flammable!).

7. Place each filter paper into a 1.5-mL Eppendorf tube and add 1 mL of the freshly prepared amyloglucosidase solution. Incubate at 37°C for 90–120 min with shaking (*see* **Note 4**).

8. Microfuge the tubes for 5–10 min to remove any fragment of filter paper, preferably at 4°C. Supernatants can be immediately assayed for glucose content or stored at –20°C for future use.

3.2. Determination of Intracellular Sodium and Lithium

1. Grow 100-mL cultures in YPD till $OD_{660} = 1$ (*see* **Note 1**).

2. Add from a stock solution (usually 5 *M*) the appropriate volume of LiCl or NaCl (or water for controls) and resume growth (*see* **Note 5**).

3. Take 20–30 mL samples at the appropriate time. (In most cases steady-state is reached after 90–120 min of culture, although it is advisable to test this for a given strain.). Filter immediately using nitrocellulose 0.45 μm membranes (as described in **Section 3.1.**). It is recommended to take duplicate samples. Try to work as quickly as possible.

4. Wash cell cake with about 3 vol of cold 1.5 *M* sorbitol plus 20 m*M* $MgCl_2$ solution. Rinse with cold deionized water.

5. Place the filter into a 10-mL capped plastic tube containing 3 mL of cold deionized water. Vortex for 20 s to resuspend the cells. Spin at 4°C for 4 min at 1500*g*. Decant the supernatant and try to eliminate remaining drops as much as possible.

6. Resuspend the cells in 0.4 mL of deionized water and transfer to an 1.5-mL Eppendorf tube. Extract by incubation at 95°C for 30 min.

7. Microfuge samples for 1 min at 10,000*g* and save supernatants for ion determination.

8. Determination of Na^+ and Li^+ concentration can be made using a flame photometer (*see* **Note 5**), following the manufacturer directions. The apparatus should be calibrated using the appropriate standards every time that it is used.

3.3. Monitoring Cell Lysis
by Using Alkaline Phosphatase as Marker

A qualitative test for the presence of alkaline phosphatase activity can be made by growing cells in agar plates containing 40 μg/mL of 5-bromo-4-

chloro-indolyl phosphate (BCIP). Cells that undergo lysis under the experimental conditions stain blue, because of hydrolysis of BCIP by the phosphatase. Alternatively, after growing of the cells, standard plates can be overlaid with a solution containing glycine hydrochloride (pH 9.5), 1% agar, and 10 mM BCIP. Plates are incubated at 28°C for 30–60 min and cells scored for blue (positive) staining *(25,26)*.

For quantitative determinations, cells must be grown on liquid medium. The following protocol has been used in the author's laboratory to monitor the release of alkaline phosphatase to the medium after exposure of *ppz1Δ ppzΔ* mutants to caffeine *(7)*. It should be easily adapted to other circumstances.

1. Grow 5 mL of cell overnight culture till $OD_{660} = 5$.
2. Next day inoculate 0.5 mL into 9 mL of fresh YPD (*see* **Note 1**). Grow with shaking for 4 h and then add 0.5 mL of the caffeine stock solution (add the same volume of sterile water to the controls).
3. Take a 0.75 mL sample of the culture every 60–90 min. Check the OD_{660}. Repeat the sampling procedure for 4–8 h.
3. Spin the sample in a microfuge at 10,000g for 2 min. Recover 0.6 mL of the supernate medium and store it at –20°C.
4. Once sampling is finished, determine the alkaline phosphatase activity in the medium. The most common method for determining alkaline phosphatase activity is based in the hydrolysis of *p*-nitrophenylphosphate at alkaline pH (usually pH 10) to yield phosphate and *p*-nitrophenol (caution, this compound is toxic. Avoid ingestion or contact). The amount of p-nitrophenol produced can be quantified by measuring the A_{405}. It is recommended to purchase any commercial kit used in clinical biochemistry (i.e., from Boehringer Mannheim), since they are both convenient and relatively inexpensive.

4. Notes

1. Glycogen content varies greatly from strain to strain. Therefore, only comparison between very related or isogenic strains should be made. For a given strain, the amount of intracellular glycogen can be very different depending on the growth conditions. When growing in rich medium (YPD, which is 1% yeast extract, 2% peptone, and 2% glucose) glycogen begins to accumulate at medium-late exponential phase and peaks when the culture reach saturation. Therefore, all these factors must be taken into account when designing the experimental conditions.
2. As a rule, a culture giving an $OD_{660} = 1$ will yield about 0.8 mg/mL of cells (wet weight). For cells in stationary phase, a value of 0.5 mg/mL is more appropriate. When working with a specific strain, it is advisable to experimentally determine the correspondence between optical density and cell mass for a given spectrophotometer. Be aware that the relationship between optical density and cell concentration is only reasonably good for samples with an optical density not higher than 0.5–0.6. Therefore, dilutions should be made when necessary.

3. It is sometimes advisable to evaluate the amount of cells left on the filter. For this purpose, remove the filter and keep it in the cold in a beaker containing 5 mL of cold water. Once the sampling is done, shake the beaker to resuspend the cells and determine the OD_{660} of this suspension. This will indicate the amount of cells left on the filter. This value should be considered when calculating the amount of glycogen per gram of cells.

4. Alternatively, samples can be maintained at 50°C for 4 h (or even overnight, if desired).

5. Lithium is an analog of sodium with higher toxicity. Many laboratory strains can still grow in rich medium plates containing 300–800 mM NaCl or 50–200 mM LiCl. When grown in liquid culture, concentrations should be reduced two- to three-fold. The appropriate concentration of salt to be added to the cells should be determined empirically for a given strain.

6. Sodium concentration can also be determined using Na^+-selective electrodes (as Radiometer G502Na) as described by the manufacturers.

References

1. Sutton, A., Immanuel, D., and Arndt, K. T. (1991) The SIT4 protein phosphatase functions in late G1 for progression into S phase. *Mol. Cell. Biol.* **11,** 2133–2148.
2. Mann, D. J., Dombradi, V., and Cohen, P. T. W. (1993) *Drosophila* protein phosphatase V functionally complements a *SIT4* mutant in *Saccharomyces cerevisiae* and its amino-terminal region can confer this complementation to a heterologous phosphatase. *EMBO J.* **12,** 4833–4842.
3. Ronne, H., Carlberg, M., Hu, G. Z., and Nehlin, J. O. (1991) Protein phosphatase 2A in *Saccharomyces cerevisiae*: effects on cell growth and bud morphogenesis. *Mol. Cell. Biol.* **11,** 4876–4884.
4. Hoffmann, R., Jung, S., Ehrmann, M., and Hofer H. W. (1994) The *Saccharomyces cerevisiae* gene *PPH3* encodes a protein phosphatase with properties different from PPX, PP1 and PP2A.Yeast **10,** 567–578.
5. Posas, F., Clotet, J., Muns, M. T., Corominas, J., Casamayor, A., and Ariño, J. (1993) The gene *PPG* encodes a novel yeast protein phosphatase involved in glycogen accumulation. *J. Biol. Chem.* **268,** 1349–1354.
6. Posas, F., Casamayor, A., Morral, N., and Ariño, J. (1992) Molecular cloning and analysis of a yeast protein phosphatase with an unusual amino-terminal region. *J. Biol. Chem.* **267,** 11,734–11,740.
7. Posas, F., Casamayor, A., and Ariño, J. (1993) The PPZ protein phosphatases are involved in the maintenance of osmotic stability of yeast cells. *FEBS Lett.* **318,** 282–286.
8. Lee, K. S., Hines, L. K., and Levin, D. E. (1993) A pair of functionally redundant yeast genes (PPZ1 and PPZ2) encoding type 1-related protein phosphatases function within the PKC1-mediated pathway. *Mol. Cell. Biol.* **13,** 5843–5853.
9. Hughes, V., Muller. A., Stark, M. J., and Cohen, P. T. (1993) Both isoforms of protein phosphatase Z are essential for the maintenance of cell size and integrity in *Saccharomyces cerevisiae* in response to osmotic stress. *Eur. J. Biochem.* **216,** 269–279.

10. Posas, F., Camps, M., and Ariño, J. (1995) The PPZ protein phosphatases are important determinants of salt tolerance in yeast cells. *J. Biol. Chem.* **270**, 13,036–13,041.

11. Chen, M. X., Chen, Y. H., and Cohen, P. T. (1993) PPQ, a novel protein phosphatase containing a Ser + Asn-rich amino-terminal domain, is involved in the regulation of protein synthesis. *Eur. J. Biochem.* **218**, 689–699.

12. Vincent, A., Newnam, G., and Liebman, S. W. (1994) The yeast translational allosuppressor, *SAL6*: a new member of the PP1-like phosphatase family with a long serine-rich N-terminal extension. *Genetics* **138**, 597–608.

13. Chen, M. X., McPartlin, A. E., Brown, L., Chen,Y. H., Barker, H. M., and Cohen, P. T. (1994) A novel human protein serine/threonine phosphatase, which possesses four tetratricopeptide repeat motifs and localizes to the nucleus. *EMBO J.* **13**, 4278–4290.

14. Thevelein, J. M. (1994) Signal transduction in yeast. *Yeast* **10**, 1753–1790.

15 Peng, Z- Y., Trumbly, R. J., and Reimann, E. M. (1990) Purification and characterization of a glycogen synthase from a deficient-deficient strain of *Saccharomyces cerevisiae*. *J. Biol. Chem.* **265**, 13,871–13,877.

16. Clotet, J., Posas, F., Casamayor, A., Schaaf-Gerstenschläger, I., and Ariño, J. (1991) The gene *DIS2S1* is essential in *Saccharomyces cerevisiae* and is involved in glycogen phosphorylase activation. *Curr. Genet.* **19**, 339–342.

17. Clotet, J., Posas, F., Hu, G-Z., Ronne, H., and Ariño, J. (1995) Role of protein phosphatase 2A in the control of glycogen metabolism in yeast. *Eur. J. Biochem.* **229**, 207–214.

18. Nakamura, T., Liu, Y., Hirata, D., Namba, H., Harada, S., Hirokawa, T., and Miyakawa, T. (1993) Protein phosphatase type 2B (calcineurin)-mediated, FK506-sensitive regulation of intracellular ions in yeast is an important determinant for adaptation to high stress salt conditions. *EMBO J.* **12**, 4063–4071.

19. Mendoza, I., Rubio, F., Rodriguez-Navarro, A., and Pardo, J. M. (1994) The protein phosphatase calcineurin is essential for salt tolerance in *Saccharomyces cerevisiae*. *J. Biol. Chem.* **269**, 8792–8796.

20. Gómez-Foix, A. M., Coats, W. S., Baqué, S., Tausif, A., Gerard, R. D., and Newgard, C. B. (1992) Adenovirus-mediated transfer of the muscle glycogen phosphorylase gene into hepatocytes confers altered regulation of glycogen metabolism. *J. Biol. Chem.* **267**, 25,129–25,134.

21. Kepler, D. and Dekler, K. (1984) Glycogen, in *Methods of Enzymatic Analysis*, 3rd. ed., vol. 6, (Bergmeyer, H. U., ed.), Chemie, Verlag, pp. 11–18.

22. Fernández-Bañares, I., Clotet, J., Ariño, J., and Guinovart, J. J. (1991) Glycogen hyperaccumulation in *Saccharomyces cerevisiae ras2* mutant: a biochemical study. *FEBS Lett.* **290**, 38–42.

23. Kunts, A., Draeger, B., and Ziegenhorn, J. (1984) D-Glucose: UV methods with hexokinase and glucose GP dehydrogenase, in *Methods of Enzymatic Analysis*, 3rd. ed., vol. 6, (Bergmeyer, H. U., ed.), Chemie, Verlag, pp. 163–172.

24. Kunts, A., Draeger, B., and Ziegenhorn, J. (1984) UV-method with glucose dehydrogenase, in *Methods of Enzymatic Analysis*, 3rd. ed., vol. 6, (Bergmeyer, H. U., ed.), Chemie, Verlag, pp. 178–185.

25. Cabib, E. And Duran, A. (1975) Simple and sensitive procedure for screening yeast mutants that lyse at nonpermisive temperatures. *J. Bacteriol.* **124,** 1604–1606.

26. Watanabe, Y., Irie, K., and Matsumoto, K. (1995) Yeast RLM1 encodes a serum response factor-like protein that may function downstream of the Mpk1 (Slt2) mitogen-activated protein kinase pathway. *Mol. Cell. Biol.* **15,** 5740–5749.

Index

Printed in the United States
by Baker & Taylor Publisher Services

Printed in the United States
by Baker & Taylor Publisher Services